LAWS OF EXPONENTS

$x^m x^n = x^{m+n}$

$(x^m)^n = x^{mn}$

$(xy)^m = x^m y^m$

$\left(\frac{x}{y}\right)^m = \frac{x^m}{y^m}, \quad y \neq 0$

$\frac{x^m}{x^n} = x^{m-n}, \quad x \neq 0$

RADICALS: $\sqrt[n]{x} = x^{1/n}$

$\sqrt[n]{x^n} = x, \quad (\sqrt[n]{x})^n = x, \quad x \geq 0$

$\sqrt[n]{\sqrt[m]{x}} = \sqrt[mn]{x}$

$\sqrt[m]{xy} = \sqrt[m]{x}\ \sqrt[m]{y}$

$\sqrt[m]{\frac{x}{y}} = \frac{\sqrt[m]{x}}{\sqrt[m]{y}}, \quad y \neq 0$

FACTORING FORMULAS

$au + av = a(u + v)$

$u^2 - v^2 = (u + v)(u - v)$

$u^2 + 2uv + v^2 = (u + v)^2$

$u^2 - 2uv + v^2 = (u - v)^2$

$u^2 + (c + d)u + cd = (u + c)(u + d)$

$abu^2 + (ad + bc)u + cd = (au + c)(bu + d)$

$u^3 + v^3 = (u + v)(u^2 - uv + v^2)$

$u^3 - v^3 = (u - v)(u^2 + uv + v^2)$

QUADRATIC FORMULA

The roots to $ax^2 + bx + c = 0,\ a \neq 0$ are $x = \frac{-b \pm \sqrt{b^2 - 4ac}}{2a}$

ABSOLUTE VALUE ($d > 0$)

$|u| = d$ is equivalent to $u = d$ or $u = -d$

$|u| < d$ is equivalent to $-d < u < d$

$|u| > d$ is equivalent to $u < -d$ or $u > d$

LOGARITHMS ($a > 0,\ a \neq 1$)

$\log_a x = y$ if and only if $a^y = x$

$\log_a (uw) = \log_a u + \log_a w, \quad u, w > 0$

$\log_a \left(\frac{u}{w}\right) = \log_a u - \log_a w, \quad u, w > 0$

$\log_a (u^c) = c \log_a u, \quad u > 0$

$\log_a \left(\frac{1}{x}\right) = -\log_a x, \quad x > 0$

$y^x = e^{x \ln y}, \quad y > 0$

$\log_b x = \frac{\log_a x}{\log_a b}, \quad x, b > 0,\ b \neq 1$

SECOND EDITION

TRIGONOMETRY

Marshall D. Hestenes
Michigan State University

Richard O. Hill, Jr.
Michigan State University

Prentice-Hall, Englewood Cliffs, New Jersey 07632

Library of Congress Cataloging-in-Publication Data

Hestenes, Marshall D.
Trigonometry.
Rev. ed. of: Trigonometry with calculators. c1982.
Includes index.
1. Trigonometry—Data processing. 2. Calculators.
I. Hill, Richard O. II. Hestenes, Marshall D.
Trigonometry with calculators. III. Title.
QA531.H6 1986 516.2′4 85-24187
ISBN 0-13-930744-3

Published by Prentice-Hall
A Division of Simon & Schuster, Inc.
Englewood Cliffs, New Jersey 07632

Previously published as *Trigonometry with Calculators*

Printed in the United States of America

10 9 8 7 6 5 4 3 2 1

The photographs at the beginning of each chapter were supplied by the Bettman Archive.

Editorial/production supervision: Paul Spencer
Cover and interior design: Judith A. Matz-Coniglio
Cover photo: ICON—Communication/FPG
Manufacturing buyer: John Hall

ISBN 0-13-930744-3 01

Prentice-Hall International (UK) Limited, *London*
Prentice-Hall of Australia Pty. Limited, *Sydney*
Prentice-Hall Canada Inc., *Toronto*
Prentice-Hall Hispanoamericana, S.A., *Mexico*
Prentice-Hall of India Private Limited, *New Delhi*
Prentice-Hall of Japan, Inc., *Tokyo*
Prentice-Hall of Southeast Asia Pte. Ltd., *Singapore*
Editora Prentice-Hall do Brasil, Ltda., *Rio de Janeiro*
Whitehall Books Limited, *Wellington, New Zealand*

Contents

Preface

The first edition of this text was written to provide a highly readable, student-oriented presentation, with an emphasis on such topics as word problems and applications, graphing, estimation, and a functional approach to the material. The scientific calculator was fully integrated into the subject matter; it was used as a pedagogical device as well as a computational aid. These features are retained in the second edition.

A big advantage of the second edition is that it is much more flexible with regard to the use of calculators. The first edition required that calculators be used, but we found that for pedagogical reasons a significant number of instructors still prefer to teach the subject matter using tables. Consequently, in this second edition, tables are included and the use of a scientific calculator is optional. The text has been designed so that individual instructors will be able to tailor calculator involvement in the subject matter to the level they prefer. Instructors may have the students use the calculator extensively, only for messy computations, or not at all.

In many places throughout the text, references are made to the calculator to help enhance the understanding of the material. For instance, viewing a calculator as the traditional "black box" helps explain functions, while pushing function buttons in different orders increases the understanding of composite functions. Such descriptions involving calculators are beneficial to all students, regardless of how much they are using calculators with the text, since certainly they have used a calculator before this (although not necessarily a scientific one).

Those who wish to use a calculator with this book will find that they can easily use *any scientific calculator on the market.* Chapter 0 sets the stage for using calculators. It includes a discussion of round-off error and estimation together with a diagnostic test on calculator usage. This test has references to Appendix B, which provides a brief but thorough explanation of the use of calculators, sufficient to prepare anyone using calculators with the text to begin. Chapter 0 can be read by students on their own and skipped by students not using calculators. Examples and exercises which require a calculator are clearly marked with a [c].

For those who wish to use tables with this book, plentiful explanations and examples on using tables are included with the tables in the appendices. Students will find these are readily understandable when reading them on their own.

Here are some other changes in the second edition:

- Many of the exercise sets have been expanded to include additional easy problems while retaining the challenging exercises. Moreover, several of the exercise sets have also been restructured to provide a more graded selection.
- Chapter 2 has been rewritten to include the wrapping function. This will benefit those students continuing on to calculus, as many calculus texts use the wrapping function.
- The section on trigonometric identities has been reworked and expanded into two sections.
- The material on the Law of Sines has been expanded with the ambiguous case split off into a separate section.
- The chapter on logarithms and several sections in other chapters have been restructured for clarity.

Earlier we mentioned that the new edition retains the key features of the first edition. In the following, we give additional information on these features:

- **Word problems and applications.** Wherever possible, we have included word problems and/or applications to illustrate the usefulness of the mathematics presented. The application sections in trigonometry (Secs. 1.5 and 2.9) and exponentials (Secs. 5.5 and 5.6) are particular examples of this. It is also exemplified by the fact that there are 179 word problems among the 1672 exercises in the text. In fact, 16 out of 39 sections have word problems in their exercise sets, and these word problems are quite varied and interesting.
- **Graphing.** There is a heavy emphasis on graphing throughout the text.
- **Estimation.** Estimations are used throughout the text, both to counteract simple errors (such as trivial computational errors or pushing the wrong buttons on a calculator) and to emphasize mathematical relationships.

- □ **Functional Approach.** A functional approach is emphasized, both because it facilitates explaining many relationships and because it is necessary for those students continuing on to calculus.
- □ **Informal Approach.** Our approach is highly readable and informal, containing 149 worked-out examples. (This averages about 4.5 examples per section.)
- □ **Right Triangle Trigonometry.** The right triangle approach to trigonometry is used. This is quite intuitive and has an appeal to many students for certain applications; we have found it works very well for students using scientific calculators. This is expanded into general trigonometry using both generalized angles and the wrapping function.
- □ **Inverse Trigonometric Functions.** This topic is discussed twice, first with right triangles, where the concept can be developed without the added confusion of domains, and then in general trigonometry, where the full treatment is discussed. We have found that this two-stage development greatly enhances understanding of this often-confusing topic.

Another feature of the text that enhances student interest is that each chapter begins with a *historical note*. These introductions are designed to increase interest by providing some historical perspective to the ensuing material. The main sources of information for these notes were Carl B. Boyer, *A History of Mathematics* (John Wiley and Sons, Inc., New York, 1968); W. W. Rouse Ball, *A Short Account of the History of Mathematics* (Dover Publications, Inc., New York, 1960); and *Historical Topics for the Mathematics Classroom* (NCTM, Washington, D.C., 1969).

The level of the second edition is the same as that of the first. It assumes the student has had the equivalent of two and a half years of high school algebra or geometry. The text is appropriate for a one-term course.

This particular book is a spinoff from a combined algebra and trigonometry text, *Algebra and Trigonometry, Second Edition,* also published in 1986 by Prentice-Hall. We are especially grateful to the reviewers of both the combined textbook and this separate trigonometry volume for their helpful criticism and the time they took in sharing their ideas and often their enthusiasm.

Reviewers of the trigonometry textbook:

Duane Deal, Ball State University
Gus Mavrigian, Youngstown State University
Phyllis J. Meckstroth, San Diego State University
Leonard M. Wapner, El Camino College

Reviewers of the combined algebra and trigonometry textbook:

C. Patric Collier, The University of Wisconsin–Oshkosh
Joseph R. Foote, The University of New Orleans

Louise Hasty, Austin Community College
Herb Johnson, Winona State University
Clayton Knoshaug, Bemidji State University
Roland Lentz, Winona State University
R. Marshall, Eastern Michigan University
Michael Mays, West Virginia University
Kenneth Ostrom, Oregon Institute of Technology
William Petry, Lansing Community College
Jack Porter, The University of Kansas
Emilio Roxin, University of Rhode Island
William Smith, University of North Carolina

We also thank the editorial and production staff at Prentice-Hall for their contribution. We especially would like to thank Paul Spencer for a second outstanding performance as production editor.

Most importantly, we express our extreme gratitude to our families, particularly our wives, for their support, encouragement, and tolerance during the preparation of this edition.

Marshall D. Hestenes
Richard O. Hill, Jr.

Michigan State University
East Lansing, MI 48824

CHAPTER 0

Calculators

This book may be used either with or without calculators. Those of you who have a scientific calculator will find that not only does the calculator help enormously with routine calculations, but it can also assist you in learning and applying the mathematics. Consequently, for those of you who wish to use a calculator with this text, it is extremely important that you know how to use your calculator and understand how it works. Although this is not a text on how to use a calculator, there are instructions regarding calculators throughout the text. These instructions are guidelines and aids to help ensure that you are using your machine correctly, particularly when new functions are introduced. Of necessity the instructions are general, because calculators vary considerably.

The purpose of this preliminary chapter is to set the stage for those using calculators with this text. However, even those who seldom use calculators should be aware of some of the pointers on accuracy and estimations and on problem solving mentioned in the first two subsections.

For those of you planning to use calculators extensively with this text, we next discuss the properties your calculator should have to be used most effectively, some things to be aware of if you are about to buy a calculator, and the use of a calculator on examinations. Finally, there is a diagnostic test to determine your proficiency at using the calculator for basic calculations. If you cannot do all the problems, there are references to Appendix B, which contains detailed instructions and exercises to help you learn to use your calculator. In addition, you may wish to refer to Appendix B later if you are having difficulty.

ACCURACY AND ESTIMATION

While calculators usually do calculations with far greater accuracy than humans, the user should be aware that they have some inaccuracies. Most machine inaccuracies are due to either round off [e.g., $1 \div 3$ is rounded to 0.33333333, so $(1 \div 3) \times 3 = 0.99999999$*] or the algorithm used by the calculator to compute the special functions (e.g., on one calculator if 2^3 is computed using the y^x button, the answer is given as 7.999998). Moreover, different calculators may give slightly different answers to the same problem because they treat round off differently or use different algorithms to compute the functions.

While the preceding inaccuracies do not usually have a great effect on the answer, there are circumstances where they can, such as when the problem has numbers which differ greatly in magnitude. For example, clearly $\dfrac{1}{10^{15} + 10^{-2} - 10^{15}} = \dfrac{1}{10^{-2}} = 100$. But if you used your calculator and first computed $10^{15} + 10^{-2}$, this rounds to 10^{15}, so $(10^{15} + 10^{-2}) - 10^{15}$ would be computed as 0. Then trying to compute its reciprocal would produce an error message.

There is another question which arises concerning accuracy. In a problem that states "the length is 5 meters," how accurate is the number 5? After all, if the length were measured by some means, there would be some error or *degree of accuracy*. The accuracy of the numbers given is important because the result of any calculation is generally no more accurate than the least accurate number used in the calculation. We shall make the following assumption:

> **Assumption** Unless otherwise indicated in a particular situation, all numerical data will be assumed to be accurate to at least eight significant figures.

* *Note:* Some calculators are tricky, here, in that they will round off this number to 1. However, internally the machine thinks the number is not 1. You can see this by now subtracting 1, for the answer will not be zero.

Note: Significant figures are discussed fully in Sec. 1.3.

If you are using a calculator and this assumption applies, you should not round off any data or intermediate results to fewer than eight places. Then usually the answers will be correct to at least six places (even with most machine inaccuracies), and **we shall give the answers to six places.** For problems in which the accuracy of the data is given, you should still *not round off any intermediate results to fewer than eight places,* but the answer should be given to only the number of places in the least accurate number used.

If you are not using a calculator, you should round off data or intermediate results only when the numbers make it unreasonable to do the calculation by hand. However, be sure to ask your instructor what his or her policy is regarding accuracy.

The reason for the prohibition against unnecessary rounding off is that accuracy can be decreased even under the simple operations of addition and multiplication. For example, if we round 21.4 to 21 and 32.3 to 32, then $21 + 32 = 53$, but $21.4 + 32.3 = 53.7$, which rounds to 54. Multiplication is worse: $21 \times 32 = 672$, but $21.4 \times 32.3 = 691.22$. Thus if you have rounded off when doing a calculation, your answer may differ from the answer in the back of the book, rounded off to the same number of places.

Notation: When rounding off, we shall use $\approx$ for *approximately equals*. For example, $\frac{2}{3} \approx 0.666667$.

We shall be rounding off numbers for two different reasons. One is to give the answer to an appropriate number of places, as we have just discussed. The other is to give a very rough approximation to a computation, correct to only one or two significant figures. This is usually done in your head in order to provide a rough check to that computation (for example, to help check that you have not pushed a wrong button). We have found that it is easier for students if we distinguish between these two uses, so we shall use the following terminology:

Estimate: a rough educated guess as to the size of a number or computation.
Approximation: a finite decimal round off of a number or computation.

For example, an estimate of $\sqrt{5}$ done in your head, $\sqrt{5} \approx 2.2$, is a check to an approximation to $\sqrt{5}$ done on your calculator, $\sqrt{5} \approx 2.23607$. Of course, most answers in the back of the text which are given to six significant figures are approximations rather than exact answers.

PROBLEM SOLVING WITH A CALCULATOR

There are two things to keep in mind when using a calculator to help you solve problems.

1. Whenever you are about to use your calculator, always estimate the answer if it is at all reasonable to do so.
2. Try to save the calculator for one computation rather than using it for several intermediate computations.

Before illustrating this, recall that when solving problems with "nice" numbers, you do all computations as you go along. For example,

$$2x + 5 = 7x + 8$$

$$2x - 7x = 8 - 5$$

$$-5x = 3$$

$$x = -\tfrac{3}{5} = -0.6$$

Now suppose you have a similar problem in which numbers are "messy," so that you might want to use a calculator. While it is possible to do the computations as you go along, it is much easier in the long run to do all the algebra first, and *save the calculator for one computation at the end*. For example,

$$2.187x + 4.918 = 7.419x + 8.213$$

$$2.187x - 7.419x = 8.213 - 4.918$$

$$(2.187 - 7.419)x = 8.213 - 4.918$$

$$x = \frac{8.213 - 4.918}{2.187 - 7.419} \quad (1)$$

Now *estimate first:*

$$x \approx \frac{8 - 5}{2 - 7} = \frac{3}{-5} = -0.6$$

Using a calculator, $x \approx -0.629778$.

Note that when the numbers are "messy" the algebra is very similar to the case where the constants are letters, and this is why it is important to understand how to do the problems in this form. For example, solve for x:

$$ax + b = cx + d, \qquad a \neq c$$

$$ax - cx = d - b$$

$$(a - c)x = d - b$$

$$x = \frac{d - b}{a - c}$$

Compare this with equation (1).

CALCULATORS FOR THIS TEXT

If you plan to use a calculator extensively with this text, virtually any calculator that is presently being sold as a *scientific calculator* is adequate. However, some older models have deficiencies. Your calculator is satisfactory for this text if it has the following capabilities:

1. Add [+], subtract [−], multiply [×], divide [÷]
2. Natural logarithm [ln x]
3. Exponentiation [y^x] or [a^x] and [e^x] (e^x may be [inv] [ln x])
4. Scientific notation and a display of at least either eight significant digits or five significant digits plus the exponent
5. Trigonometric functions [sin], [cos], [tan]
6. Pi [π]
7. Inverse trigonometric functions [inv] or [$\sin^{-1}$], [$\cos^{-1}$], [$\tan^{-1}$]

The following are optional but useful features:

1. On algebraic logic machines, parentheses. These can be highly useful.
2. (Trigonometry) Calculators vary as to the values of x for which they will compute $\sin x$, $\cos x$, and $\tan x$. It is strongly recommended that your calculator *should* compute these functions *at least* for $0° \leq x \leq 360°$ (x in degrees) or $0 \leq x \leq 2\pi$ (x in radians).
3. A factorial button [$x!$].
4. A noticeable click when an entry is made.

If you are about to buy a calculator, you should consider the logic used on the machine. There are two basic kinds of machine logic, *algebraic* and *reverse Polish notation* (RPN). In addition, there are two kinds of algebraic calculators, standard and algebraic hierarchy. See Sec. B.1 in Appendix B for a discussion of these. Our experience seems to indicate that if you have trouble with algebra, you probably should get an algebraic calculator and that a calculator with algebraic hierarchy is more advantageous than a standard algebraic calculator. On the other hand, if you are technically oriented, you might also consider an RPN calculator. Using such a calculator requires keeping track of a *stack* but requires slightly fewer keystrokes for many computations.

Finally, if you are taking a course which requires calculators for examinations, you probably should follow these guidelines:

1. If your machine has replaceable batteries, you should probably carry an extra set to tests. If it has rechargeable batteries, make sure you fully charge them the night before a test.
2. If you have a new calculator, make sure it has had some use before

walking into a test with it. (If something is wrong with a machine, it usually appears early in the calculator's life.)

3. Different models work differently. If you have to suddenly use a different machine (for example, if you have to borrow a friend's for a test), make sure you understand its idiosyncrasies.

DIAGNOSTIC TEST

The following diagnostic test is designed to determine if you are sufficiently proficient at using your calculator if you intend to use it extensively with this text. As you do the problems, remember to *estimate first*. If you have no trouble, you are ready. Otherwise, refer to the appropriate parts of Appendix B:

- □ Go to Sec. B.1 if you have difficulty with problems 1–6.
- □ Go to Sec. B.2 if you have difficulty with problems 7–9.
- □ Go to Sec. B.3 if you have difficulty with problems 10–13.
- □ Go to Sec. B.4 if you have difficulty with problems 14–17.

Problem	*Estimate (Done in Your Head as Much as Possible)*	*Answer (To Six Significant Figures)*
1. $672.81 + 919.07$	$\approx 700 + 900 = 1600$	1591.88
2. $0.0013412 - 0.0079492$	$\approx 0.001 - 0.008 = -0.007$	-0.00660800
3. $5.932 \times (-411.8)$	$\approx 6 \times (-400) = -2400$	-2442.80
4. $0.7735 \div (-0.02134)$	$\approx 0.8 \div (-0.02) = 80 \div (-2) = -40$	-36.2465
5. $5.813 + 2.114(7.812)$	$\approx 6 + 2(8) = 22$	22.3276
6. $4.819 - 20.81 \div (2.119)$	$\approx 5 - 20 \div (2) = -5$	-5.00167
7. $2.134 \times 4.691 + 3.814 \times 4.192 + 1.992 \times 6.874$	$\approx 2(5) + 4(4) + 2(7) = 40$	39.6919
8. $(2.134 + 4.691)(3.814 + 4.192)(1.992 + 6.874)$	$\approx 7(8)(9) \approx 500$	484.447
9. $\dfrac{2.819 - 1.414}{7.196 + 3.192}$	$\approx \dfrac{3-1}{7+3} = \dfrac{2}{10} = 0.2$	0.135252
10. $5.213 + \sqrt{4.115}$	$\approx 5 + \sqrt{4} = 7$	7.24155
11. $\sqrt{5.213 + 4.115}$	$\approx \sqrt{5 + 4} = 3$	3.05418
12. $\left(1 + \dfrac{4.1}{8}\right)^2$	$\approx (1 + 0.5)^2 \approx 2$	2.28766
13. $\dfrac{1}{2.19\sqrt{4.12} - \frac{1}{9}}$	$\approx \dfrac{1}{2(2) - 0.1} \approx \dfrac{1}{4} = 0.25$	0.230728
14. $(5.128 \times 10^{21})(471.2 \times 10^{-18})$	$\approx 5 \times 10^{21} \times 5 \times 10^{-16} \approx 25 \times 10^5$	2.41631×10^6
15. $3.121^{-1.94}$	$\approx 3^{-2} = \dfrac{1}{9} \approx 0.11$	0.109918
16. $30.13^{4/3}$	$\approx (27^{1/3})^4 = 3^4 = 81$	93.7560
17. How many significant figures have the numbers (a) 173.078 (b) 4.3001×10^{18} (c) 0.00145?		(a) 6 (b) 5 (c) 3

CHAPTER 1

Right Triangle Trigonometry

The exact origins of trigonometry are lost in prehistory. All that we know is that early people became interested in astronomy for its relation to religion (and astrology), to predict the seasons and planting time, and as an aid to navigation and geography. The mathematics they developed to describe their observations in astronomy formed the beginnings of trigonometry. Indeed, trigonometry remained an appendage to astronomy until 1200–1400 A.D.

The early Babylonians divided the circle into 360 equal parts, giving us degrees, perhaps because they thought there were 360 days in a year. The number system they developed was based on the number 60 (a sexagesimal system rather than our decimal system), and because of this, they divided a degree into 60 minutes and a minute into 60 seconds. (They also divided the day into hours, the hour into 60 minutes, and the minute into 60 seconds.)

Ancient Egyptian and Babylonian scholars knew theorems about the ratios of sides of similar triangles. Both cultures used rudimentary trigonometric computations by 2000 B.C. However, the Greeks were the first to systematically study trigonometric relationships. The first significant trigonometry book was written by Ptolemy of Alexandria around the second century A.D. This book was later called the *Almagest* by the Arabs for "the greatest" and was to be the definitive text on astronomy and trigonometry until the sixteenth century.

The sine function was invented in India, perhaps around 300 to 400 A.D. However, it was really a modification of a slightly different function used by the Greeks. By the end of the ninth century A.D., all six trigonometric functions and the identities relating them were known to the Arabs.

George Rheticus (1514–1577) was the first to define trigonometric functions completely in terms of right triangles. Before him, they were defined in terms of half chords in circles.

1.1 Angles

Most people reading this already have an intuitive feeling for an angle. For completeness, we give a definition of an angle. It is one of many possible definitions of an angle and is more general than is needed for this chapter, but it will be useful to have this definition for later work.

> **(1) Definition** Suppose l and m are two half lines with a common endpoint O, L is a point on l, and M is a point on m. See Fig. 1.
>
> 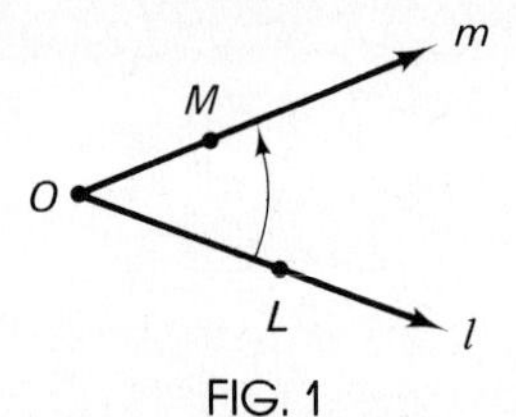
>
>
> FIG. 1
>
> Then an **angle** from OL to OM (denoted $\angle LOM$) is the amount of rotation of l about O (in the plane containing l and m) required so that l will coincide with m. The side being rotated, OL, is called the **initial side,** the other side, OM, is the **terminal side,** and the common point O is the **vertex.**

Note that the definition says "an" angle. This is because for a given configuration, the initial side may be rotated clockwise or counterclockwise, and it may be rotated completely around several times before stopping at the terminal side. See Fig. 2 for a few of the angles determined by the same configuration.

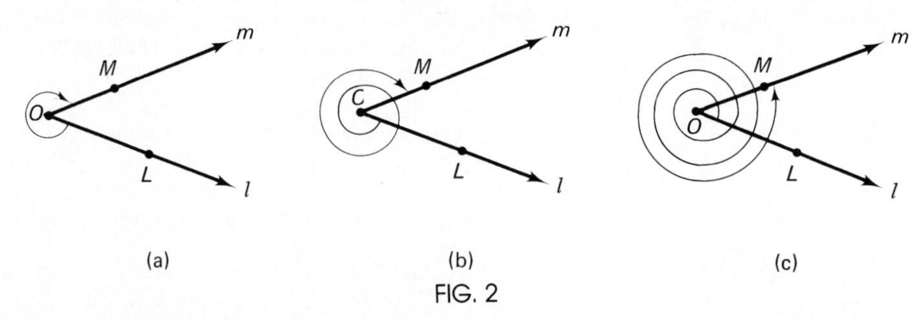

FIG. 2

If the rotation is counterclockwise as in Fig. 2(c), we say that the angle is **positive,** while if the rotation is clockwise as in Fig. 2(a) and (b), we say that the angle is **negative.**

Given an angle, we often find it convenient to measure it, so that we can compare it with other angles or do computations involving the angle. There are two widely used units for measuring angles, called degrees and radians.* People have been measuring angles in degrees since the Babylonians, but in many modern applications the radian is a more useful unit.

> **(2) Definition** One **degree,** denoted 1°, is $\frac{1}{360}$ of a complete rotation counterclockwise. If $x \geq 0$ and θ is $\frac{x}{360}$ of a complete rotation counterclockwise, we say that the degree measurement of θ is $x°$ and write $\theta = x°$. If the rotation is clockwise, $\theta = -x°$.

In other words, if a complete rotation is divided into 360 equal parts, each measuring 1°, the degree measurement of θ is the number of these parts (or fractions thereof) in θ, with the sign assigned according to our sign convention: plus for counterclockwise and minus for clockwise rotations.

EXAMPLE 1 (Rotations are counterclockwise unless stated otherwise.)

(a) If θ is $\frac{1}{2} = \frac{180}{360}$ of a complete rotation, $\theta = 180°$. This is called a **straight angle.**

(b) If θ is $\frac{1}{4} = \frac{90}{360}$ of a complete rotation, $\theta = 90°$. This is called a **right angle.**

(c) If θ is $\frac{63.21}{360}$ of a complete rotation, $\theta = 63.21°$.

(d) If θ is $\frac{6}{5} = \frac{432}{360}$ of a complete rotation, $\theta = 432°$.

(e) If θ is $\frac{142}{360}$ of a complete rotation clockwise, $\theta = -142°$. ■

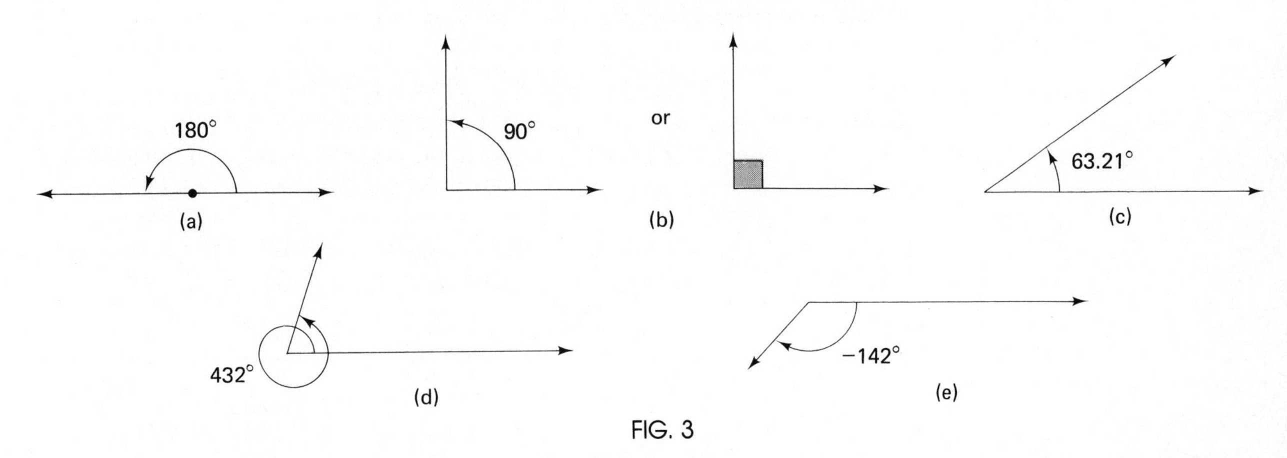

FIG. 3

*There are other units for measuring angles. For example, another unit which is used in Europe, mainly in surveying, is a **gradian** or **grad.** One grad is 1/100 of a right angle.

The degree can be further divided into minutes and seconds. One **minute,** denoted $1'$, is $\frac{1}{60}$ degree and one **second,** denoted $1''$, is $\frac{1}{60}$ minute. Thus $1°$ is broken up into minutes and seconds in exactly the same way that an hour is. With the advent of computers and hand-held calculators which work with decimal numbers, it is usually best to have degrees in decimal notation. If an angle is given in degrees, minutes, and seconds, this can easily be converted to degrees in decimal notation and vice versa.

EXAMPLE 2 Convert $6°13'18''$ to degrees in decimal notation.

Solution We know

$$1' = \left(\frac{1}{60}\right)^{\circ} \quad \text{and} \quad 1'' = \left(\frac{1}{60}\right)'$$

Combining these, we have

$$1'' = \left(\frac{1}{60}\right)' = \left(\frac{1}{60}\right)\left(\frac{1}{60}\right)^{\circ} = \left(\frac{1}{3600}\right)^{\circ}$$

Thus $$6°13'18'' = 6° + \frac{13°}{60} + \frac{18°}{3600} \approx 6.22167° \quad \blacksquare$$

EXAMPLE 3 Convert $87.4129°$ to degrees, minutes, and seconds.

Solution The $87.4129°$ is $87°$ plus a part of a degree, namely, $0.4129°$. We must convert $0.4129°$ to minutes and seconds. Since $1° = 60'$, we have

$$0.4129° = 0.4129(60') = 24.774'$$

The $24.774'$ is $24'$ plus a part of a minute, namely, $0.774'$. Since $1' = 60''$, we have

$$0.774' = 0.774(60'') = 46.44''$$

Thus $87.4129° = 87°24'46.44''$. $\blacksquare$

A second unit of measurement of angles is radians (rad).

> **(3) Definition** On the initial side l of an angle θ, let P be the point one unit from the vertex. If θ is positive or 0, let s be the length of the circular arc traced out by P as l rotates through θ; if θ is negative, let s be the negative of that length. Then the **radian measure** of θ is s, and we write
>
> $$\theta = s \text{ rad}$$

In Fig. 4(a) and (b), the arc lengths along the circle of radius 1 are 1 and s, respectively. Thus Fig. 4(a) illustrates an angle of 1 rad, while Fig. 4(b) illustrates an angle of s rad.

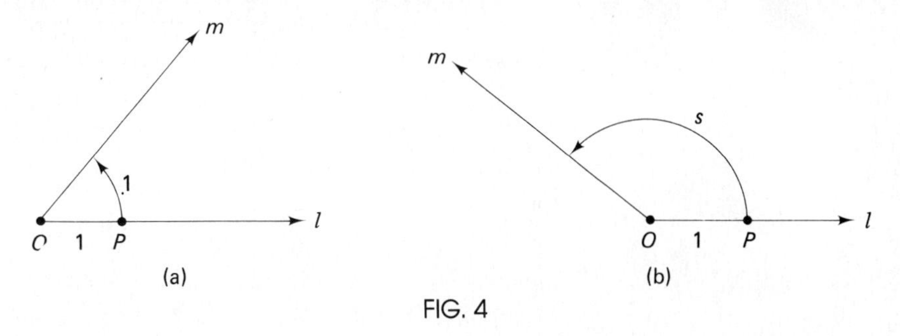

FIG. 4

Since a circle of radius r has a circumference of $2\pi r$, a circle of radius 1 (which is referred to as a unit circle) has circumference 2π. When θ is a complete rotation, P travels all around the circumference of a unit circle. This yields the following:

> If θ is a complete rotation (counterclockwise), $\theta = 2\pi$ rad.

On the other hand, we already know that one complete rotation (counterclockwise) is 360°. Consequently, $360° = 2\pi$ rad, or

(4) $$180° = \pi \text{ rad}$$

From (4), it follows that

(5) $$1° = \frac{\pi}{180} \text{ rad} \qquad \text{and} \qquad 1 \text{ rad} = \frac{180°}{\pi}$$

These are what we use to change from degrees to radians or radians to degrees. When doing conversions or any other calculations involving π, it is important to realize that π is an irrational number and hence may be expressed as an infinite nonrepeating decimal number. On a calculator, the $\boxed{\pi}$ button gives π rounded off, usually to 8 or 10 significant digits and displaying 3.1415927. You must remember that this is only an approximation.

We can use the conversion process to find out what 1 rad is in degrees and vice versa:

(6) $$1 \text{ rad} = \frac{180°}{\pi} \approx 57.2958°$$

$$\text{and } 1° = \frac{\pi}{180} \text{ rad} \approx 0.0174533 \text{ rad}$$

From this, we see that as a rough estimate 1 rad $\approx 60°$. Hence

> **(7)** For rough estimates for conversion, use either
>
> $$1 \text{ rad} \approx 60^\circ \qquad \text{or} \qquad 1^\circ \approx \frac{1}{60} \text{ rad}$$

EXAMPLE 4 Convert (a) 8.3 rad to degrees and (b) 37° to radians by first (i) estimating, using (7) and any other reasonable approximations that would allow mental arithmetic and then (ii) using (5) or (6) and multiplying.

Solution (a)

(i) $8.3 \text{ rad} \approx 8(60^\circ) = 480^\circ$

(ii) $8.3 \text{ rad} = 8.3\left(\frac{180}{\pi}\right)^\circ \approx 475.555^\circ$

(b)

(i) $37^\circ \approx 37\left(\frac{1}{60}\right) \text{ rad} \approx \frac{6}{10} \text{ rad} = 0.6 \text{ rad}$

(ii) $37^\circ = 37\left(\frac{\pi}{180}\right) \text{ rad} \approx 0.645772 \text{ rad}$ ■

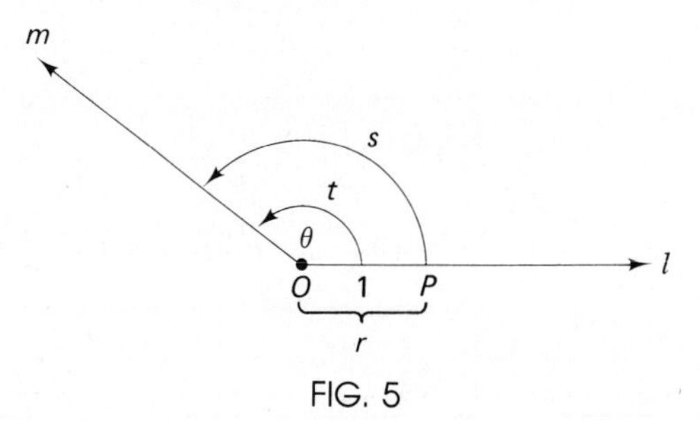

FIG. 5

Let us examine radian measure more closely. A theorem in geometry says that for a fixed angle θ, the ratio of arc length to radius is constant. Thus in Fig. 5, $\frac{t}{1} = \frac{s}{r}$. Since θ has measure t rad, we see the following:

> **(8) Theorem** Let P be a point on the initial side of an angle θ. Suppose P is r units from the vertex O. As the initial side of θ is rotated to the terminal side, suppose P traces out an arc of length s. Then
>
> $$\theta = \frac{s}{r} \text{ rad}$$

Theorem (8) leads to an interesting observation about radians. Suppose, for instance, that $s = 8$ centimeters and $r = 2$ centimeters. Then

$$\frac{s}{r} = \frac{8 \text{ cm}}{2 \text{ cm}} = 4$$

That is, the units cancel, leaving no units whatsoever. Consequently, the radian measure of an angle is usually regarded as a real number. For this reason, henceforth we shall use the following:

> **Convention** We shall write $\theta = s$ instead of $\theta = s$ rad except in places where "rad" is needed for clarity or emphasis.

The formula $\theta = \frac{s}{r}$ from Theorem (8) is very useful in solving many applied problems.

EXAMPLE 5 Find the approximate diameter of the moon if an observer on the earth measures an angle of 0.5° from the top to the bottom of the disk when the moon is 240,000 miles away.

Solution Referring to Fig. 6, let O be the eye of the observer, C be the center of the moon, r be the distance from O to C, and DE be a diameter of the moon perpendicular to OC. Finally, let A and B be the points where the circle of

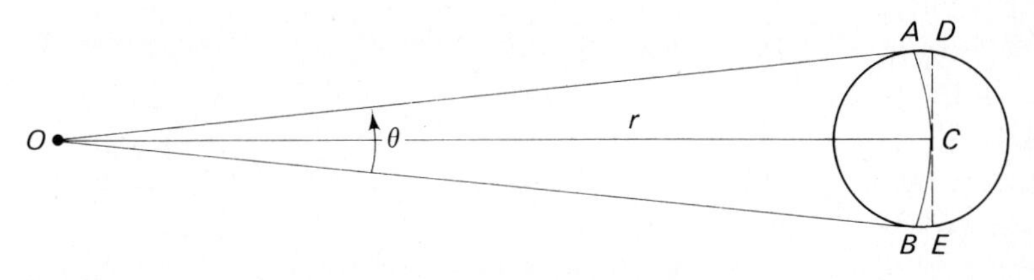

FIG. 6

radius r centered at O intersects the moon. Then r is approximately the distance from the earth to the moon, and the length of DE is approximately s, where s is the length of the arc ACB. From $\frac{s}{r} = \theta$ (rad) and since $\theta = 0.5° = 0.5\left(\frac{\pi}{180}\right)$,

$$s = r\theta = 240{,}000(0.5)\left(\frac{\pi}{180}\right)\text{mi}$$

Estimating

$$s = 240{,}000(0.5)\left(\frac{\pi}{180}\right) \approx 240{,}000\left(\frac{1}{2}\right)\left(\frac{1}{60}\right) = 2000 \text{ mi}$$

With a calculator, $s \approx 2094.395$ miles, but the distance $r = 240{,}000$ is probably accurate to only two significant figures. Thus the most we can say is that the diameter ≈ 2100 miles. ■

EXERCISES

In Exercises 1–8, find the degree and radian measure of θ.

1. θ is $\frac{4}{3}$ of a complete rotation (counterclockwise).
2. θ is $\frac{3}{4}$ of a complete rotation (counterclockwise).
3. θ is $\frac{3}{2}$ of a complete rotation (counterclockwise).
4. θ is $\frac{5}{6}$ of a complete rotation (counterclockwise).
5. θ is $\frac{1}{3}$ of a complete rotation (clockwise).
6. θ is $\frac{1}{5}$ of a complete rotation (clockwise).
7. θ is $\frac{8}{9}$ of a complete rotation (clockwise).
8. θ is $\frac{8}{3}$ of a complete rotation (clockwise).

In Exercises 9–20, convert the given angle to degrees in decimal notation.

9. $14°25'12''$
10. $434°43'48''$
11. $-24°15'$
12. $27°43'12''$
13. $231°32'$
14. $-38°11'$
Ⓒ**15.** $84°6'7''$
Ⓒ**16.** $5°5'5''$
Ⓒ**17.** $-132°41'40''$
Ⓒ**18.** $189°17''$
Ⓒ**19.** $1227°4'38''$
Ⓒ**20.** $-625°10'53''$

In Exercises 21–26, convert the given angle to degrees, minutes, and seconds.

21. $9.46°$
22. $106.53°$
23. $-75.6135°$
24. $804.06°$
25. $431.98°$
26. $-147.317°$

In Exercises 27–42, convert the angle from degrees to radians. First estimate your answer.

27. $135°$
28. $15°$
29. $75°$
30. $-30°$
31. $120°$
32. $270°$
33. $180°$
34. $-45°$
35. $-85.4°$
36. $18°$
Ⓒ**37.** $311°14'19''$
Ⓒ**38.** $-160.41°$
Ⓒ**39.** $-46°12'$
Ⓒ**40.** $52°28'48''$
Ⓒ**41.** $861.428°$
Ⓒ**42.** $74°26'24''$

In Exercises 43–58, convert the angle from radians to degrees in decimal notation. Estimate first.

43. $\frac{7\pi}{6}$ rad
44. $\frac{\pi}{6}$ rad
45. $\frac{5\pi}{4}$ rad
46. $-\frac{\pi}{4}$ rad
47. 9π rad
48. $\frac{\pi}{3}$ rad

49. $\frac{5\pi}{2}$ rad

50. $-\frac{\pi}{2}$ rad

51. 3 rad

52. 2 rad

53. 0.8 rad

54. -1 rad

55. 8.2 rad

56. 0.4 rad

57. 0.349 rad

58. $-\frac{2\pi}{3}$ rad

In Exercises 59–66, convert the angle from radians to degrees, minutes, and seconds.

59. 2.5 rad

60. 4.13 rad

61. 0.682 rad

62. 13.9 rad

63. -1.6 rad

64. -3.07 rad

65. -0.091 rad

66. -22 rad

67. The length of an arc of a circle is 7.00 centimeters. Find the angle (in degrees and radians) it makes at the center of the circle if the radius is 4.00 centimeters. Give the answer to three significant digits.

68. The arc of a circle 10.00 centimeters long makes an angle of 36° at the center. Find the radius to three significant digits.

69. An arc of a circle makes an angle of 170° at the center. If the radius of the circle is 16 meters, find the length of the arc to three significant digits.

70. A wheel makes 120 revolutions in 1 minute. In 1 second, through how many radians and how many degrees does it turn?

71. The large hand on a wall clock is 1 foot, 7 inches long. In 20 minutes
a. Through what angle (in both degrees and radians) does the hand move?
b. How many inches does its tip move?

72. The hour hand on a clock is 10 inches long. In a 50-minute class period,
a. Through what angle (in both degrees and radians) does the hand move?
b. How far does its tip move?

73. The second hand on a wristwatch is $\frac{3}{4}$ inch long. In one second,
a. Through what angle (in both degrees and radians) does the hand move?
b. How far does its tip move?

74. Your horse on a merry-go-round is 18.3 meters from the center. Suppose the merry-go-round makes 12.4 revolutions per ride. If you take three rides,
a. Through what angle do you move?
b. How many meters do you travel?

75. The height of a mountain 10 miles away makes an angle of 3.2° at the eye of an observer. How high is the mountain to the nearest 10 feet?

76. Suppose that the sun is 93 million miles away from the earth and that its diameter makes an angle of 32′ at a point on the earth. What is the approximate diameter of the sun?

77. A satellite is in a circular orbit 1000 miles above the surface of the earth. The radius of the earth is about 4000 miles. How many miles (to two significant figures) does the satellite travel while sweeping an angle of 40° with the center of the earth?

78. If the satellite in Exercise 77 makes 1 revolution every 80 minutes, how fast (in miles per hour) is it traveling?

79. Suppose a bike wheel has a radius of 27 inches. If the bike is rolling at 10 miles per hour, through what angle does a spoke turn in 1 minute?

80. If the bike wheel in Exercise 79 made 200 revolutions in a minute, how fast was the bike traveling?

1.2 The Trigonometric Functions

The trigonometric functions were originally defined as functions of angles in right triangles and then extended to more general situations. To follow this approach, we need to review briefly some facts about triangles. In working with triangles, we use a standard labeling, with side a opposite angle A, etc. See Fig. 7.

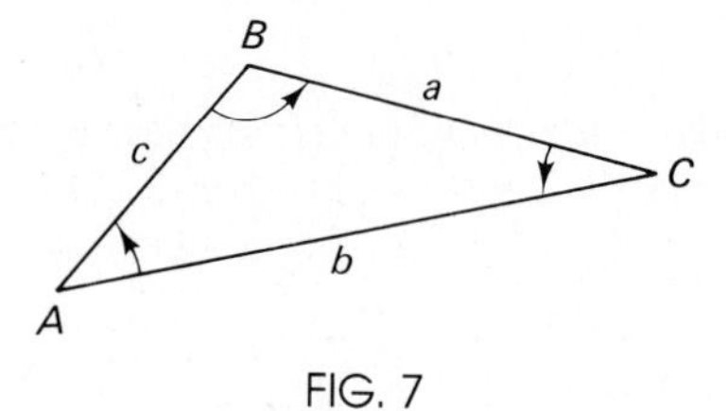

FIG. 7

We must clarify what we mean by an angle in a triangle. An angle in a triangle is obtained by rotating the initial side counterclockwise through the interior of the triangle until the first time it coincides with the terminal side. Thus the three angles are each positive. See Fig. 7. Moreover, the following is true (see Exercise 35 for a proof):

> The sum of the three angles of any triangle is $180° = \pi$ rad.

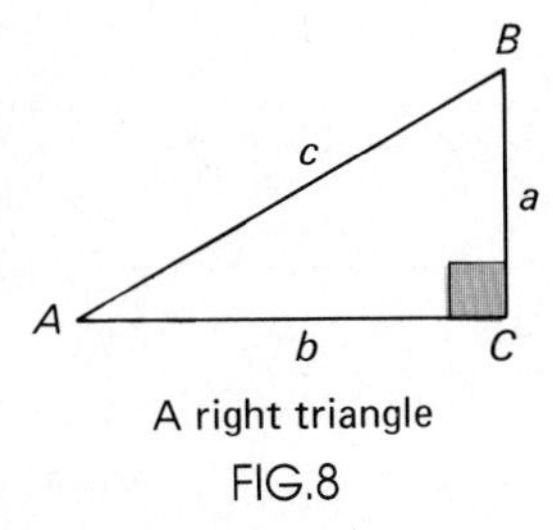

A right triangle

FIG.8

If one of the angles of a triangle is a right angle, the triangle is called a **right triangle.** See Fig. 8. The sides of a right triangle adjacent to the right angle are called **legs** of the triangle, while the side opposite the right angle is called the **hypotenuse** of the triangle. In any right triangle, the Pythagorean theorem holds.*

> **Pythagorean Theorem** In a right triangle with legs of lengths a and b and hypotenuse of length c (as in Fig. 8),
>
> $$a^2 + b^2 = c^2$$

Now let us consider the two right triangles in Fig. 9. Suppose $\angle A = \angle A'$. Then since $\angle C = \angle C' = \frac{\pi}{2}$ (rad) and $\angle A + \angle B + \angle C = \pi = \angle A' + \angle B' + \angle C'$ (the sum of the angles of any triangle is π), it follows that $\angle B = \angle B'$. Thus if $\angle A = \angle A'$, the two right triangles have precisely the

*See Exercise 34 for a proof. Also, see the *Pythagorean Proposition* by E. S. Loomis, NCTM, Ann Arbor, MI, 1968, which contains 367 different proofs of the Pythagorean theorem.

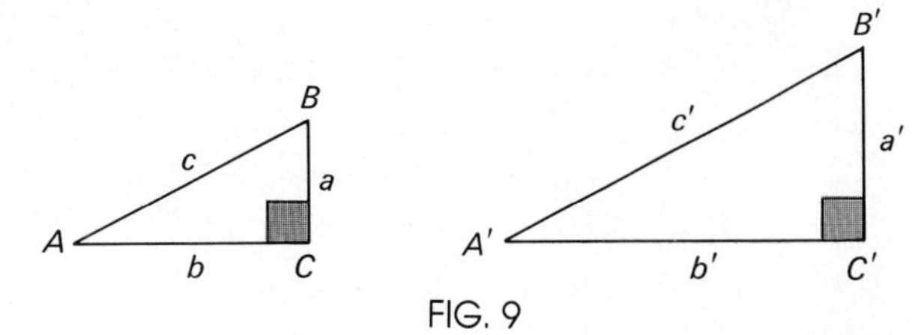

FIG. 9

same three angles. Two triangles with the same three angles are called **similar,** and it is known that in similar triangles the ratios of the lengths of corresponding sides are the same. Thus in our example

$$\frac{a}{c} = \frac{a'}{c'}, \qquad \frac{b}{c} = \frac{b'}{c'}, \qquad \frac{a}{b} = \frac{a'}{b'}, \qquad \text{etc.}$$

If we know the sides of one triangle and one side of a second triangle which is similar to the first, we can use the ratios of corresponding sides to find the remaining sides of the second triangle.

EXAMPLE 1 Find the coordinates of P if P is one unit from the origin along the line from the origin to $(-4, 3)$. See Fig. 10(a).

Solution If we drop perpendiculars from $(-4, 3)$ and P to the x-axis, we obtain right triangles with the *lengths* of the sides indicated in Fig. 10(b). By the Pythag-

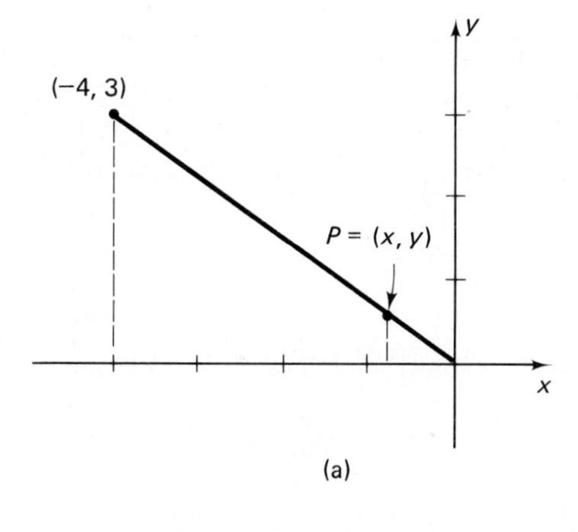

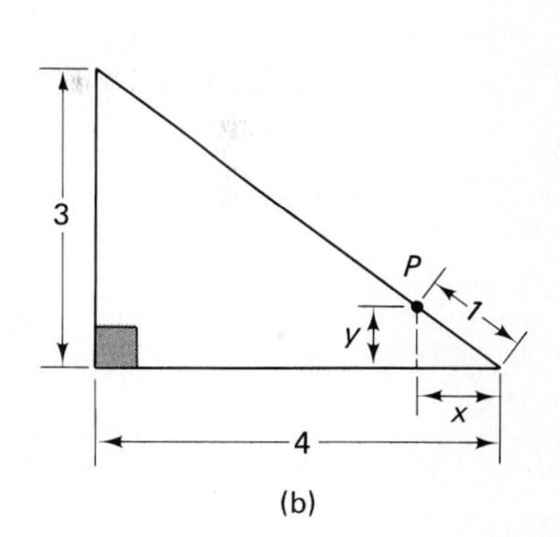

FIG. 10

orean theorem, the larger triangle has hypotenuse of length 5. Since the two triangles are similar,

$$\frac{|y|}{1} = \frac{3}{5} \quad \text{or} \quad |y| = \frac{3}{5} \qquad \text{and} \qquad \frac{|x|}{1} = \frac{4}{5} \quad \text{or} \quad |x| = \frac{4}{5}$$

Since (x, y) is in the second quadrant, $(x, y) = (-\frac{4}{5}, \frac{3}{5})$. ■

We are now ready to define the trigonometric functions of any acute angle θ. Let A be the vertex of θ. Pick any point B other than A on the terminal side of θ and drop a perpendicular from B to the initial side of θ, labeling the point of intersection as C. See Fig. 11(a). We now have a right triangle ABC with sides a, b, c. By the discussion preceding Example 1, we know that the six ratios $\frac{a}{c}, \frac{b}{c}, \frac{a}{b}, \frac{b}{a}, \frac{c}{b}$, and $\frac{c}{a}$ are independent of what point B we picked on the terminal side of θ. Thus the ratios depend only on the size of θ; i.e., they are a function of θ alone. These six functions are the **trigonometric functions** and are called the **sine** of θ (denoted by $\sin\theta$), the **cosine** of θ (denoted by $\cos\theta$), the **tangent** of θ (denoted by $\tan\theta$), the **cotangent** of θ (denoted by $\cot\theta$), the **secant** of θ (denoted by $\sec\theta$), and the **cosecant** of θ (denoted by $\csc\theta$).

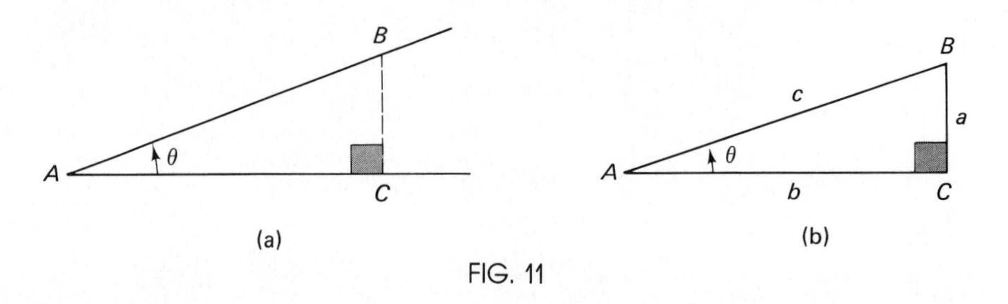

FIG. 11

The trigonometric functions are defined as follows [referring to Fig. 11(b)]:

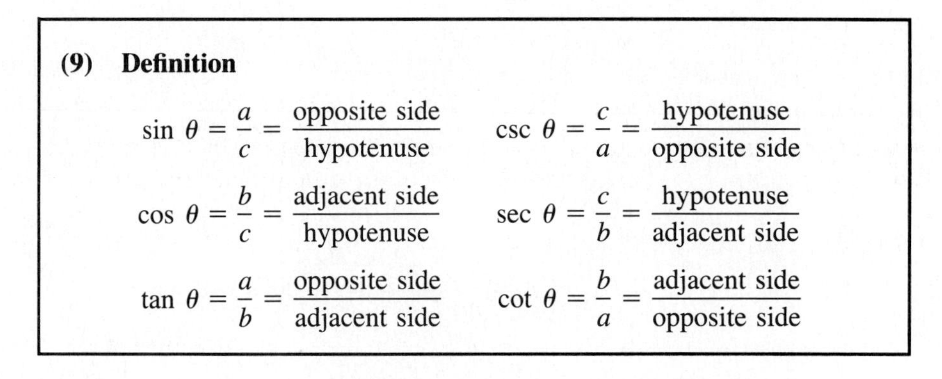

(9) Definition

$$\sin\theta = \frac{a}{c} = \frac{\text{opposite side}}{\text{hypotenuse}} \qquad \csc\theta = \frac{c}{a} = \frac{\text{hypotenuse}}{\text{opposite side}}$$

$$\cos\theta = \frac{b}{c} = \frac{\text{adjacent side}}{\text{hypotenuse}} \qquad \sec\theta = \frac{c}{b} = \frac{\text{hypotenuse}}{\text{adjacent side}}$$

$$\tan\theta = \frac{a}{b} = \frac{\text{opposite side}}{\text{adjacent side}} \qquad \cot\theta = \frac{b}{a} = \frac{\text{adjacent side}}{\text{opposite side}}$$

EXAMPLE 2 In each of the following, find the six trigonometric functions of θ:

(a) (b)

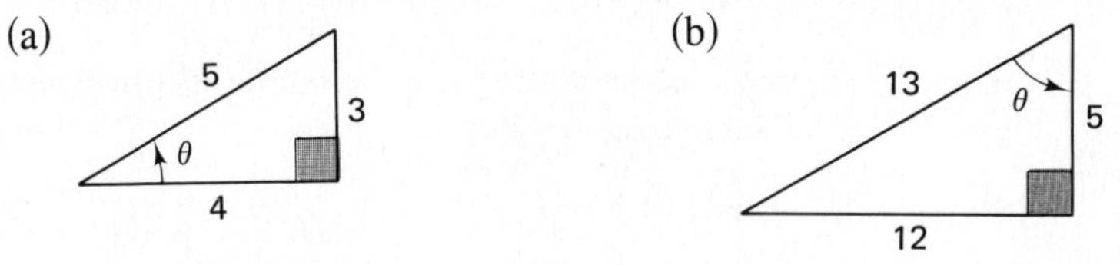

Solution We apply the definitions (9):

(a) $\sin\theta = \frac{3}{5}$ $\csc\theta = \frac{5}{3}$ (b) $\sin\theta = \frac{12}{13}$ $\csc\theta = \frac{13}{12}$

$\cos\theta = \frac{4}{5}$ $\sec\theta = \frac{5}{4}$ $\cos\theta = \frac{5}{13}$ $\sec\theta = \frac{13}{5}$

$\tan\theta = \frac{3}{4}$ $\cot\theta = \frac{4}{3}$ $\tan\theta = \frac{12}{5}$ $\cot\theta = \frac{5}{12}$ ■

From the definitions (9), it is immediate that there are relationships among the trigonometric functions. For instance, each of the functions defined on the same line of (9) is clearly the reciprocal of the other. Thus

(10) $$\csc\theta = \frac{1}{\sin\theta}, \qquad \sec\theta = \frac{1}{\cos\theta}, \qquad \text{and} \qquad \cot\theta = \frac{1}{\tan\theta}$$

Moreover, $\dfrac{\sin\theta}{\cos\theta} = \dfrac{a/c}{b/c} = \dfrac{a}{b} = \tan\theta$. Since in addition $\cot\theta = \dfrac{1}{\tan\theta}$, we have the following:

(11) $$\tan\theta = \frac{\sin\theta}{\cos\theta} \qquad \text{and} \qquad \cot\theta = \frac{\cos\theta}{\sin\theta}$$

One further relation follows directly from the Pythagorean theorem: $a^2 + b^2 = c^2$. If both sides are divided by c^2, one obtains $\left(\dfrac{a}{c}\right)^2 + \left(\dfrac{b}{c}\right)^2 = 1$. By using $\sin\theta = \dfrac{a}{c}$ and $\cos\theta = \dfrac{b}{c}$, the following holds:

(12) $$\sin^2\theta + \cos^2\theta = 1$$

Note: In the last equation we have used a convention regarding powers of trigonometric functions. When raising a trigonometric function, say $\sin\theta$, to a power n, we write $\sin^n\theta$ instead of $(\sin\theta)^n$ whenever $n \neq -1$. [We shall see later that the notations $\sin^{-1}\theta$, $\cos^{-1}\theta$, etc., have a special meaning. To denote $\dfrac{1}{\sin\theta}$, $\dfrac{1}{\cos\theta}$, etc., using exponents, write $(\sin\theta)^{-1}$, $(\cos\theta)^{-1}$, etc.]

Equations (10), (11), and (12) are called the **elementary trigonometric identities.** You can sometimes use these identities to find the remaining trigonometric functions from one of them.

EXAMPLE 3 If θ is an angle in a right triangle and $\cos\theta = \frac{3}{5}$, find the remaining trigonometric functions of θ.

Solution Since $\cos\theta = \frac{3}{5}$, from (10) we have

$$\sec\theta = \frac{1}{\cos\theta} = \frac{5}{3}$$

From $\sin^2\theta + \cos^2\theta = 1$, identity (12), we have

$$\sin^2\theta + \tfrac{9}{25} = 1 \quad \text{or} \quad \sin^2\theta = \tfrac{16}{25}, \qquad \sin\theta = \pm\tfrac{4}{5}$$

Since θ is in a right triangle, $\sin\theta$ is positive, so $\sin\theta = \frac{4}{5}$. Then from (10),

$$\csc\theta = \frac{1}{\sin\theta} = \frac{5}{4}$$

Finally, from (11),

$$\tan\theta = \frac{\sin\theta}{\cos\theta} = \frac{4/5}{3/5} = \frac{4}{3} \quad \text{and} \quad \cot\theta = \frac{\cos\theta}{\sin\theta} = \frac{3}{4}$$

Alternate Solution The same problem may be done directly from the definitions of the trigonometric functions and the Pythagorean theorem. Since θ is an angle in a right triangle and $\cos\theta = \frac{3}{5}$, we can look at any right triangle with one angle θ in which the ratio $\frac{b}{c} = \frac{3}{5}$. The easiest is to choose $b = 3$ and $c = 5$. Then $a^2 + b^2 = c^2$, or $a^2 + 9 = 25$, or $a^2 = 16$. Thus, $a = 4$ (not $a = \pm 4$ since a length is positive). Consequently,

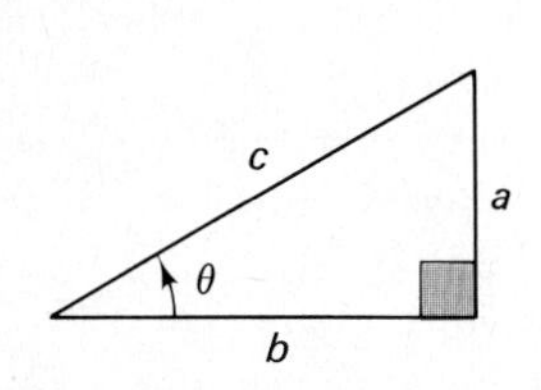

$$\sin\theta = \frac{a}{c} = \frac{4}{5}, \qquad \tan\theta = \frac{a}{b} = \frac{4}{3}, \qquad \csc\theta = \frac{c}{a} = \frac{5}{4},$$

$$\sec\theta = \frac{c}{b} = \frac{5}{3}, \qquad \text{and} \qquad \cot\theta = \frac{b}{a} = \frac{3}{4} \qquad ■$$

Trigonometry can be extremely useful in solving word problems which involve triangles. Let us look at an example.

EXAMPLE 4 From a point 80 feet from the base of a smokestack along level ground, it is observed that the angle between the ground and a line drawn from the point to the top of the smokestack has tangent equal to 0.635. How tall is the smokestack?

Solution First we draw a picture as shown. We are given that $\tan\theta = 0.635$. However, since $\tan\theta = \frac{h}{80}$,

$$h = 80\tan\theta = 80(0.635) = 50.8 \text{ feet}$$

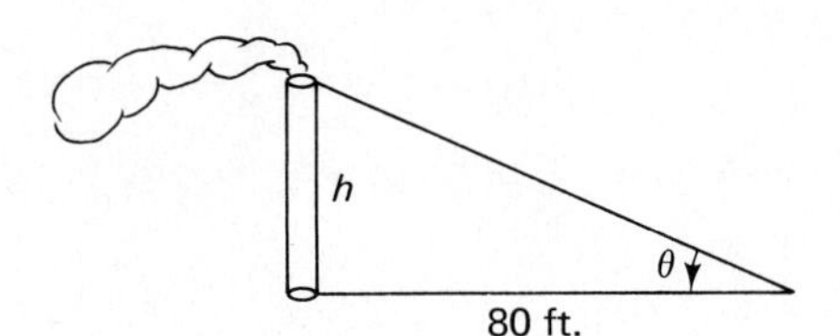

■

EXERCISES

In Exercises 1–6, assume the following triangles are similar:

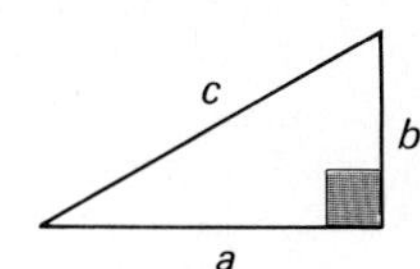

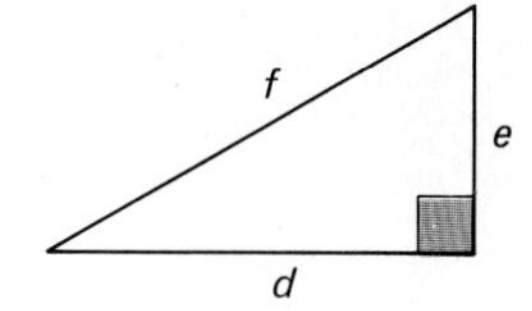

1. If $a = 4$, $b = 3$, and $d = 8$, find c, e, f, and then find $\frac{a}{c}$ and $\frac{d}{f}$.

2. If $a = 3$, $c = 5$, and $e = 10$, find b, d, f, and then find $\frac{b}{c}$ and $\frac{e}{f}$.

3. If $d = 36$, $e = 15$, and $a = 12$, find f, b, c, and then find $\frac{b}{a}$ and $\frac{e}{d}$.

4. If $a = 24$, $c = 25$, and $d = 36$, find b, e, f, and then find $\frac{a}{c}$ and $\frac{d}{f}$.

5. If $a = 2$, $b = 3$, and $f = 12$, find c, d, e, and then find $\frac{a}{b}$ and $\frac{d}{e}$.

6. If $e = 3$, $f = 4$, and $c = 96$, find a, b, d, and then find $\frac{f}{e}$ and $\frac{c}{b}$.

7. Find the coordinates of a point P if P is one unit from the origin along the line from the origin to $Q = (-3, 4)$.

8. Find the coordinates of a point P if P is one unit from $Q = (-5, -12)$ along the line from the origin to Q.

9. Find the coordinates of a point P if P is along the line from the origin through $Q = (2, -4)$ and P is twice as far from the origin as Q is.

10. If P has x-coordinate -7 and is along the line from the origin through $Q = (-\frac{4}{5}, \frac{3}{5})$, how far is P from the origin?

In Exercises 11–18, find the six trigonometric functions of θ.

11.

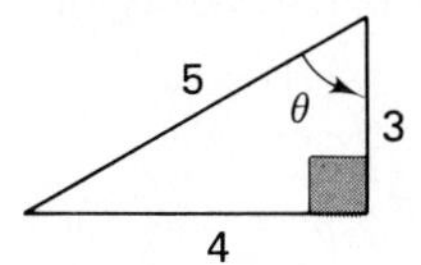

12.

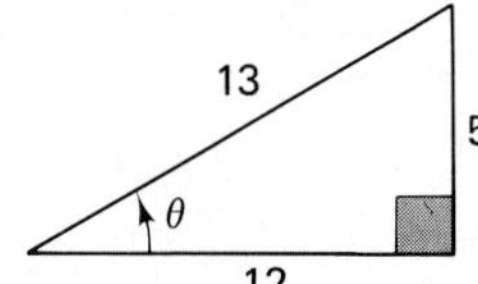

13.

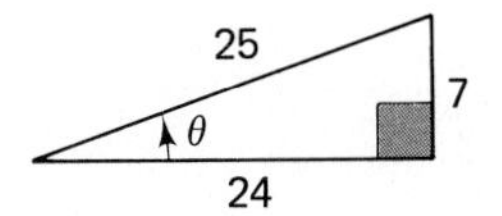

14.

15.

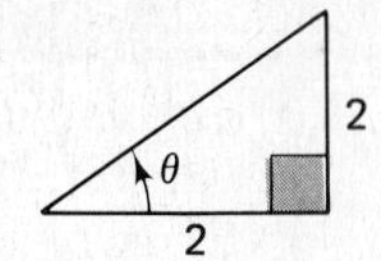

16.

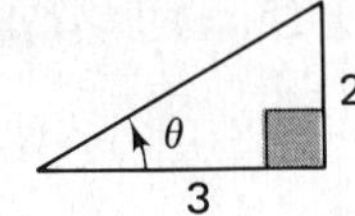

17.

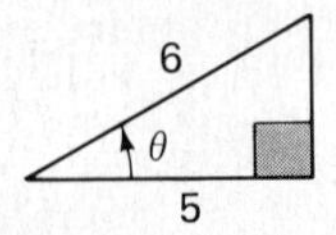

18.

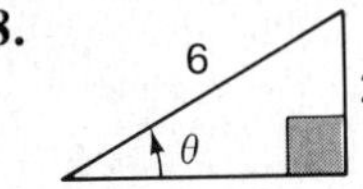

Find the trigonometric functions of θ in Exercises 19–26. θ is an angle in a right triangle.

19. $\sin \theta = \frac{12}{13}$

20. $\tan \theta = \frac{3}{4}$

21. $\cot \theta = 1$

22. $\cos \theta = \frac{3}{4}$

23. $\cos \theta = \frac{1}{2}$

24. $\sin \theta = \frac{1}{2}$

25. $\sin \theta = \frac{2}{3}$

26. $\tan \theta = \frac{3}{2}$

27. Find two further relationships like equation (12) among the trigonometric functions by dividing the Pythagorean relation $a^2 + b^2 = c^2$ first by a^2 and then by b^2.

In Exercises 28– 33, find the answers to three significant digits.

28. Suppose a 12-meter ladder is resting against a wall with the top of the ladder at the top of the wall. Suppose the ground, the wall, and the ladder form a right triangle. If θ is the angle between the ground and the ladder and $\sin \theta = 0.623$, how tall is the wall, and how far from the wall is the base of the ladder?

29. One leg of a triangle is 9.2 feet long. The tangent of the angle between the other leg and the hypotenuse is 1.56. How long is the other leg?

30. John and Diane are on the edges of opposite sides of a canyon. Nancy is on the same edge as Diane, 20 meters from her. A line from John to Diane would be perpendicular to a line from Diane to Nancy. The angle between the line from Nancy to Diane and the line from Nancy to John has tangent 1.437. How far apart are John and Diane?

31. A telephone pole is perpendicular to the ground. A taut anchoring cable 60 feet long runs from the ground to the top of the pole (so that the ground, cable, and pole form a right triangle). If the tangent of the angle between the ground and the cable is 2.6, how tall is the pole (to the nearest foot)?

32. A 3-meter ramp is put from a level driveway to a loading dock. If the cosine of the angle between the ramp and the driveway is 0.95, how high is the loading dock?

33. A 6-foot electric cord is stretched taut from the bottom of a clock on a wall to a socket on the same wall. The socket is 1 foot above the floor. If you drew a line from the clock straight down to the floor, it would make an angle with the cord whose cosine is 0.9. How high above the floor is the bottom of the clock?

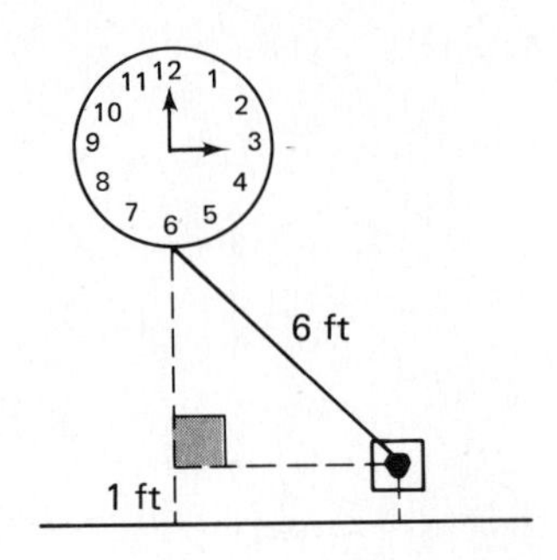

34. Prove the Pythagorean theorem which says that in a right triangle labeled as shown, $a^2 + b^2 = c^2$. Do so by computing the area of the accompanying large square in two ways: first in the normal way and then as the sum of the areas of the four copies of the triangle and the inner square. Set the two equal, and see what happens.

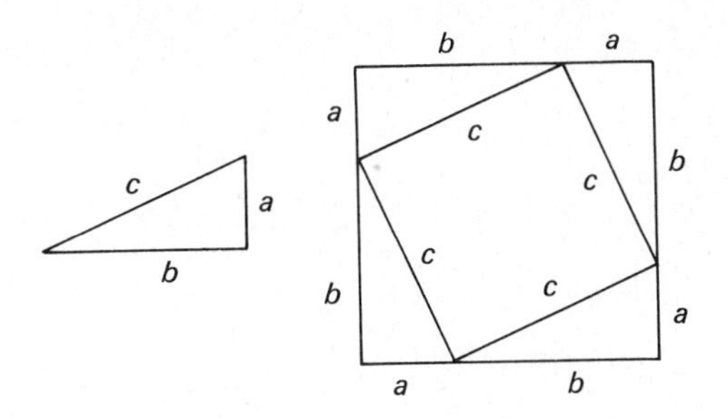

35. Give a proof that the sum of the angles in a triangle is 180° as follows. Given any triangle ABC, draw a line through the vertex C parallel to side AB and label as shown. State the sum of $\theta_5 + \theta_3 + \theta_4$, giving a reason. Give a reason why $\theta_5 = \theta_1$. Give a reason why $\theta_4 = \theta_2$. What can you conclude about $\theta_1 + \theta_2 + \theta_3$?

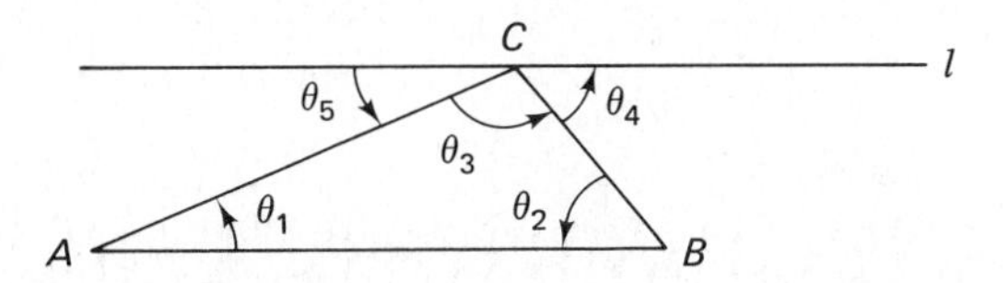

1.3 Evaluation of Trigonometric Functions

In this section we consider how to compute the values of the trigonometric functions of an acute angle θ when you are given just the angle. This is relatively easy if $\theta = \frac{\pi}{6} = 30°$, $\theta = \frac{\pi}{4} = 45°$, or $\theta = \frac{\pi}{3} = 60°$. Because of this and the fact that they arise fairly often, these angles are called **special angles.** (Observe that we are following our convention of not writing "rad" if radians is the measure but *always writing the symbol* ° if degrees is the measure.)

TRIGONOMETRIC FUNCTIONS OF $\frac{\pi}{4} = 45°$

If one acute angle of a right triangle has measure $\frac{\pi}{4} = 45°$, then so does the other since the sum of the angles in any triangle is $\pi = 180°$. Any triangle having two equal angles is called **isosceles;** the sides opposite the equal angles are also equal. In the present case, denote the equal sides by a. Then the hypotenuse c can be found by the Pythagorean theorem, $c^2 = a^2 + a^2 = 2a^2$, or $c = \sqrt{2}\,a$. We could now compute the trigonometric ratios, but for any two choices of a positive number for a, the resulting triangles would be similar and the corresponding trigonometric ratios the same. Consequently, for convenience we choose $a = 1$ (which means $c = \sqrt{2}$). See Fig. 12. Then

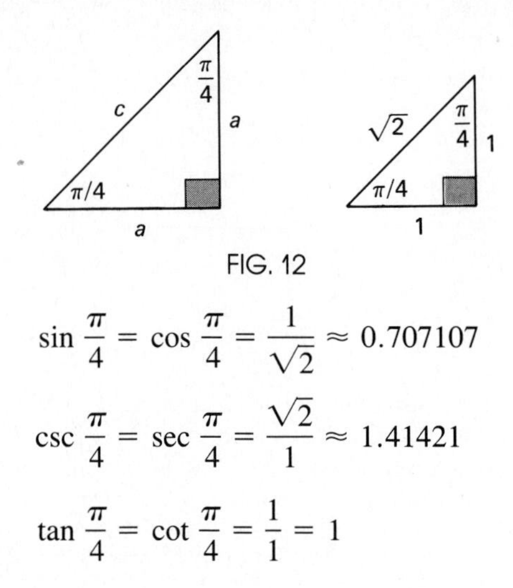

FIG. 12

$$\sin\frac{\pi}{4} = \cos\frac{\pi}{4} = \frac{1}{\sqrt{2}} \approx 0.707107$$

$$\csc\frac{\pi}{4} = \sec\frac{\pi}{4} = \frac{\sqrt{2}}{1} \approx 1.41421$$

$$\tan\frac{\pi}{4} = \cot\frac{\pi}{4} = \frac{1}{1} = 1$$

Remark: Although most of you have been taught to rationalize denominators, we shall not emphasize that here. This is because when evaluating on the calculator, it is quicker to evaluate $\frac{1}{\sqrt{2}}$ (press [2] [√] [1/x]) than to evaluate $\frac{\sqrt{2}}{2}$ (press on alg.: [2] [√] [÷] [2] [=]).

TRIGONOMETRIC FUNCTIONS OF $\frac{\pi}{6}$ = 30° AND $\frac{\pi}{3}$ = 60°

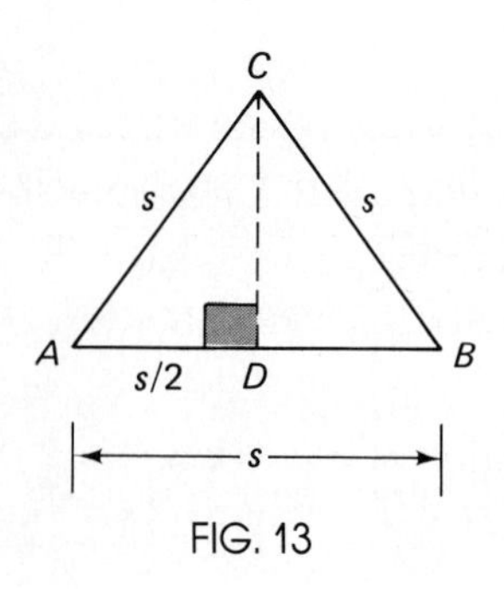

FIG. 13

If one acute angle of a right triangle has measure $\frac{\pi}{6}$, then the other acute angle has measure $\frac{\pi}{3}$. Consequently, we can use the same right triangle to compute the trigonometric functions of both angles. To see how the sides of such a triangle are related, we start with an equilateral triangle. See Fig. 13. Each of the angles of triangle ABC measures $\frac{\pi}{3} = 60°$, and each of the sides has equal length s. The altitude CD bisects the side AB and the angle $\angle ACB$. Thus triangle CAD is a $\frac{\pi}{6}, \frac{\pi}{3}, \frac{\pi}{2}$ right triangle (30°, 60°, 90° right triangle). To compute the trigonometric functions, we may assign any nonzero value to s, since for all such values the triangles are similar, and hence corresponding ratios are the same. For convenience, we assume $s = 2$. Then the length of AD, which we denote by $|AD|$, is $|AD| = \frac{s}{2} = 1$ (the side opposite the 30° angle in a 30°, 60°, 90° right triangle is half as long as the hypotenuse), and $|CD| = \sqrt{2^2 - 1^2} = \sqrt{3}$. See Fig. 14.

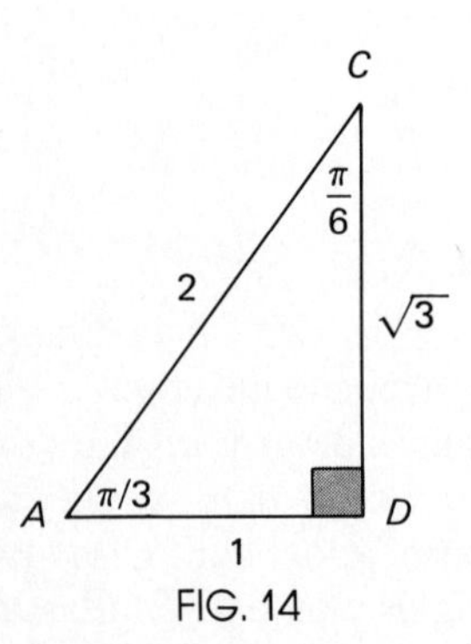

FIG. 14

Thus

$$\sin\frac{\pi}{3} = \frac{\sqrt{3}}{2} \approx 0.866025 \qquad \csc\frac{\pi}{3} = \frac{2}{\sqrt{3}} \approx 1.15470$$

$$\cos\frac{\pi}{3} = \frac{1}{2} = 0.5 \qquad \sec\frac{\pi}{3} = 2$$

$$\tan\frac{\pi}{3} = \sqrt{3} \approx 1.73205 \qquad \cot\frac{\pi}{3} = \frac{1}{\sqrt{3}} \approx 0.577350$$

and

$$\sin\frac{\pi}{6} = \frac{1}{2} = 0.5 \qquad \csc\frac{\pi}{6} = 2$$

$$\cos\frac{\pi}{6} = \frac{\sqrt{3}}{2} \approx 0.866025 \qquad \sec\frac{\pi}{6} = \frac{2}{\sqrt{3}} \approx 1.15470$$

$$\tan\frac{\pi}{6} = \frac{1}{\sqrt{3}} \approx 0.577350 \qquad \cot\frac{\pi}{6} = \sqrt{3} \approx 1.73205$$

We now know how to compute the trigonometric functions of the special angles $\frac{\pi}{4}$, $\frac{\pi}{3}$, and $\frac{\pi}{6}$. These occur so frequently that you should either memorize them or, better yet, *memorize the triangles from which they are obtained* (Figs. 12 and 14).

EVALUATION OF TRIGONOMETRIC FUNCTIONS

Unfortunately, in general if you are given an angle θ, there is no easy way to compute the trigonometric functions of θ directly, but there are sophisticated techniques for doing this that require many calculations. For many years, extensive tables of the trigonometric functions of various angles have been provided so that it would be practical to use the functions. Nowadays these functions are built into scientific calculators. It is much easier and more accurate to evaluate trigonometric functions using calculators than using tables. However, there are pedagogical reasons why some instructors prefer to teach trigonometry using tables. Consequently, this text is written so that you can use either calculators or tables.

USING CALCULATORS

Most scientific calculators have a switch or a button that allows you to place the calculator in degree or radian mode. If the calculator is in degree mode, all calculations involving trigonometric functions are performed with the assumption all angles are in degrees. Similarly, in radian mode, the same calculations would be performed with the assumption all angles are in radians. *Make sure you have read the calculator manual and know how to put the calculator into both radian and degree mode.*

> **Warning** A very common error that occurs when using calculators in trigonometry is to have the calculator in the wrong mode. If you are getting incorrect answers, the first thing to check is that your calculator is in the correct mode.

Once the calculator is in the appropriate mode, the value of sin θ, cos θ, or tan θ is found in the same way on most machines, whether RPN or algebraic: You enter θ ***first,*** and then press the [sin], [cos], or [tan] button, ***not*** the other way around.

EXAMPLE 1 ⓒIn (a)–(c), the angles are measured in degrees, so the calculator should be in degree mode.

(a) Find sin 23°. Press [23] [sin], obtaining 0.390731.

(b) Find tan 54°. Press [54] [tan], obtaining 1.37638.

(c) Find $\sin^2 18° + 2 \tan 37° \cos 8°$. Using algebraic without parentheses, press [8] [cos] [×] [37] [tan] [×] [2] [+] [18] [sin] [x^2] [=], obtaining 1.58793.

In (d) and (e), the angles are measured in radians, so your calculator should be in radian mode.

(d) Find sin 1. Press [1] [sin], obtaining 0.841471.

(e) Find $3 \cos^4 0.83$. Using algebraic, press [.83] [cos] [x^y] [4] [×] [3] [=], obtaining 0.622324. ■

Calculators do not have buttons for all the trigonometric functions. Instead they have only [sin], [cos], and [tan] buttons. This is because these are the most commonly used trigonometric functions, and we have seen that the remaining trigonometric functions can be obtained by using the identities (10):

$$\csc \theta = \frac{1}{\sin \theta}, \quad \sec \theta = \frac{1}{\cos \theta}, \quad \text{and } \cot \theta = \frac{1}{\tan \theta}.$$

EXAMPLE 2 ⓒFind csc 31°. Since $\csc 31° = \dfrac{1}{\sin 31°}$, in degree mode press [31] [sin] [$1/x$], obtaining 1.94160. ■

USING TABLES

Appendix D contains tables of trigonometric functions evaluated in degrees or in radians. There are also ample instructions as to how to use these tables. If you intend to use tables (rather than, or in addition to, calculators), this would be an appropriate time to read Appendix D. Before continuing, you should be able to use the tables to determine

$\tan 54° \approx 1.376$	$\csc 31° \approx 1.942$
$\sin 18.2° \approx 0.3123$	$\cos 1.12 \approx 0.4357$
$\tan 0.918 \approx 1.308$	$\cot 1.293 \approx 0.2852$

When you are using tables, do not forget they are in general accurate to at most four significant figures. Hence the answers to your computations in examples and exercises may differ somewhat from the book's answers, which were computed using a calculator.

AN APPLICATION

Let us look at an example to begin to see how knowledge of the trigonometric functions of the angles of a right triangle can be useful. First we define some terms that occur frequently. If an observer looks at an object that is higher than eye level, it is necessary for the observer to *elevate* his or her eyes to look straight at the object. Suppose we draw a horizontal line through the eyes of the observer and another line through the eyes of the observer and the object (i.e., along the line of sight of the observer). Then the angle between the two lines is the **angle of elevation** of the object from the observer. See Fig. 15. Similarly, if the object is below the observer, it is necessary for the observer to *depress* his or her eyes to look straight at the object. If we draw the same two lines (the horizontal line and the line of sight), then the angle between the two lines is the **angle of depression** between the observer and the object. See Fig. 15. Remember, *elevation* means that the object is higher than the observation point, while *depression* means the object is lower.

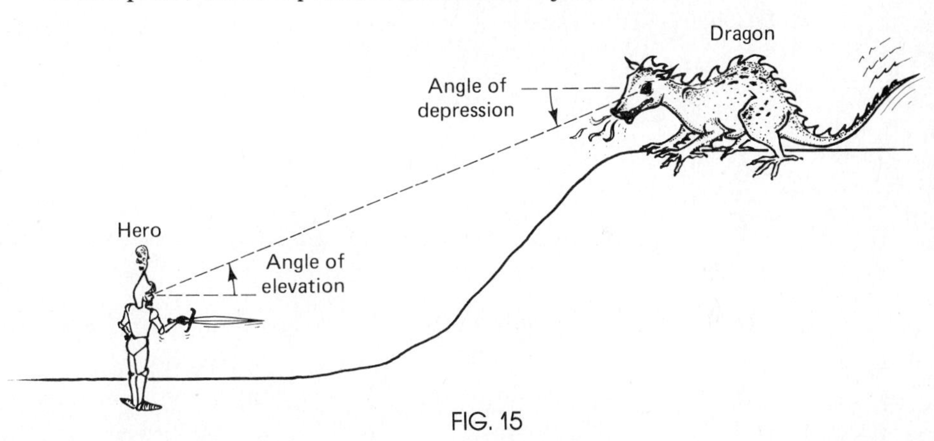

FIG. 15

EXAMPLE 3 The angle of elevation of the top of a 40-foot tree from a point on the ground is 30° (the ground is level). How far is the point from the base of the tree?

Solution We have the accompanying picture. Since $\tan 30^\circ = \dfrac{40}{b}$,

$$b = \frac{40}{\tan 30^\circ} = \frac{40}{1/\sqrt{3}} = 40\sqrt{3} \text{ ft} \approx 69.2820 \text{ ft}$$

■

EXERCISES

1. Copy triangle (a) on a separate piece of paper. *Without* looking back in the text, label the sides with appropriate lengths, and then compute the six trigonometric functions of 30° and of 60°.

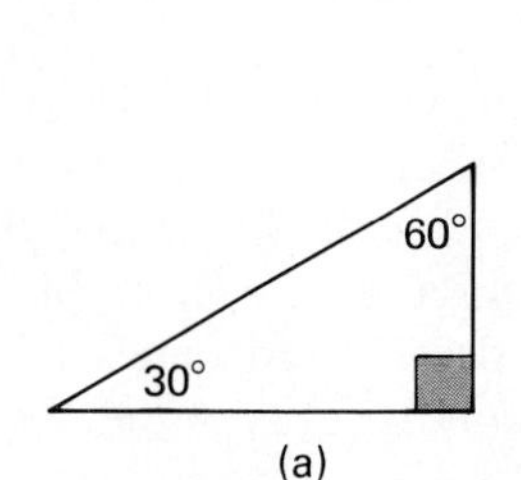

(a)

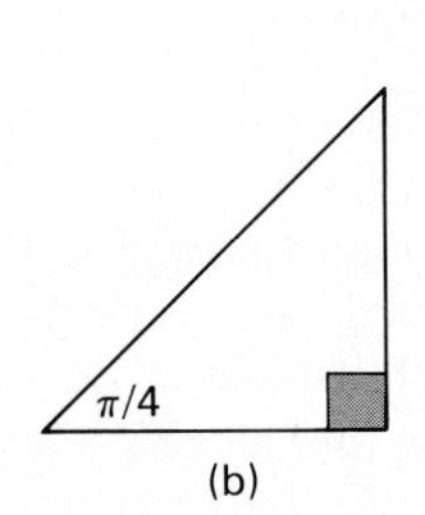

(b)

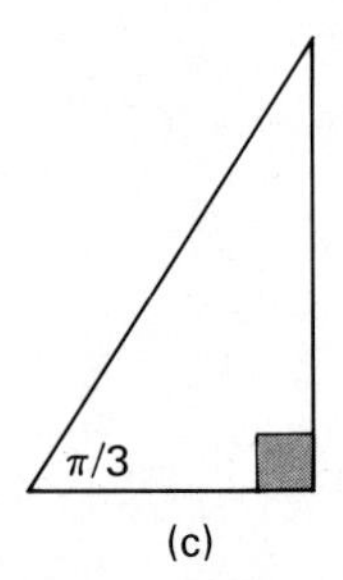

(c)

2. Copy triangle (b) on a separate piece of paper. *Without* looking back in the text, label the sides with appropriate lengths, and then compute the six trigonometric functions of $\frac{\pi}{4}$.
3. Copy triangle (c) on a separate piece of paper. *Without* looking back in the text, label the sides with appropriate lengths, and then compute the six trigonometric functions of $\frac{\pi}{3}$ and $\frac{\pi}{6}$.
4. Fill in the following chart without using a calculator:

	$\sin\theta$	$\cos\theta$	$\tan\theta$	$\csc\theta$	$\sec\theta$	$\cot\theta$
30°						
45°						
60°						

In Exercises 5–31, evaluate the trigonometric functions, using a calculator or tables only where necessary

5. $\tan 60°$
6. $\cos 12.3°$
7. $\tan 57°$
8. $\csc 22°$
9. $\cos 34° 15'$
10. $\cot 71.36°$
11. $\csc 63.5°$
12. $\sec 82°$
13. $\sin 45°$
14. $\tan 30°$
15. $\cos 30°$
16. $\sin 60°$
17. $\sin 1$
18. $\cos 0.5$
19. $\cos \frac{\sqrt{3}}{2}$
20. $\tan \frac{\pi}{4}$
21. $\tan \frac{\pi}{5}$
22. $\sin \frac{\pi}{6}$
23. $\csc 1.1$
24. $\sec 0.75$
25. $\cot \frac{2}{5}$
26. $\csc \frac{\pi}{3}$

27. $\sec \dfrac{\pi}{5}$

28. $\cot \dfrac{\pi}{5}$

29. $\cos 12° 17'31''$

30. $\sin 58° 46'6''$

31. $\cot 26° 32''$

In Exercises 32–39, evaluate the given expressions

32. $3 \sin 26.2° + \tan 74°$

33. $8 \sin^3 \dfrac{\pi}{4}$

34. $\sec 14° \cos 41° + 5 \csc 19.4°$

35. $6.753 + 4 \tan 1.14 - \dfrac{3}{\csc 0.75}$

Ⓒ36. $\cot^2 0.86 + \dfrac{\sin 0.48}{1 + \sec 0.3}$

Ⓒ37. $\sin^3 32° + \sin 24° \tan 14°$

Ⓒ38. $\dfrac{\cos 0.42 + \tan 1.03}{\sin 0.17 - \cos 0.68}$

Ⓒ39. $(\tan 32° + 1)(\cos 84° - \sin 38°)\csc^3 51°$

Ⓒ40. Verify the identities $\sin^2 \theta + \cos^2 \theta = 1$ and $1 + \tan^2 \theta = \sec^2 \theta$ for $\theta = 15°, 42°, \dfrac{\pi}{3}, 1$. (If they do not work, explain why.)

In Exercises 41–47, find the answer to three significant digits.

41. From a point 36 feet from the base of a tower along level ground, the angle of elevation of the top of the tower is 45°. How tall is the tower?

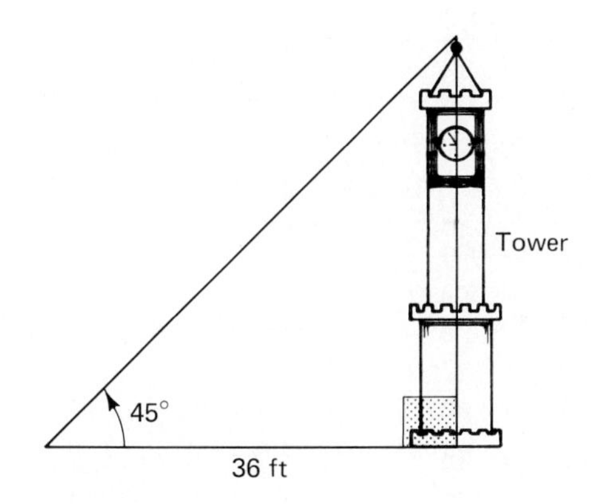

42. The angle of elevation of the top of a tower from the tip of its shadow (along level ground) is $\dfrac{\pi}{6}$. If the tower is 60 feet tall, how long is its shadow?

43. A field is in the shape of a right triangle. One leg is 521 yards long, and that leg makes an angle of 60° with the hypotenuse. Find the area of the field. [Recall that the area of a triangle is $\frac{1}{2}$(base)(height).]

44. A partially full oil storage tank is a cylinder 50 feet tall and 30 feet in diameter. From a point on the top edge, the angle of depression of the surface of the oil on the opposite side of the tank is 30°. What is the height of the oil in the tank? How many cubic feet of oil are in the tank?

45. On a baseball diamond, what is the sine of the angle between a line from home to first and a line from first to third?

46. The angle of depression of a car from a balloon is 45°. If the balloon is 2000 feet from the car, how high is the balloon?

47. A building (with straight sides) is 43.6 meters high. From a point on the edge of the roof it is observed that if you look directly across the street, the angle of depression to the near edge of the base of the building there is 60°. How far apart are the buildings?

48. A plane is 6 miles directly over city A. If the plane flew on a path that makes an angle of 30° with the ground, it would land at airport B. How far is it from A to B?

1.4 Inverse Trigonometric Functions

Suppose the right triangle shown in Fig. 16 is given. We would like to determine the angle θ. It seems likely that we could use our knowledge of the trigonometric functions of θ, say the fact that $\sin \theta = \frac{3}{5}$, to find θ. This is the case, for whenever θ is an acute angle in a right triangle $\left(0 < \theta < \frac{\pi}{2}\right)$, the trigonometric functions each have an inverse function.

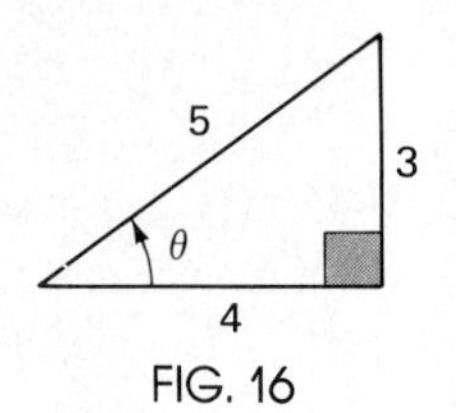

FIG. 16

INVERSE SINE AND COSINE

We first show that sine and cosine have inverse functions. In Fig. 17, let OA have length 1, let l be a half line originating at O, let AB be perpendicular to l, and let θ be angle $\angle BOA$. Then since $|OA| = 1$, $\sin \theta = |AB|$ and $\cos \theta = |OB|$.

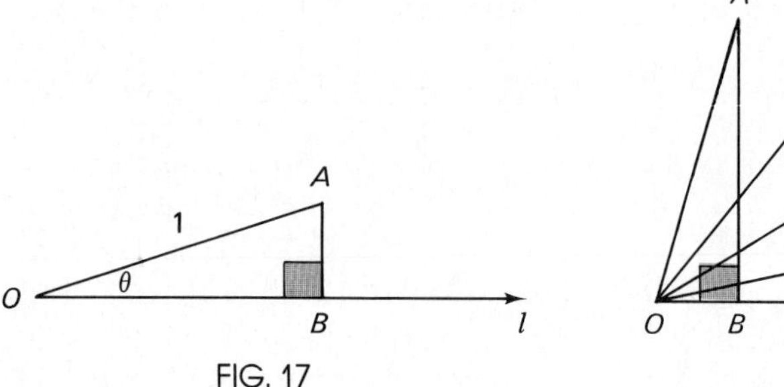

FIG. 17

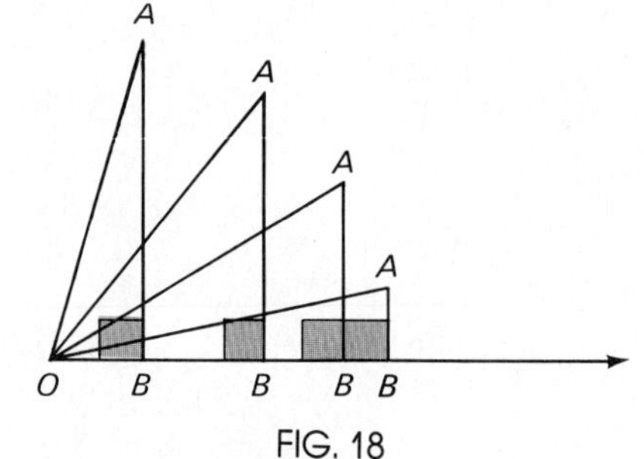

FIG. 18

Now consider what happens when θ (in radians) increases from "nearly" 0 to "nearly" $\frac{\pi}{2}$. We see from Fig. 18 that $|AB| = \sin \theta$ increases from "nearly" 0 to "nearly" 1, and for no two different values of θ is $|AB|$ the same length.

Thus we have seen the following:

> **(13)**
>
> **i.** If $0 < \theta < \frac{\pi}{2}$ and $u = \sin \theta$, then $0 < u < 1$.
>
> **ii.** For each u, $0 < u < 1$, there is a unique θ, $0 < \theta < \frac{\pi}{2}$, such that $\sin \theta = u$.

Similarly, the corresponding result holds for $\cos\theta$. For as θ increases from "nearly" 0 to "nearly" $\frac{\pi}{2}$, we see from Fig. 18 that $|OB| = \cos\theta$ decreases from "nearly" 1 to "nearly" 0, and for no two different values of θ is $|OB|$ the same length.

(14)

i. If $0 < \theta < \frac{\pi}{2}$ and $u = \cos\theta$, then $0 < u < 1$.

ii. For each u, $0 < u < 1$, there is a unique θ, $0 < \theta < \frac{\pi}{2}$, such that $\cos\theta = u$.

The statements (13) and (14) are exactly what we need for the sine and cosine functions to have inverse functions. The inverse sine function is usually denoted by either $\sin^{-1}$ (like f^{-1}) or arcsin. The other inverse trigonometric functions are denoted similarly.

(15) Definition

i. Suppose $0 < u < 1$. Then $\sin^{-1} u = \theta$ (or $\arcsin u = \theta$) if and only if $0 < \theta < \frac{\pi}{2}$ and $\sin\theta = u$.

ii. Suppose $0 < u < 1$. Then $\cos^{-1} u = \theta$ (or $\arccos u = \theta$) if and only if $0 < \theta < \frac{\pi}{2}$ and $\cos\theta = u$.

Of course θ may be measured in degrees; then the restriction on θ in (15i) and (15ii) would be $0° < \theta < 90°$. (However, if we want $\theta = \sin^{-1} u$ or $\theta = \cos^{-1} u$ in degrees, we shall explicitly state this. Otherwise they will be measured in radians.)

You can think of the equation $\theta = \sin^{-1} u$ as "θ is the angle, $0 < \theta < \frac{\pi}{2}$, whose sine is u." In particular, note the following:

Warning: $\sin^{-1} u$ *does not mean* $\frac{1}{\sin u}$. For example, $\sin^{-1} 0.5 = \frac{\pi}{6} \approx 0.523599$ (since $\sin\frac{\pi}{6} = 0.5$), whereas $\frac{1}{\sin 0.5} \approx \frac{1}{0.47942554} \approx 2.08583$.

EXAMPLE 1 Find $\sin^{-1}\left(\frac{\sqrt{2}}{2}\right)$ in both radians and degrees.

Solution Since $\sin\frac{\pi}{4} = \frac{\sqrt{2}}{2}$, we have $\sin^{-1}\left(\frac{\sqrt{2}}{2}\right) = \frac{\pi}{4} = 45°$. ■

COMPUTING INVERSES

If you are using tables, inverse trigonometric functions can be evaluated simply by going "backward," i.e., using the relationship $\theta = \sin^{-1} u$ exactly when $\sin \theta = u, 0 < \theta < \frac{\pi}{2}$, and similarly for $\cos^{-1} u$. This is explained in Appendix D, and it would be appropriate to read this part now if you have not already done so.

If you are using a calculator, $\sin^{-1} u$ and $\cos^{-1} u$ are evaluated using a combination of buttons.

First make sure the calculator is in the proper mode.

Then the process is basically the same on most machines, whether RPN or algebraic. There usually is a button marked [INV] or [ARC] or [F]. To find $\sin^{-1} u$, you enter u *first*, next press that button marked [INV] or [ARC] or [F], and then press [sin]. The computation of $\cos^{-1}$ is similar. However, *there are variations. Consult the manual* if you have any problems.

EXAMPLE 2 Find $\cos^{-1} 0.23$ in both degrees and radians.

Solution Using a calculator, put it into the proper mode and then press [.23] [INV] [cos], obtaining 76.7029° or 1.33872 (rad).

Using Table D1, we obtain 76°42′ or 1.3387 (rad). ■

INVERSE TANGENT

We now turn to the tangent function, which also has an inverse if $0 < \theta < \frac{\pi}{2}$.

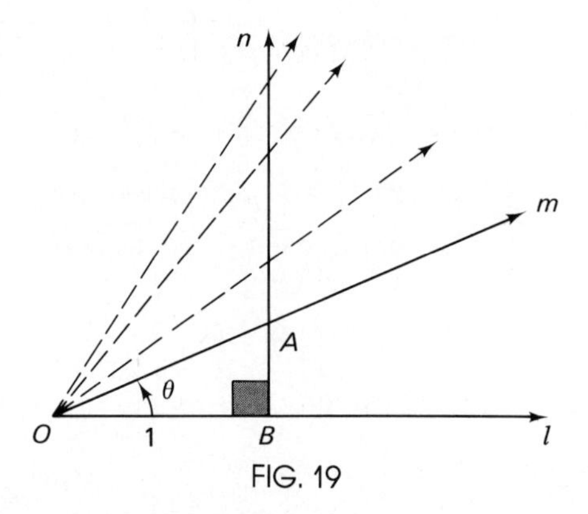

FIG. 19

In Fig. 19, let θ be an angle with vertex O, let l be the half line originating at O which is the initial side of θ, and let m be the terminal side of θ. Further, let B be the point on l one unit from O, let n be the line perpendicular to l at

B, and let A be the intersection of m and n. Then θ is angle $\angle BOA$. Since AOB is a right triangle,

$$\tan \theta = \frac{|AB|}{|OB|} = \frac{|AB|}{1} = |AB|$$

Consider what happens as θ increases from "nearly" 0 to "nearly" $\frac{\pi}{2}$. We see that m gets rotated and that $|AB|$ increases from "nearly" 0 and becomes arbitrarily large. In addition, for no two different values of θ is $|AB|$ the same length, and $|AB|$ may be any positive real number. Thus since $|AB| = \tan \theta$, we have seen the following:

> **1.** If $0 < \theta < \frac{\pi}{2}$ and $u = \tan \theta$, then $u > 0$.
>
> **2.** For each $u > 0$, there is a unique θ, $0 < \theta < \frac{\pi}{2}$, such that $\tan \theta = u$.

As with sine and cosine, this is exactly what we need for tangent to have an inverse function.

> **(16) Definition** Suppose $u > 0$. Then $\tan^{-1} u = \theta$ (or $\arctan u = \theta$) if and only if $0 < \theta < \frac{\pi}{2}$ and $\tan \theta = u$.

Again, you can think of the equation $\theta = \tan^{-1} u$ as "θ is the angle, $0 < \theta < \frac{\pi}{2}$, whose tangent is u." Also, θ may be in degrees, in which case the restriction on θ is $0° < \theta < 90°$.

EXAMPLE 3 Find $\tan^{-1} 1$ in both radians and degrees.

Solution Since $\tan \frac{\pi}{4} = 1$, we have $\tan^{-1} 1 = \frac{\pi}{4} = 45°$. ■

Whether using a calculator or tables, computing $\tan^{-1} u$ is similar to computing $\sin^{-1} u$ or $\cos^{-1} u$.

EXAMPLE 4 Find arctan 10 in both radians and degrees.

Solution Using a calculator, put it in the proper mode and then press [10] [INV] [tan], obtaining 1.47113 (rad) or 84.2894°.

Using Table D1, you obtain 1.4711 (rad) or 84.28°. ■

The remaining trigonometric functions have inverses, but we shall not use them here.

EXERCISES

In Exercises 1–6, evaluate in both degrees and radians without using tables or a calculator.

1. $\sin^{-1} 0.5$ **2.** $\cos^{-1} \dfrac{1}{\sqrt{2}}$ **3.** $\cos^{-1} 0.5$

4. $\tan^{-1} \dfrac{1}{\sqrt{3}}$ **5.** $\sin^{-1} \dfrac{\sqrt{3}}{2}$ **6.** $\tan^{-1} \sqrt{3}$

In Exercises 7–30, evaluate in both degrees and radians, using tables or a calculator only where necessary.

7. $\sin^{-1} \dfrac{\sqrt{2}}{2}$ **8.** $\cos^{-1} 0.5$ **9.** $\tan^{-1} 1$

10. $\tan^{-1} 5.3$ **11.** $\sin^{-1} \dfrac{\pi}{4}$ **12.** $\cos^{-1} \dfrac{\sqrt{3}}{2}$

13. $\cos^{-1} 0.1$ **14.** $\sin^{-1} 0.9$ **15.** $\sin^{-1} 0.8$

16. $\sin^{-1} 0.7$ **17.** $\sin^{-1} 0.6$ **18.** $\sin^{-1} 0.5$

19. $\sin^{-1} 0.4$ **20.** $\sin^{-1} 0.3$ **21.** $\sin^{-1} 0.2$

22. $\sin^{-1} 0.1$ **23.** $\tan^{-1} 0.2$ **24.** $\tan^{-1} 0.5$

25. $\tan^{-1} 0.8$ **26.** $\tan^{-1} \frac{3}{2}$ **27.** $\tan^{-1} 3$

28. $\tan^{-1} 10$ **29.** $\tan^{-1} 100$ **30.** $\tan^{-1} 1000$

Exercises 31–36 refer to the following triangle:

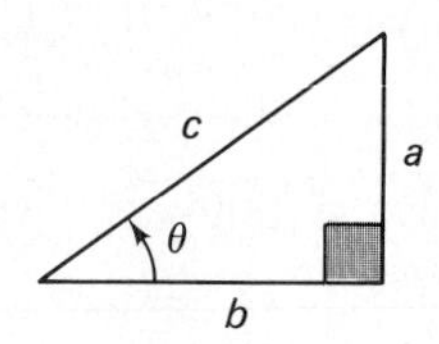

31. If $a = 3$ and $c = 5$, find $\sin \theta$, and then find θ (in degrees).

32. If $b = 11$ and $c = 20$, find $\cos \theta$, and then find θ (in radians).

33. If $a = 0.608$ and $b = 0.128$, find $\tan \theta$, and then find θ (in degrees).

34. If $a = 643$ and $c = 700$, find $\sin \theta$, and then find θ (in radians).

35. If $b = 1.89$ and $c = 10$, find $\cos \theta$, and then find θ (in degrees).

36. If $a = 37.2$ and $b = 120$, find $\tan \theta$, and then find θ (in radians).

37. The roads joining three villages form a right triangle, with the right angle at village X. It is 4 kilometers from village X to village Y and 9 kilometers from village Y to village Z. At what angle (in whole degrees) do the roads meet at village Z?

38. A small rocket is to be launched from the ground aimed toward a big tree on the other side of a brick wall. If it is 12 feet from the launching position to the base of the wall and the wall is 10 feet tall, what must be the minimum angle in *whole* degrees between the rocket and the ground so that when launched, the rocket will clear the wall?

39. A 12-foot ladder is leaning against the wall in a room. The top of the ladder is 7 feet, 4 inches from the floor. What (acute) angle (in whole degrees) does the ladder make with the floor?

40. A tree is 17.2 meters tall. What is the angle of elevation (in radians to three significant figures) of the top of the tree from a point that is 40 meters from the base of the tree along level ground?

41. Two flagpoles are 50 feet apart. One pole is 80 feet tall, while the other is 60 feet tall. Find the angle of depression (in radians to three significant figures) of the top of the shorter pole from the top of the taller pole.

42. A golfer is about to hit his golf ball. It is 183 yards straight west from his ball to the hole. He hits the ball straight but slightly off-line. It ends up 24 yards straight north of the hole. What is the angle (in degrees, to the nearest $\frac{1}{10}$ degree) between the intended line of flight and the actual line of flight of the ball?

43. It is 153 meters from an observation point on the ground to a hang-glider flying in the air. It is 47 meters along level ground from that point to a point directly under the glider. What is the angle of depression (in whole degrees) of the observation point from the glider?

44. The cross section of a roof (and the beams across at the base) is in the shape of an isosceles triangle. It is 36 feet across at the base, and each of the slant edges is 26 feet long. What is the pitch of the roof (in radians, to two decimal places)? (The pitch is the angle between the base and a slanted side of the triangle.)

1.5 Applications of Right Triangles

There are many instances where you can use right triangles to find the length or width or height of some object which would otherwise be very difficult to measure. The principle involved is that if you are given one acute angle and one side of a right triangle or two sides of a right triangle, you can use trigonometry to find the remaining sides and angles.

EXAMPLE 1 In Fig. 20, suppose $\theta = 1$ and $a = 3.2$ centimeters. Find b and c.

Solution Since $\tan\ \theta = \dfrac{a}{b}$, we have $b = \dfrac{a}{\tan\ \theta} = \dfrac{3.2}{\tan 1} \approx 2.05470$ cm. Since $\sin\ \theta = \dfrac{a}{c}$, we have $c = \dfrac{a}{\sin\ \theta} = \dfrac{3.2}{\sin 1} \approx 3.80286$ cm. ■

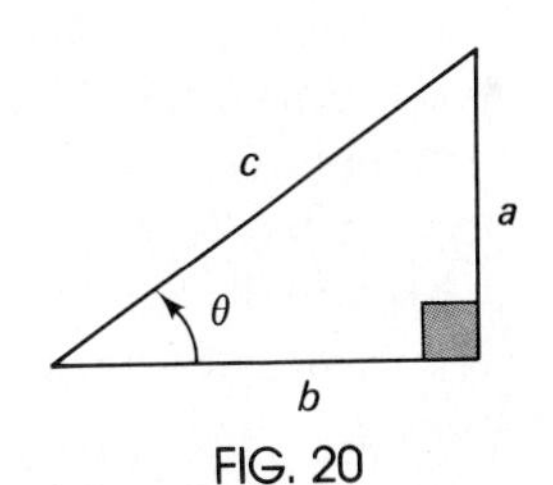

FIG. 20

EXAMPLE 2 In Fig. 20, suppose $b = 4.64$ kilometers and $c = 16$ kilometers. Find a and θ (in radians).

Solution Since $\cos\ \theta = \dfrac{b}{c} = \dfrac{4.64}{16} = 0.29$, $\theta = \cos^{-1} 0.29 \approx 1.27657$. To find a, you can either use the Pythagorean theorem,

$$a = \sqrt{c^2 - b^2} = \sqrt{16^2 - 4.64^2} \approx 15.3124 \text{ km}$$

or trigonometry, say the fact that $\sin\ \theta = \dfrac{a}{c}$, to obtain

$$a = c \sin\ \theta = 16 \sin\ (\cos^{-1} 0.29) \approx 15.3124 \text{ km}$$ ■

Examples 1 and 2 illustrate how, given sufficient information about a right triangle, you can find "missing" information. When doing word problems, you should draw pictures if at all possible and look for triangles, particularly right triangles. If a right triangle is involved, then after you label the parts, you can proceed as in Examples 1 and 2 to find the desired information.

EXAMPLE 3 Suppose you are at the base of a vertical cliff and you wish to know its height. You walk out along level ground to a point 100 feet from the base of the cliff, and from that point you measure the angle of elevation of the top of the cliff to be 68.1°. What is its height?

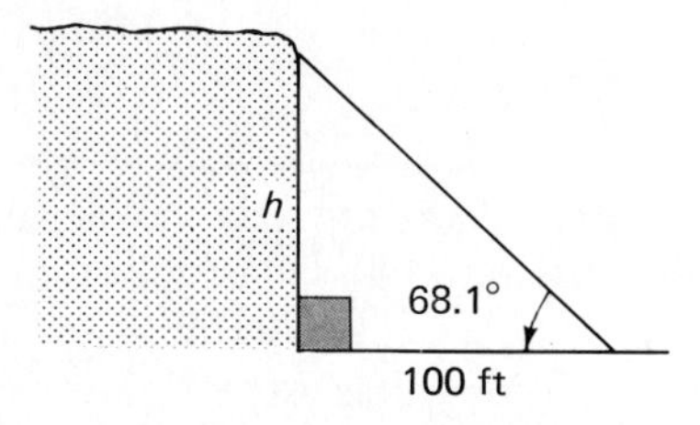

Solution First draw a picture, letting h be the unknown height. Then $\tan 68.1° = \dfrac{h}{100}$, or $h = 100 \tan 68.1° \approx 248.758$ feet.

If in the original problem we had stated that the data are accurate to three significant figures, we would now round the answer to $h \approx 249$ feet. ■

In many cases the information given makes it natural to draw a triangle, but the triangle is not a right triangle. In that case, the best thing to do in order to use the techniques we have developed is to drop an altitude from one of the vertices, thereby forming two right triangles. In Fig. 21, we illustrate how dropping an altitude from vertex C of triangle ABC forms two right triangles, ACD and BCD, whether the base D of the altitude falls within the original triangle or not.

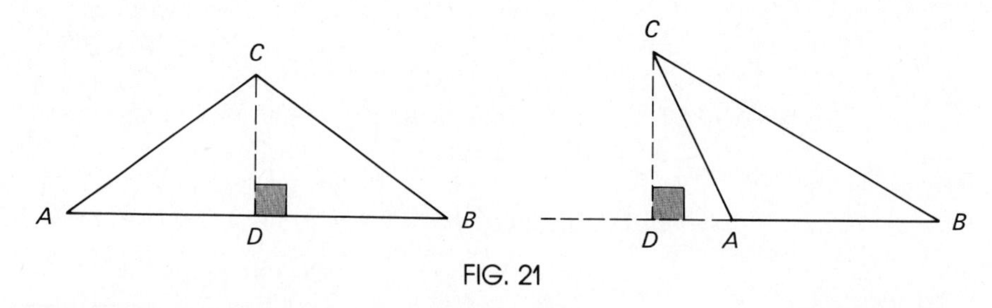

FIG. 21

EXAMPLE 4 Suppose you are on level ground and you wish to measure the height of a mountain above ground level. From some point A along the level ground you determine that the angle of elevation of the peak is 0.95, while from a point B which is 200 feet farther from the mountain, the angle of elevation of the peak is 0.92. Find the height.

Solution First we draw a picture:

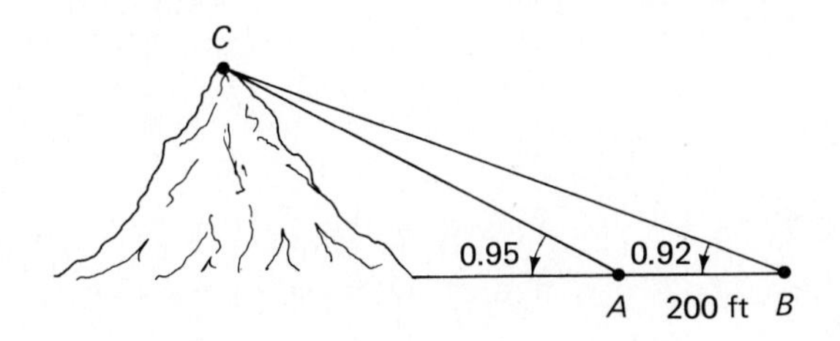

We see triangle ABC, but the height h is the length of an altitude from vertex C. We drop this altitude, forming right triangles ACD and BCD. In each right triangle we know an acute angle, but we do not know any of the sides.

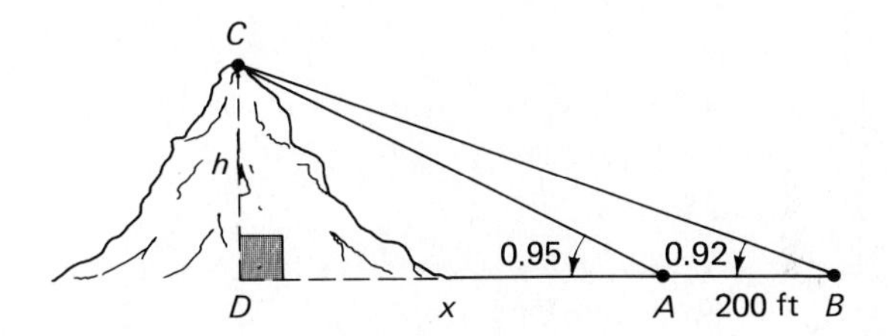

We must use some algebra to solve this problem. Let $x = |AD|$. Then $|BD| = x + 200$. From triangle ACD,

$$\tan 0.95 = \frac{h}{x}$$

while from triangle BCD,

$$\tan 0.92 = \frac{h}{x + 200}$$

These are two equations in two unknowns, and we can solve them for h as follows: From the first equation, $x \tan 0.95 = h$, or $x = \dfrac{h}{\tan 0.95}$. From the second equation,

$$h = (x + 200) \tan 0.92 = x \tan 0.92 + 200 \tan 0.92$$

Substituting $x = \dfrac{h}{\tan 0.95}$ into this equation yields

$$h = \frac{h}{\tan 0.95} \tan 0.92 + 200 \tan 0.92$$

$$h - h\frac{\tan 0.92}{\tan 0.95} = 200 \tan 0.92$$

$$h\left(1 - \frac{\tan 0.92}{\tan 0.95}\right) = 200 \tan 0.92$$

$$h = \frac{200 \tan 0.92}{1 - \dfrac{\tan 0.92}{\tan 0.95}} \approx 4315.01 \text{ ft}$$

If the problem had stated that the data are accurate to two significant figures, we would now round the answer to $h \approx 4300$ feet. ■

EXERCISES

In Exercises 1–19, use the given quantities to find the quantities indicated. (See the triangle below.)

1. $\theta = 15°$, $b = 12.4$ feet. Find a, c.
2. $\theta = 0.6$, $a = 38$ feet. Find b, c.
3. $\theta = 0.432$, $c = 29.2$ meters. Find a, b.
4. $\theta = 78°24'$, $a = 6.1$ kilometers. Find b, c.
5. $\phi = 1.12$, $c = 3.6$ inches. Find a, b.
6. $\phi = 0.58$, $a = 13.1$ centimeters. Find b, c.
7. $\tan \theta = 1.1$, $c = 19.3$ feet. Find a, b.
8. $\tan \theta = 0.43$, $a = 6.5$ inches. Find b, c.
9. $\cos \theta = 0.125$, $a = 48.7$ centimeters. Find b, c.
10. $\cos \theta = 0.64$, $b = 183$ inches. Find a, c.
11. $\sin \theta = 0.312$, $b = 245$ inches. Find a, c.
12. $\sin \theta = 0.8743$, $c = 12$ inches. Find a, b.
13. $\sin \theta = 0.56$, $a = 93$ feet. Find b, c.
14. $a = 3$ meters, $b = 4$ meters. Find c, θ, ϕ.
15. $a = 5.23$ feet, $c = 12.81$ feet. Find θ, ϕ.
16. $b = 4.635$ kilometers, $c = 21.289$ kilometers. Find θ, ϕ.
17. $a = 976$ miles, $b = 1329$ miles. Find θ, ϕ.
18. $a = 0.0115$ millimeter, $c = 0.0925$ millimeter. Find θ, ϕ.
19. $b = 47{,}531$ yards, $c = 51{,}982$ yards. Find θ, ϕ.

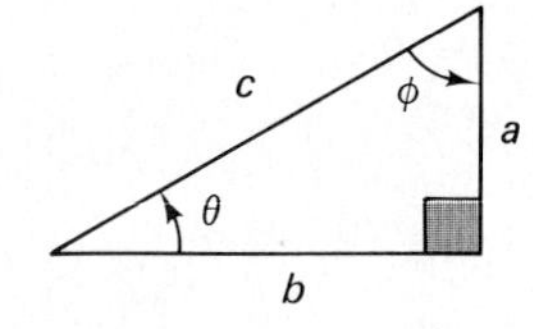

20. A mountain on the moon makes a shadow, which we can measure fairly accurately from earth to be 5 kilometers. We can also measure the angle of the sun's rays to be 0.23 rad. Approximately how tall is the mountain (in meters)? Assume two-significant-figure accuracy.

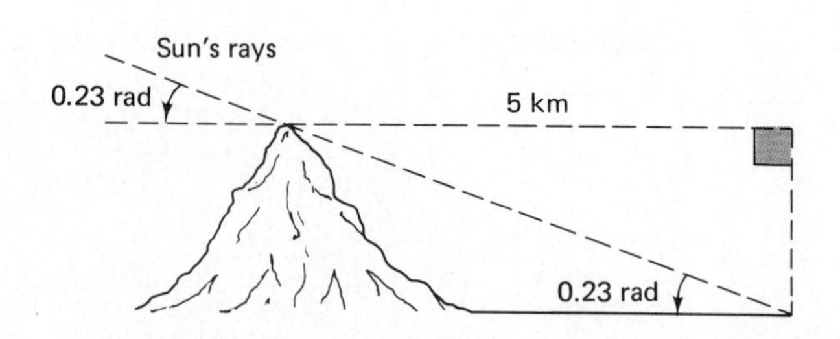

21. Steps to the entrance to a building rise a total of 2 feet. They are to be torn out and replaced by a ramp for wheelchairs, inclined at 10°. How long (along the slant) is the ramp? Assume two-significant-figure accuracy.

22. A tree casts a shadow (along level ground) that is 82 feet long. If the angle of elevation of the top of the tree from the tip of the shadow is 42°, how tall is the tree? Assume two-significant-figure accuracy.

23. A plane flying level at 10,000 feet flies directly over an observer on the ground. Two minutes later the angle of elevation of the plane from the observer is 0.1411 rad. How many miles has the plane flown in those 2 minutes? How fast was the plane flying (in miles per hour)? Assume four-significant-figure accuracy.

24. A seaplane is on an approach path for landing which is 12° below horizontal. It passes over a ship at an altitude of 1626 feet. How far from the ship will the plane touch down? Assume four-significant-figure accuracy.

25. A field is in the shape of a right triangle, with a fence along the two legs and a river (with a straight bank) along the hypotenuse. If the side along the river is 436.9 meters long and that side makes an angle of 0.53627 with one of the legs, how long is the fence (to four significant figures)?

26. A 15-foot guy wire runs from the ground (which is level) to the middle of a telephone pole. The guy wire makes an angle of 39°18′12″ with the pole. How tall is the pole?

27. An oil well driller has just struck oil, and a geyser is spouting forth. From a point along level ground 62 feet from the base of the geyser the driller observes that the angle of elevation of the top is about 37°. To the nearest foot, how tall is the geyser?

28. A straight railroad track is rising at a constant rate. At the end of 1 mile, it is 138.3 feet higher than at the start. What is the angle between the rail and a horizontal (to the nearest $\frac{1}{100}$ degree)?

29. A laser is on level ground 6.570 meters from a mark at the base of a wall. At what angle with the ground should the laser be inclined in order to point at an object 8.831 meters up the wall directly over the mark? Assume four-significant-figure accuracy.

30. The ground around a smokestack is level. From the ground at a distance of 126 feet from the base, the angle of elevation of the top is 34.31509°. What is the angle of elevation of the top from a point on the ground 205 feet from the base?

31. In the regular hexagon shown, find the length of AC and AD if each side has length 6.23. Use the fact that in a regular n-gon, each angle is $\left(\frac{n-2}{n}\right)180°$.

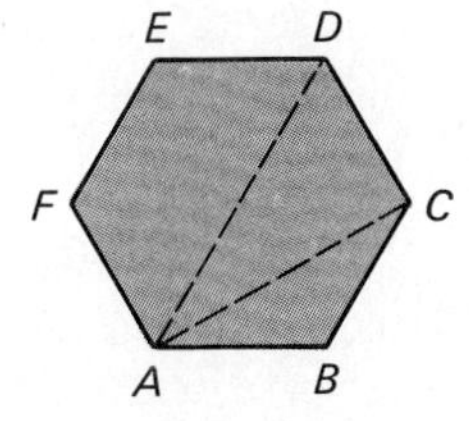

32. In the regular octagon shown, find the length of AC if each side has length 7.

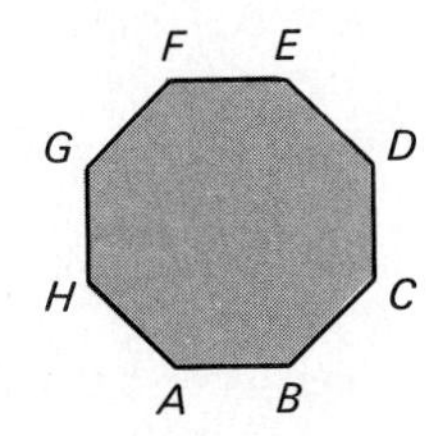

33. The angle of elevation of a mountain peak from a point on level ground is 65°, while from a point 100 feet farther from the peak it is 61°. How high is the mountain to the nearest foot?

34. You wish to determine how far it is across a river with straight, parallel banks. As you stand on one bank, you notice that your line of sight to a rock farther up on the opposite bank makes an angle of 12° (with your bank). After you walk 30 feet along the bank, you notice the angle is now 28°30′. How wide is the river to the nearest foot?

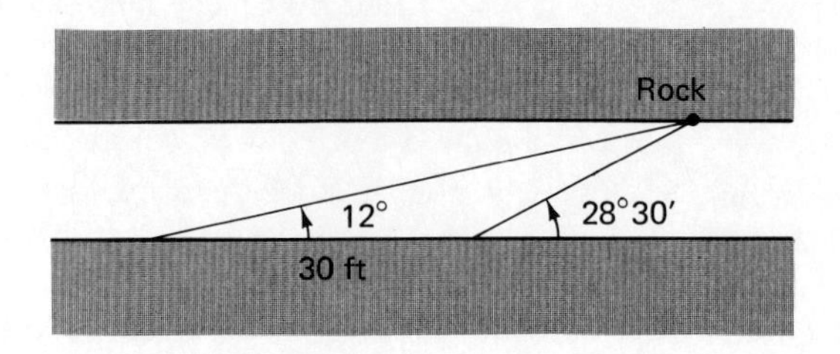

35. As you stand on a bridge that is 100 feet above the water, you are looking at an approaching barge. If the angle of depression of the front of the barge is 0.507 rad and the angle of depression of the rear of the barge is 0.303 rad, find the length of the barge to the nearest foot.

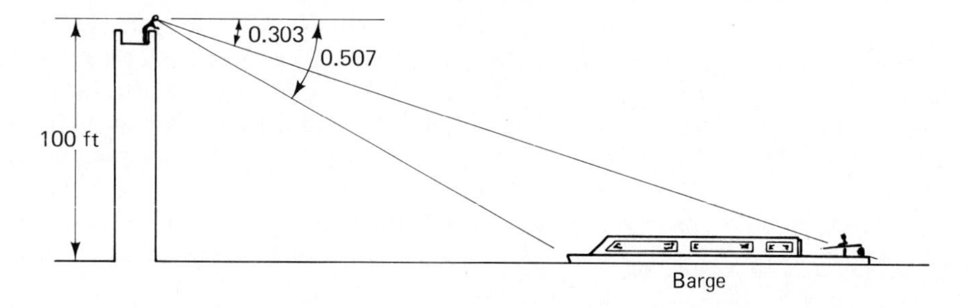

36. You are at the window of a building looking out from a point which is 93.1 feet above ground level (the ground nearby is flat). As you look at a nearby building, you notice that the angle of depression to its base is 72° and the angle of elevation to its peak is 81.6°. How far apart are the buildings, and how tall is the second building (to the nearest $\frac{1}{10}$ foot)?

37. A plane is flying at 480 miles per hour on a level course directly away from you. At one time you observe that its angle of elevation is 72°, while 5 minutes later it is 5°. How much above the ground level where you are is the plane flying? Assume two-significant-figure accuracy.

38. Three stars (label them A, B, and C) are the vertices of a right triangle with right angle at A. The angle $\angle ACB$ is known to be 0.7364. A spaceship traveling along the line from A to C observes that the angle between its path and a line from it to B is 0.8946. If it is 1.64 light-years from A to B (a light-year is the distance a ray of light would travel in a year), how far is it from the spaceship to C? Assume four-significant-figure accuracy.

39. A field is in the shape of a right triangle with hypotenuse of length 7250 yards and one angle 0.632. Suppose we label the vertices A, B, and C, where the right angle is at A and $\angle ABC = 0.632$. Then BC is the hypotenuse, so $|BC| = 7250$. A path starts at vertex C and proceeds in a straight line through the field until it meets side AB in a point D and then proceeds from there along side AB to vertex B. The angle $\angle ACD = 0.237$. How long is the path to the nearest yard? (*Hint:* First find $|AC|$ and $|AB|$, and then find $|AD|$ and $|CD|$.)

40. A ship is sailing on a straight course. A lighthouse is sighted ahead, and it is noted that the angle between a line to the lighthouse and the ship's course is 27°. After traveling 2.65 kilometers, the ship has gone past the lighthouse. The angle between the ship's course and a line to the lighthouse is now 42°. How close (to the nearest meter) did the ship come to the lighthouse?

41. Two radio towers are standing in a field. One is 123.2 feet tall. From a point partway up the second, it is observed that the angle of depression of the base of the first tower is $34°13'$, while the angle of elevation of its top is $51°35'$. How far apart are the towers (to the nearest $\frac{1}{10}$ foot)?

42. The roads between cities A, B, and C form a right triangle, with right angle at city A. City D is located along the road from A to B and is 6.4 kilometers from A. There is a straight road from C to D, and road CD meets road AD in an angle of 53°. Road CB meets road AB in an angle of 39°. Find the remaining distances between the towns (to the nearest $\frac{1}{10}$ kilometer).

43. A plane is flying directly away from you at a height of 6996 feet. It is flying at a constant speed on a level course. At one time, you observe that the angle of elevation of the plane is 47°, while 2 minutes later it is 10°. How fast is the plane flying (in miles per hour, to the nearest whole mph)?

REVIEW EXERCISES

In Exercises 1–6, convert to degrees in decimal notation.

1. $62°18'18''$
2. $-41°13'27''$
3. $4817°52'$
4. $3°36''$
5. $-187°35'43''$
6. $95°42'$

In Exercises 7–12, convert to radians.

7. 60°
8. −240°
9. 84°
10. 154.286°
11. −471.32°
12. $321°14'51''$

In Exercises 13–18, convert to degrees in decimal notation.

13. $\frac{2\pi}{3}$
14. $-\frac{\pi}{6}$
15. 2.3
16. −3.4172
17. −0.047
18. 631.2

19. A right triangle has legs of lengths 1 and 2. Find the six trigonometric functions of the angle θ opposite the side of length 1.

20. A right triangle has hypotenuse of length 3 and one leg of length 1. Find the six trigonometric functions of the angle θ opposite the side of length 1.

21. If θ is an angle in a right triangle and $\sin \theta = \frac{1}{4}$, find the remaining trigonometric functions of θ.

22. If θ is an angle in a right triangle and $\cot \theta = 2$, find the remaining trigonometric functions of θ.

In Exercises 23–32, evaluate using a calculator or tables.

23. $\sin 1°$
24. $\sin 1$
25. $\csc 61.4°$
26. $\cos 40°10'20''$
27. $\sec 27.2°$
28. $\sec 0.912$
29. $\tan 14.1°$
30. $\cot 1.1°$
31. $\cot 1.1$
32. $\cos 0.82$

Ⓒ33. Find $(\sin 0.294 - 3 \cos 0.82)^3$.

Ⓒ34. Find $\dfrac{\cot^4 18° - \tan 41.2°}{3 \sec^2 52° + 1}$.

35. A balloon is rising straight up above a point A on the ground. An observation point B is 600 feet from A along the (level) ground. At one time, it is noticed that the angle of elevation of the balloon is 30°. A short while later the angle of elevation is 60°. How much did the balloon rise in the meantime? Assume three-significant-figure accuracy.

36. Three small islands, A, B, and C, are the vertices of a right triangle with right angle at C. If it is 28.2 miles from A to B and the angle between AB and AC is 30°, how far is it from A to C and from B to C? If a plane went down somewhere inside the triangle, how much area would have to be searched? Assume three-significant-figure accuracy.

In Exercises 37–44, evaluate in both degrees and radians. Use tables or a calculator only when necessary.

37. $\sin^{-1} 0.5$ **38.** $\cos^{-1} \frac{\sqrt{2}}{2}$ **39.** $\sin^{-1} 0.48$

40. $\tan^{-1} 0.5$ **41.** $\tan^{-1} 1$ **42.** $\cos^{-1} 0.91$

43. $\sin^{-1} \frac{\sqrt{3}}{2}$ **44.** $\tan^{-1} \sqrt{2}$

In Exercises 45–51, give answers to three significant digits.

45. In a right triangle, one angle is 39.2°, and the hypotenuse has length 6.3. Find the length of the legs.

46. In a right triangle, one angle is 0.517 rad, and the adjacent leg has length 350. Find the length of the other leg.

47. One angle θ in a right triangle has $\sin \theta = 0.234$. If the side opposite θ has length 0.2897, find the length of the adjacent side.

48. One angle θ in a right triangle has $\cos \theta = 0.234$. If the side opposite θ has length 31.2, find the length of the adjacent side.

49. The side of a hill makes an angle of 12° with a horizontal. A horizontal mine shaft goes into the hill. A vertical air shaft is to be drilled into the mine shaft starting from a point 150 feet up along the side of the hill from the mine entrance. How long will the air shaft be?

50. A ship is proceeding on a straight course. At one time, the angle between the ship's course and a line from the ship to a small island is 18.2°. After the ship proceeds a mile, the angle has changed to 37.43°. How close will the ship pass to the island?

51. You are at the top of a building 400 feet tall. As you look at a nearby smaller building, you observe that the angle of depression of the base is 81° and the angle of depression of the top is 49.3°. Find the height of the nearby building.

CHAPTER 2

General Trigonometry

The Arabs assimilated mathematics from the neighbors they conquered, organized and expanded it somewhat, and then passed it along to the Europeans. The origin of the word *sine* illustrates this. When the Hindus invented the sine function, they called it *jiva* for "half chord," which the Arabs adopted as *jiba*. However, around 1150, an Englishman, Robert of Chester, translated the Arabic into Latin (the scientific language of Europe until the nineteenth century). He confused *jiba* with the Arabic word *jaib*, which means "bay" or "inlet," and so he used the Latin word meaning "bay" or "inlet," which is *sinus*. Later, sinus was shortened to sine.

As mathematics spread from Arabia to Europe, it very slowly evolved. An Arab, Nasir Eddin al-Tusi (1201–1274), began to treat trigonometry as a field of study separate from astronomy. But this approach was not adopted in Europe until Regiomontanus (1436–1476), who also began to use decimals in trigonometric tables, rather than sexagesimal fractions.

Francois Viete (1540–1603) developed a generalized approach to trigonometry. However, he was hesitant to use negative numbers, and he did not know about periodicity.

The Bernoulli brothers Jean (1667–1748) and Jacques (1654–1705) were the first to consider trigonometric functions as functions of numbers, not just angles.

Gilles Persone de Roberval (1602–1675), a contemporary of Descartes, was the first to graph the sine curve, though he only did it for $0° \leq x \leq 90°$.

2.1 The Trigonometric Functions

In this section, we shall extend the definitions of the trigonometric functions to arbitrary angles.

First, suppose θ is any angle. We say θ is in **standard position** if its vertex is the origin and its initial side is the positive x-axis. Figure 1 illustrates some angles in standard position.

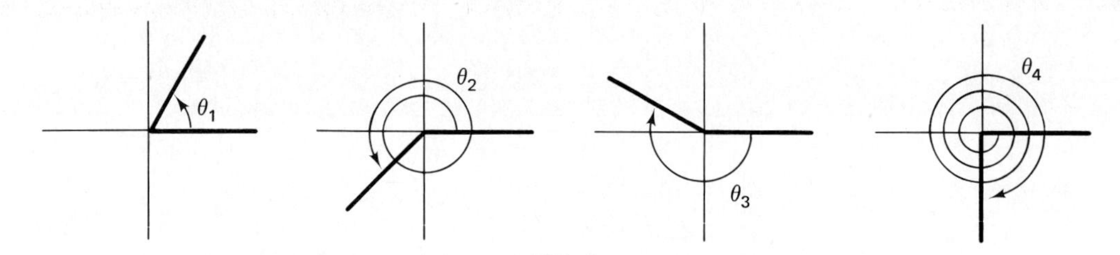

FIG. 1

We shall use the phrase "θ is in Q_3" to mean "θ is an angle in standard position, and its terminal side is in the third quadrant." We shall use the corresponding phrase when the terminal side is in any other quadrant. For example, in Fig. 1, θ_1 is in Q_1, θ_2 is in Q_3, and θ_3 is in Q_2. An angle in standard position with its terminal side on an axis is not in a quadrant; such an angle is called a **quadrantal angle** or a **between quadrant angle.** In Fig. 1, θ_4 is a quadrantal angle.

If θ is in Q_1 and is acute, we can apply the techniques of the previous chapter to find the trigonometric functions of θ. First, let $P = (x, y)$ be any point (different from the origin) on the terminal side of θ. Then $r = \sqrt{x^2 + y^2}$ is the distance from P to the origin. See Fig. 2(a). The definitions of the trigonometric functions state that

$$\sin \theta = \frac{y}{r} \qquad \csc \theta = \frac{r}{y}$$

$$\cos \theta = \frac{x}{r} \qquad \sec \theta = \frac{r}{x}$$

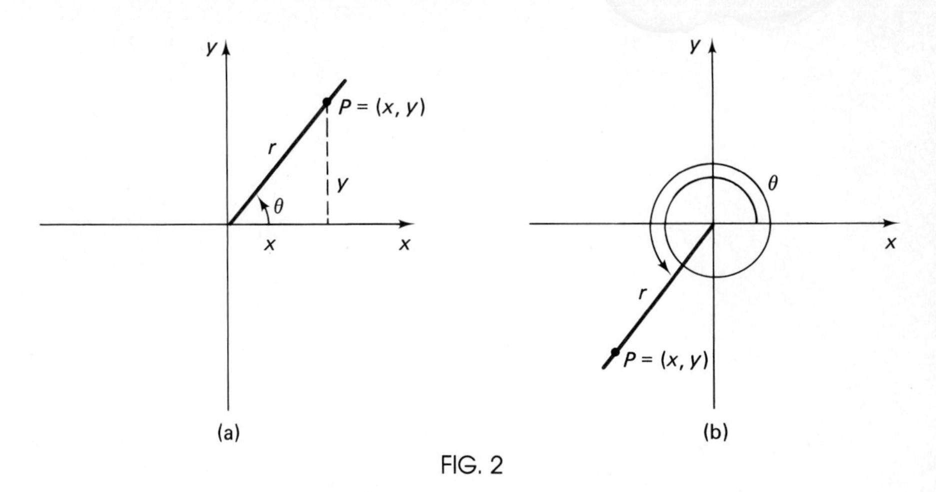

FIG. 2

$$\tan\theta = \frac{y}{x} \qquad \cot\theta = \frac{x}{y}$$

In this form, the six trigonometric functions generalize immediately to arbitrary angles.

> **(1) Definition** Let θ be an angle in standard position, let $P = (x, y)$ be any point (different from the origin) on the terminal side of θ, let $r = \sqrt{x^2 + y^2}$ be the distance from P to the origin. See Fig. 2(b). Then
>
> $$\sin\theta = \frac{y}{r} \qquad \csc\theta = \frac{r}{y},\quad y \neq 0$$
>
> $$\cos\theta = \frac{x}{r} \qquad \sec\theta = \frac{r}{x},\quad x \neq 0$$
>
> $$\tan\theta = \frac{y}{x},\quad x \neq 0 \qquad \cot\theta = \frac{x}{y},\quad y \neq 0$$

It can easily be shown that the trigonometric functions of θ as defined in (1) do not depend on which point P (different from the origin) is chosen on the terminal side. We express this by saying that the trigonometric functions are **well defined**. The proof uses straight calculations if θ is a quadrantal angle and similar triangles otherwise. (See Exercises 48–51.)

From Definition (1), observe that $\tan\theta$ and $\sec\theta$ are undefined if $x = 0$, while $\cot\theta$ and $\csc\theta$ are undefined if $y = 0$.

Let us see how we can use the definitions.

EXAMPLE 1 Suppose $(-2, -3)$ is a point on the terminal side of θ. See Fig. 3. Find the trigonometric functions of θ.

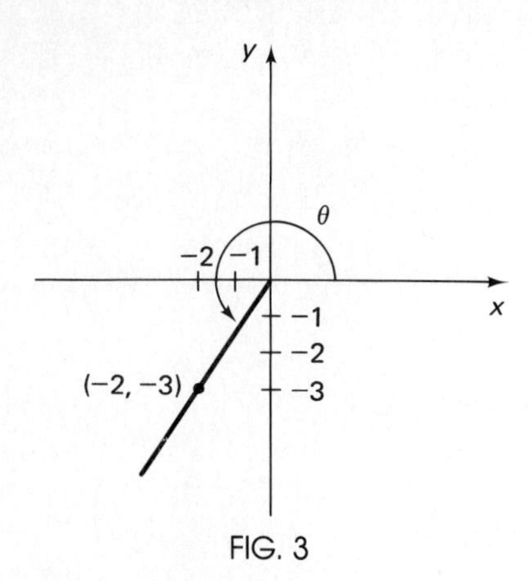

FIG. 3

Solution Set $r = \sqrt{(-2)^2 + (-3)^2} = \sqrt{4 + 9} = \sqrt{13}$. From Definition (1),

$$\sin\theta = \frac{y}{r} = \frac{-3}{\sqrt{13}} \approx -0.832050 \qquad \csc\theta = \frac{r}{y} = \frac{\sqrt{13}}{-3} \approx -1.20185$$

$$\cos\theta = \frac{x}{r} = \frac{-2}{\sqrt{13}} \approx -0.554700 \qquad \sec\theta = \frac{r}{x} = \frac{\sqrt{13}}{-2} \approx -1.80278$$

$$\tan\theta = \frac{y}{x} = \frac{-3}{-2} = \frac{3}{2} \qquad \cot\theta = \frac{x}{y} = \frac{-2}{-3} = \frac{2}{3}$$ ■

As Example 1 illustrates, trigonometric functions may be negative. For instance, since r is always positive, $\sin\theta = \frac{y}{r}$ and $\csc\theta = \frac{r}{y}$ have the same sign as y. Thus $\sin\theta$ and $\csc\theta$ are positive when θ is in Q_1 or Q_2 and negative when θ is in Q_3 or Q_4. The signs of the other trigonometric functions can be found similarly. The charts in Fig. 4 indicate the signs according to where θ lies. These can be combined as in the single chart in Fig. 5. This is remembered by saying "All Students Take Courses" (or something similar). In this chart the letters indicate the positive functions: A denotes all, S denotes sine (and its reciprocal cosecant), T denotes tangent (and cotangent), while C denotes cosine (and secant).

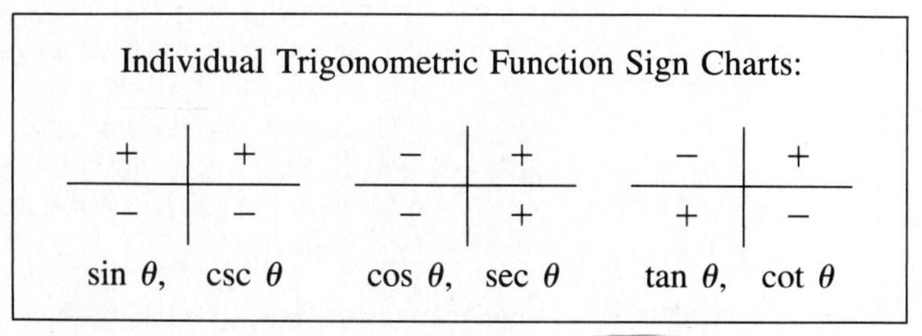

FIG. 4

Combined Trigonometric Function Sign Chart:

S	A
T	C

FIG. 5

In the previous chapter, we always assumed that θ was an acute angle, which meant that the terminal side would be in the first quadrant. Thus all the trigonometric functions were positive (we just saw this in our sign charts). Now when we are given one of the trigonometric functions and asked to find the others, we may not be able to determine the signs unless we have further information.

EXAMPLE 2 Suppose $\cos\theta = \frac{3}{5}$. Find the remaining trigonometric functions of θ.

Solution First, observe from the sign charts that the terminal side of θ is in Q_1 or Q_4 since $\cos\theta$ is positive. Next we know that we may choose any convenient point $P = (x, y)$ on the terminal side when computing the trigonometric functions. Since $\cos\theta = \frac{3}{5} = \frac{x}{r}$, we choose the point P with $x = 3$ and y such that $r = \sqrt{x^2 + y^2} = 5$. In other words, $y^2 = r^2 - x^2 = 25 - 9 = 16$, or $y = \pm 4$. See Fig. 6. Then from the definitions,

$$\sin\theta = \pm\frac{4}{5},\ \csc\theta = \pm\frac{5}{4},\ \sec\theta = \frac{5}{3},\ \tan\theta = \pm\frac{4}{3},\ \text{and } \cot\theta = \pm\frac{3}{4} \quad \blacksquare$$

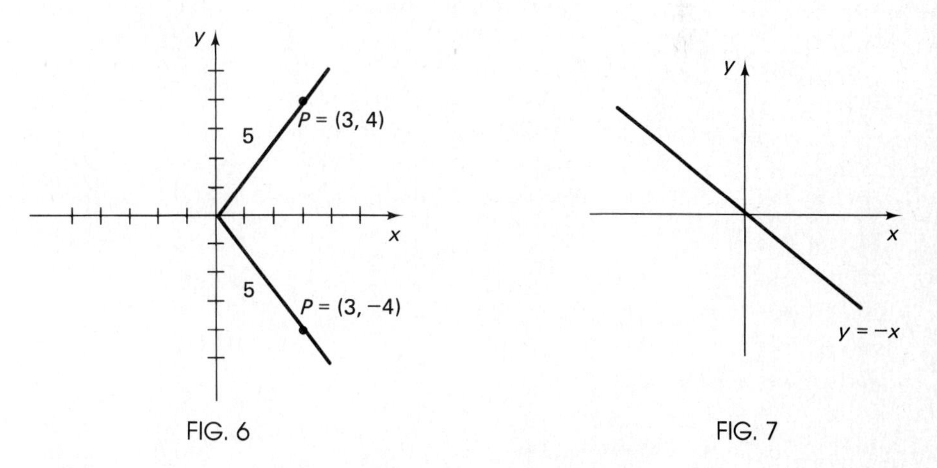

FIG. 6

FIG. 7

EXAMPLE 3 Suppose that for every point on the terminal side of θ, $y = -x$. See Fig. 7. Find the trigonometric functions of θ.

Solution The terminal side of θ is in either Q_2 or Q_4. We pick a point on the terminal side in each quadrant and find the trigonometric functions by the definitions: $x = -1$, $y = 1$ satisfies the equation and is in $Q_2\,(r = \sqrt{2})$, and $x = 1$, $y = -1$ satisfies the equation and is in $Q_4\,(r = \sqrt{2})$. Hence

$$\sin\theta = \pm\frac{1}{\sqrt{2}}, \qquad \cos\theta = \mp\frac{1}{\sqrt{2}}, \qquad \tan\theta = -1$$

$$\csc\theta = \pm\sqrt{2}, \qquad \sec\theta = \mp\sqrt{2}, \qquad \cot\theta = -1 \qquad \blacksquare$$

EXERCISES

In Exercises 1–15, use the information given to find the values of six trigonometric functions of θ, where θ is an angle in standard position.

1. $(-4, 3)$ is on the terminal side of θ.

2. $(12, -5)$ is on the terminal side of θ.

3. $(2, 3)$ is on the terminal side of θ.

4. $(-2, -3)$ is on the terminal side of θ.

5. $(-1, -3)$ is on the terminal side of θ.

6. $(1, -2)$ is on the terminal side of θ.

7. $(3, -2)$ is on the terminal side of θ.

8. $(6, 4)$ is on the terminal side of θ.

9. For some point on the terminal side of θ, $r = 7$, $x = -2$.

10. For some point on the terminal side of θ, $r = 6$, $y = -3$.

11. For some point on the terminal side of θ, $r = 4$, $x = 2$.

12. For some point on the terminal side of θ, $r = 3$, $y = 1$.

13. For every point on the terminal side of θ, $y = 3x$.

14. For every point on the terminal side of θ, $y = -2x$.

15. For every point on the terminal side of θ, $y = -4x$.

In Exercises 16–23, determine what quadrant θ is in and the signs of $\sin\theta$, $\cos\theta$, and $\tan\theta$.

16. $756°$

17. $3467°$

18. $-255.6°$

19. $-1106.53°$

20. $\dfrac{11\pi}{7}$

21. 2.67

22. -8.0325

23. $-\dfrac{97\pi}{3}$

In Exercises 24–43, find the values of the trigonometric functions of θ.

24. $\sin\theta = \frac{5}{13}$

25. $\cos\theta = -\frac{7}{25}$

26. $\tan\theta = -\frac{3}{4}$

27. $\cot\theta = \frac{12}{5}$

28. $\csc\theta = -\frac{25}{7}$

29. $\sec\theta = \frac{13}{5}$

30. $\sin\theta = \frac{3}{4}$, $\cos\theta$ negative

31. $\cos\theta = -\frac{2}{3}$, $\sin\theta$ negative

32. $\sin\theta = -\frac{1}{2}$, $\tan\theta$ positive

33. $\tan\theta = 2$, $\cos\theta$ negative

34. $\cos\theta = \frac{1}{3}$, $\cot\theta$ positive

35. $\tan\theta = -\frac{3}{4}$, $\sin\theta$ negative

36. $\csc\theta = 3$, $\sec\theta$ negative

37. $\sin\theta = -\frac{1}{4}$, $\sec\theta$ positive

38. $\cos\theta = \frac{1}{5}$, θ in Q_4

39. $\sin\theta = -\frac{2}{5}$, θ in Q_3

40. $\tan\theta = \sqrt{3}$, θ in Q_3

41. $\tan\theta = \sqrt{3}$, θ in Q_1

42. $\sec\theta = -2$, θ in Q_2

43. $\sin\theta = \frac{3}{5}$, θ in Q_2

In Exercises 44–47, find the unknown coordinate of B and the distance of B from the origin if A and B are on the terminal side of the same angle.

44. $A = (4, -12)$, $B = (x, -24)$

45. $A = (-7, -24)$, $B = (-3, y)$

46. $A = (-2, 3)$, $B = (-5, y)$

47. $A = (1, 4)$, $B = (x, 27)$

48. Assume θ is in Q_3. Show that Definition (1) does not depend on which point (different from the origin) is chosen on the terminal side as follows: Let $P = (x, y)$ and $P' = (x', y')$ be any two such points, and drop perpendiculars to the x-axis. Use similar triangles and the signs of x, y, x', y' to show the values of the trigonometric functions are the same whether you use P or P'.

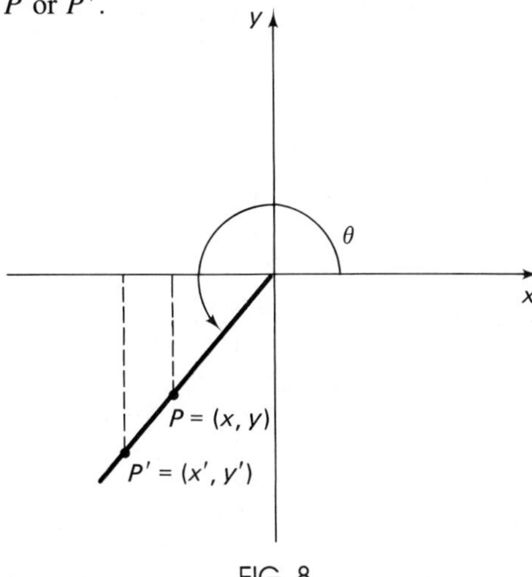

FIG. 8

49. Do Exercise 48 under the assumption θ is in Q_2.

50. Do Exercise 48 under the assumption θ is in Q_1.

51. Do Exercise 48 under the assumption θ is in Q_4.

2.2 The Wrapping Function; the Circular Functions

To use trigonometry effectively in most modern scientific applications and in calculus, it is necessary to have the concept of the trigonometric functions of real numbers. In order to discuss this concept, we must first introduce a new function, called the **wrapping function**.

We begin by drawing a circle of radius 1 centered at the origin. (A circle or radius 1 is called a **unit circle**.) See Fig. 9. Next, draw a vertical line one unit to the right of the y-axis, with the zero on the number line at the point $P = (1, 0)$. We shall call this number line the t-axis.

We now "wrap" the t-axis around the unit circle exactly as though the unit circle were a spool and the t-axis were an infinitely long thread. We wrap the positive t-axis (the part above the x-axis) around the unit circle in a counterclockwise direction. See Fig. 10(a) for the start of this wrapping. We wrap the negative t-axis around the unit circle in the clockwise direction. See Fig. 10(b) for the start of this wrapping.

If t is any point on the t-axis, then, under our wrapping operation, t goes to a corresponding point on the unit circle, which we denote by $W(t)$. This

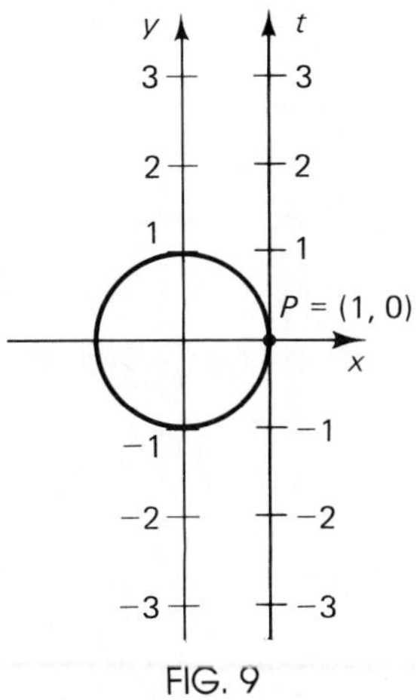

FIG. 9

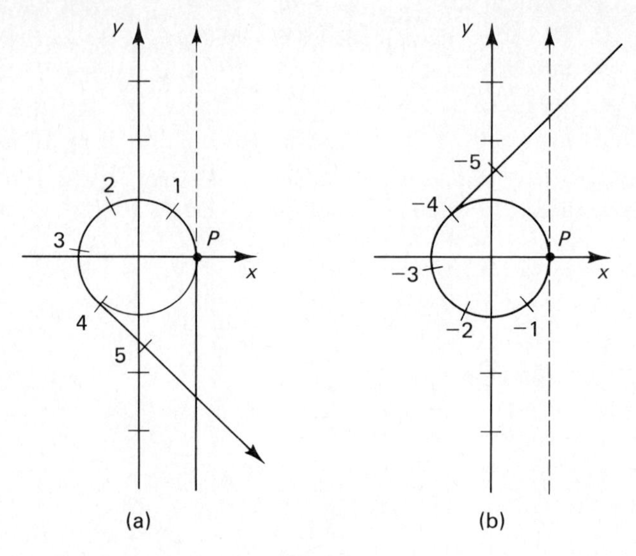

FIG. 10

correspondence, W, is a function, called the **wrapping function**. From the construction, we see that for any real number t, we locate $W(t)$ on the unit circle as follows:

1. If $t = 0$, then $W(t) = (1, 0)$, the point we labeled P.
2. If t is positive, then start at P and go t units along the circumference of the unit circle *in the counterclockwise direction* to obtain the point $W(t)$.
3. If t is negative, then start at P and go $|t|$ units along the circumference of the unit circle *in the clockwise direction* to obtain the point $W(t)$.

With just this information, there is no easy way to find the coordinates of $W(t)$ for most numbers t. However, there are some values of t for which this is easy.

EXAMPLE 1 Find the coordinates of (a) $W\left(\frac{\pi}{2}\right)$, $W(\pi)$, $W\left(\frac{3\pi}{2}\right)$, $W(2\pi)$, $W\left(\frac{5\pi}{2}\right)$, and (b) $W\left(-\frac{\pi}{2}\right)$, $W(-\pi)$, $W\left(-\frac{3\pi}{2}\right)$, $W(-2\pi)$, $W\left(-\frac{5\pi}{2}\right)$.

Solution A unit circle has radius 1, so its circumference is 2π. Thus, starting at (1, 0), each increment of $\frac{\pi}{2}$ units traveled along the circumference of the unit circle takes you a quarter of the way around the unit circle, to the next place where an axis crosses the unit circle. See Fig. 11. Consequently:

(a) $W\left(\frac{\pi}{2}\right) = (0, 1)$ (b) $W\left(-\frac{\pi}{2}\right) = (0, -1)$

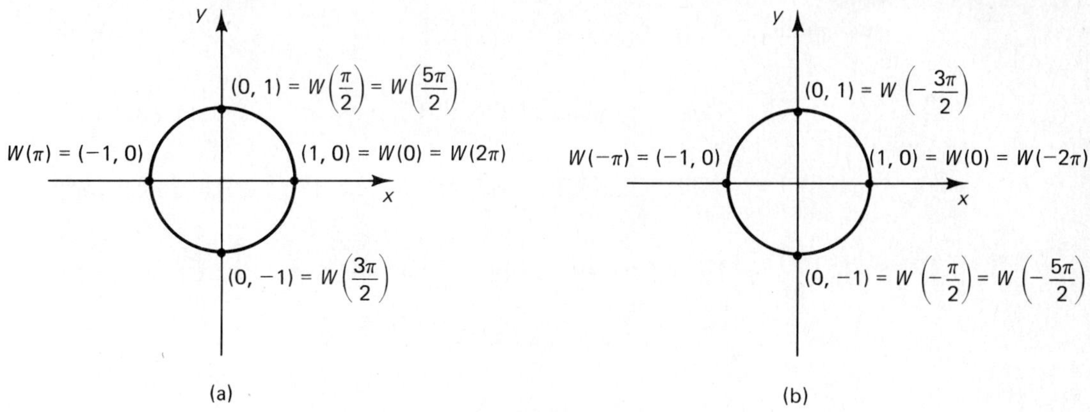

FIG. 11

$$W(\pi) = (-1, 0) \qquad W(-\pi) = (-1, 0)$$

$$W\left(\frac{3\pi}{2}\right) = (0, -1) \qquad W\left(-\frac{3\pi}{2}\right) = (0, 1)$$

$$W(2\pi) = (1, 0) \qquad W(-2\pi) = (1, 0)$$

$$W\left(\frac{5\pi}{2}\right) = (0, 1) \qquad W\left(-\frac{5\pi}{2}\right) = (0, -1) \qquad \blacksquare$$

PERIODICITY OF THE WRAPPING FUNCTION

In Example 1, a pattern is beginning to emerge. Since 2π is one complete circuit around the circumference of the unit circle, the wrapping function maps the real numbers t and $t + 2\pi$ *to the same point* on the unit circle. That is,

(2) $W(t + 2\pi) = W(t)$ for any real number t

This property is important enough to merit a name. In general, a nonconstant function f is called **periodic** if there is a real number $p \neq 0$ such that $f(x + p) = f(x)$ for all x for which $f(x)$ is defined. If p_1 is the smallest positive real number such that $f(x + p_1) = f(x)$ for all x for which $f(x)$ is defined, then p_1 is called the **period** of f. From the preceding, the wrapping function is periodic of period 2π. Note that by applying (2) twice, $W(t + 4\pi) = W[(t + 2\pi) + 2\pi] = W(t + 2\pi) = W(t)$. More generally,

(3) $W(t + 2n\pi) = W(t)$ for any real number t and any integer n

EXAMPLE 2 Find $W\left(\frac{73\pi}{2}\right)$.

Solution Since $\frac{73\pi}{2} = \frac{\pi}{2} + 18(2\pi)$, $\quad W\left(\frac{73\pi}{2}\right) = W\left(\frac{\pi}{2}\right) = (0, 1)$. ■

This periodicity of W will be exploited in the next section.

RELATIONSHIP BETWEEN THE WRAPPING FUNCTION AND RADIAN MEASURE

The reason the wrapping function is so important is the following very nice relationship.

> **(4)** If t is any real number and θ is the angle in standard position with radian measure t, then $W(t)$ is the point on the terminal side of θ one unit from the origin.

To see this, we need to recall the definition of radian measure.

> **(5) Definition of Radian Measure** On the initial side l of an angle θ, let P be the point one unit from the vertex. If θ is positive or 0, let t be the length of the circular arc traced out by P as l rotates through θ; if θ is negative, let t be the negative of that length. Then the radian measure of θ is t, and we write
>
> $$\theta = t \text{ rad}$$

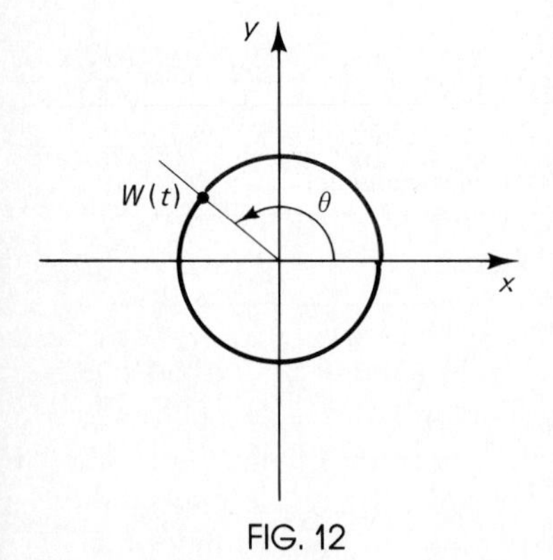

FIG. 12

Now let t be any real number, and plot $W(t)$. See Fig. 12. On the same drawing, form the angle θ in standard position with radian measure t by following the procedure of the above definition. Then θ is the central angle subtended by the arc t and $W(t)$ is on the terminal side of θ. Thus we have the relationship given in (4).

To give us a few examples to work with, we now use the relationship to find the coordinates of $W(t)$ when t is a multiple of $\frac{\pi}{6}$ or $\frac{\pi}{4}$. In these cases, we can use the special triangles in Fig. 13. This is demonstrated in Example 3. (In Sec. 5 of this chapter we shall show how to find the coordinates of $W(t)$ for any value of t.)

1 $\quad \frac{\pi}{4} \quad \frac{1}{\sqrt{2}} \quad \frac{\pi}{4} \quad \frac{1}{\sqrt{2}}$

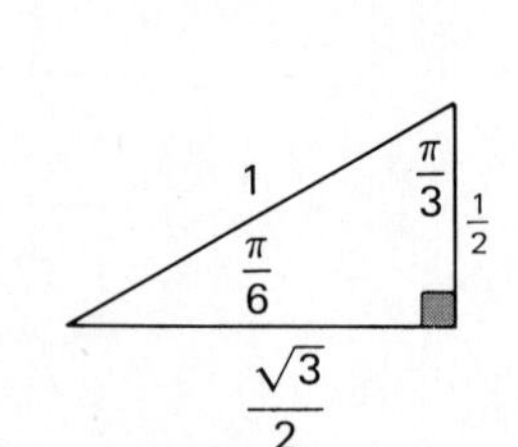

FIG. 13

EXAMPLE 3 Find the coordinates of (a) $W\left(\frac{\pi}{4}\right)$, (b) $W\left(\frac{5\pi}{6}\right)$, and (c) $W\left(\frac{4\pi}{3}\right)$.

Solution In each of these cases, form a right triangle by dropping a perpendicular from $W(t)$ to the x-axis, then connecting $W(t)$ to the origin. See Fig. 14. Let $W(t) = (x, y)$. In these right triangles, the hypotenuse has length 1, and the lengths of the horizontal and vertical legs are $|x|$ and $|y|$, respectively (a length cannot be negative). Thus the coordinates x and y are the lengths of the legs, together with the appropriate sign. The sign is determined by the quadrant in which $W(t)$ lies.

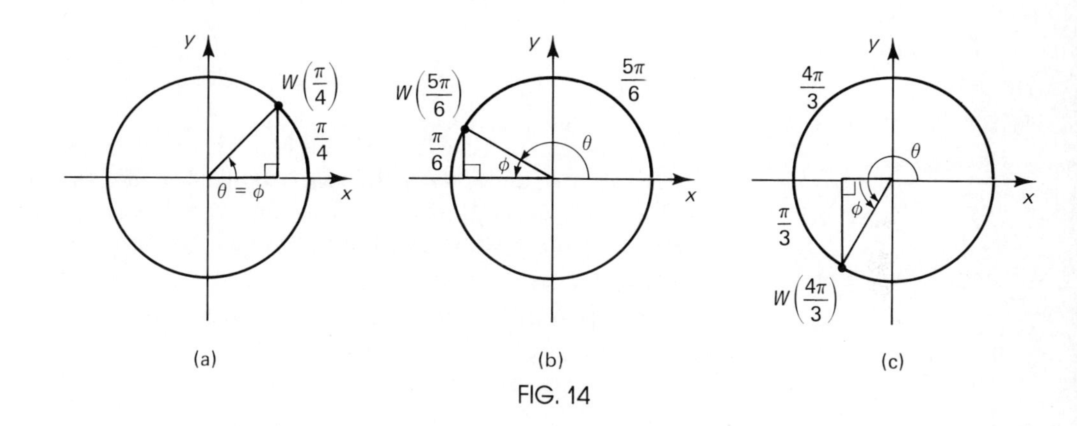

FIG. 14

From the relationship (4), in each case we know the measure of the angle θ subtended by the arc of length t. Since we know θ, we can determine the measure of the angle ϕ as indicated in Fig. 14. Then from the special triangles in Fig. 13, we can immediately determine the lengths of the legs of the triangles in Fig. 14. Adding the appropriate sign gives the coordinates of $W(t)$.

(a) $\theta = \phi = \frac{\pi}{4}$. By Fig. 13, each leg has length $\frac{1}{\sqrt{2}}$. Since $W(t)$ is in the first quadrant, both coordinates are positive. Thus $W\left(\frac{\pi}{4}\right) = \left(\frac{1}{\sqrt{2}}, \frac{1}{\sqrt{2}}\right)$.

(b) $\theta = \frac{5\pi}{6}$, and so $\phi = \frac{\pi}{6}$. By Fig. 13, the lengths of the legs are $\frac{\sqrt{3}}{2}$ and $\frac{1}{2}$. Since $W(t)$ is in the second quadrant, the x-coordinate is negative and the y-coordinate is positive. Thus $W\left(\frac{5\pi}{6}\right) = \left(-\frac{\sqrt{3}}{2}, \frac{1}{2}\right)$.

(c) $\theta = \frac{4\pi}{3}$, and so $\phi = \frac{\pi}{3}$. By Fig. 13, the lengths of the legs are $\frac{1}{2}$ and $\frac{\sqrt{3}}{2}$. Since $W(t)$ is in the third quadrant, both coordinates are negative. Thus $W\left(\frac{4\pi}{3}\right) = \left(-\frac{1}{2}, -\frac{\sqrt{3}}{2}\right)$. ■

THE CIRCULAR FUNCTIONS

We have just seen that if t is any real number, then there is an associated angle θ in standard position such that the radian measure of θ is t, and the point $W(t) = (x, y)$ is on the terminal side of θ. Of course, the distance r from $W(t)$ to the origin is 1 (since $W(t)$ is on the unit circle). Using the definition of the trigonometric functions from (1) in the previous section, we see

$$\sin\theta = \frac{y}{r} = y \qquad \csc\theta = \frac{r}{y} = \frac{1}{y} \quad (\text{if } y \neq 0)$$

$$\cos\theta = \frac{x}{r} = x \qquad \sec\theta = \frac{r}{x} = \frac{1}{x} \quad (\text{if } x \neq 0)$$

$$\tan\theta = \frac{y}{x} \quad (\text{if } x \neq 0) \qquad \cot\theta = \frac{x}{y} \quad (\text{if } y \neq 0)$$

Given the association between t and θ, it is natural to *define* the **trigonometric functions of the real number t** in terms of the coordinates (x, y) of $W(t)$ in exactly the same way. Because $W(t)$ is on the unit circle, these functions are often referred to as the **circular functions.**

(6) Definition For any real number t, suppose $W(t) = (x, y)$. Then

$$\sin t = y \qquad \csc t = \frac{1}{y} \quad (\text{if } y \neq 0)$$

$$\cos t = x \qquad \sec t = \frac{1}{x} \quad (\text{if } x \neq 0)$$

$$\tan t = \frac{y}{x} \quad (\text{if } x \neq 0) \qquad \cot t = \frac{x}{y} \quad (\text{if } y \neq 0)$$

From the motivation behind the definition, we see immediately:

(7) The trigonometric functions of a real number t are exactly the same as the trigonometric functions of the associated angle with radian measure t.

We now illustrate how to evaluate the circular functions for those values of t for which we know $W(t)$.

EXAMPLE 4 Find the values of the six circular functions of $-\dfrac{3\pi}{2}$.

Solution We know $W\left(-\dfrac{3\pi}{2}\right) = (0, 1)$ from Example 1. Hence

$$\sin\left(-\frac{3\pi}{2}\right) = y = 1 \qquad \csc\left(-\frac{3\pi}{2}\right) = \frac{1}{y} = 1$$

$$\cos\left(-\frac{3\pi}{2}\right) = x = 0 \qquad \sec\left(-\frac{3\pi}{2}\right) = \frac{1}{x} \text{ is undefined}$$

$$\tan\left(-\frac{3\pi}{2}\right) = \frac{y}{x} \text{ is undefined} \qquad \cot\left(-\frac{3\pi}{2}\right) = \frac{x}{y} = 0 \quad \blacksquare$$

EXAMPLE 5 Find the values of the six circular functions of $\frac{4\pi}{3}$.

Solution We know $W\left(\frac{4\pi}{3}\right) = \left(-\frac{1}{2}, -\frac{\sqrt{3}}{2}\right)$ from Example 3. Hence

$$\sin\frac{4\pi}{3} = -\frac{\sqrt{3}}{2} \qquad \csc\frac{4\pi}{3} = -\frac{2}{\sqrt{3}}$$

$$\cos\frac{4\pi}{3} = -\frac{1}{2} \qquad \sec\frac{4\pi}{3} = -2$$

$$\tan\frac{4\pi}{3} = \frac{-\sqrt{3}/2}{-1/2} = \sqrt{3} \qquad \cot\frac{4\pi}{3} = \frac{-1/2}{-\sqrt{3}/2} = \frac{1}{\sqrt{3}} \quad \blacksquare$$

EXERCISES

In Exercises 1–24, find the coordinates of W(t).

1. $W(3\pi)$ **2.** $W\left(\frac{7\pi}{2}\right)$ **3.** $W\left(\frac{9\pi}{2}\right)$ **4.** $W(5\pi)$

5. $W(-3\pi)$ **6.** $W\left(-\frac{7\pi}{2}\right)$ **7.** $W\left(-\frac{9\pi}{2}\right)$ **8.** $W(-5\pi)$

9. $W\left(\frac{49\pi}{2}\right)$ **10.** $W(55\pi)$ **11.** $W(86\pi)$ **12.** $W\left(\frac{91\pi}{2}\right)$

13. $W\left(-\frac{113\pi}{2}\right)$ **14.** $W\left(-\frac{83\pi}{2}\right)$ **15.** $W\left(\frac{\pi}{6}\right)$ **16.** $W\left(\frac{\pi}{3}\right)$

17. $W\left(\frac{3\pi}{4}\right)$ **18.** $W\left(\frac{2\pi}{3}\right)$ **19.** $W\left(\frac{7\pi}{6}\right)$ **20.** $W\left(\frac{11\pi}{6}\right)$

21. $W\left(-\frac{\pi}{4}\right)$ **22.** $W\left(-\frac{\pi}{6}\right)$ **23.** $W\left(-\frac{5\pi}{6}\right)$ **24.** $W\left(-\frac{\pi}{3}\right)$

In Exercises 25–40, find the values of the six circular functions of t for the given value of t.

25. $\frac{\pi}{2}$ **26.** π **27.** $-\frac{\pi}{2}$ **28.** $-\frac{7\pi}{2}$

29. $\frac{49\pi}{2}$ **30.** $-\frac{83\pi}{2}$ **31.** $\frac{\pi}{4}$ **32.** $\frac{\pi}{3}$

33. $\frac{\pi}{6}$ **34.** $\frac{2\pi}{3}$ **35.** $\frac{3\pi}{4}$ **36.** $\frac{11\pi}{6}$

37. $-\frac{\pi}{4}$ **38.** $-\frac{\pi}{6}$ **39.** $-\frac{5\pi}{6}$ **40.** $-\frac{\pi}{3}$

2.3 Properties of the Trigonometric Functions

In this section we shall describe some of the properties of the trigonometric functions. Of primary importance is that the fundamental identities we observed in the previous chapter still hold. These identities and others in this section will be stated and verified in terms of the circular functions, i.e., the trigonometric functions of real numbers. However, since the trigonometric functions of a real number t are exactly the same as the trigonometric functions of the associated angle with radian measure t, it easily follows that the identities hold for trigonometric functions of angles.

THE FUNDAMENTAL TRIGONOMETRIC IDENTITIES

(8) $$\csc t = \frac{1}{\sin t}, \qquad \sec t = \frac{1}{\cos t}, \qquad \cot t = \frac{1}{\tan t}$$

(9) $$\tan t = \frac{\sin t}{\cos t}, \qquad \cot t = \frac{\cos t}{\sin t}$$

(10) $$\sin^2 t + \cos^2 t = 1$$

It is immediate from Definition (6) that (8) and (9) are still true. We can verify (10) as follows:

If $W(t) = (x, y)$, then $x^2 + y^2 = 1$, since (x, y) is on the unit circle. Consequently,

$$\sin^2 t + \cos^2 t = y^2 + x^2 = 1 \qquad \blacksquare$$

There are two identities similar to (10) that can be obtained from (8)–(10). If (10) is divided by $\cos^2 t$, we obtain

$$\frac{\sin^2 t}{\cos^2 t} + \frac{\cos^2 t}{\cos^2 t} = \frac{1}{\cos^2 t}$$

or

$$\left(\frac{\sin t}{\cos t}\right)^2 + 1 = \left(\frac{1}{\cos t}\right)^2$$

By (8) and (9),

(11) $$\tan^2 t + 1 = \sec^2 t$$

If instead we divide (10) by $\sin^2 t$, we obtain

(12)	$1 + \cot^2 t = \csc^2 t$

There are two other identities that follow immediately from Defintion (6) and from the fact that if $W(t) = (x, y)$, then $W(-t) = (x, -y)$. See Fig. 15. These are

$$\sin(-t) = -y = -(y) = -\sin t$$

and

$$\cos(-t) = x = \cos t.$$

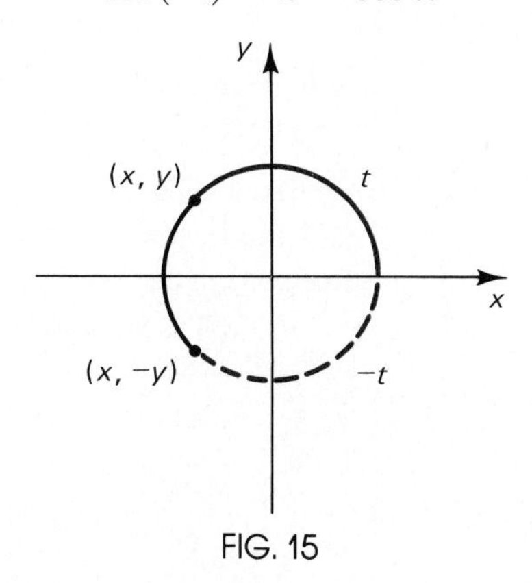

FIG. 15

Therefore we have shown the following:

(13)	$\sin(-t) = -\sin t$ and $\cos(-t) = \cos t$

In Sec. 1 of this chapter, we saw that if we are given the value of one trigonometric function, then we can find the values of the others directly from Definition (1). However, this type of problem can also be done using the fundamental identities (8)–(10), just as we did in the previous chapter. If tan t or cot t is given, it is sometimes useful to use (11) or (12) as well.

EXAMPLE 1 Suppose $\tan t = -\frac{1}{2}$ and $\sin t$ is positive. Find the remaining trigonometric functions of t.

Solution Since $\tan t$ is negative and $\sin t$ is positive, $W(t)$ must be in Q_2. Hence, $\cos t$ and $\sec t$ are negative. From (11), $\tan^2 t + 1 = \sec^2 t$, we have $(-\frac{1}{2})^2 + 1 = \sec^2 t$, or $\sec^2 t = \frac{5}{4}$. Thus, $\sec t = -\dfrac{\sqrt{5}}{2}$, and by (8), $\cos t = -\dfrac{2}{\sqrt{5}}$. From (9),

$$\tan t = \frac{\sin t}{\cos t}, \quad \text{or} \quad \sin t = \cos t \tan t = \left(-\frac{2}{\sqrt{5}}\right)\left(-\frac{1}{2}\right) = \frac{1}{\sqrt{5}}$$

Thus by (8), $\csc t = \sqrt{5}$. Also by (8), since $\tan t = -\frac{1}{2}$, $\cot t = -2$. ■

PERIODICITY OF THE TRIGONOMETRIC FUNCTIONS

In the previous section, we observed that the wrapping function was periodic of period 2π; i.e., $W(t) = W(t + 2\pi)$ for any real number t. But this means that if $W(t) = (x, y)$, then $W(t + 2\pi)$ also has the same coordinates (x, y). Since the trigonometric functions are defined only in terms of coordinates, it follows that

(14)

$$\sin t = \sin(t + 2\pi) \qquad \csc t = \csc(t + 2\pi)$$
$$\cos t = \cos(t + 2\pi) \qquad \sec t = \sec(t + 2\pi)$$
$$\tan t = \tan(t + 2\pi) \qquad \cot t = \cot(t + 2\pi)$$

whenever the functions are defined.

In other words, the trigonometric functions are all periodic. We shall see later that sine, cosine, cosecant, and secant each have period 2π, while tangent and cotangent each have period π.

By repeated applications of (14), we have

(15)

$$\sin t = \sin(t + n \cdot 2\pi) \qquad \csc t = \csc(t + n \cdot 2\pi)$$
$$\cos t = \cos(t + n \cdot 2\pi) \qquad \sec t = \sec(t + n \cdot 2\pi)$$
$$\tan t = \tan(t + n \cdot 2\pi) \qquad \cot t = \cot(t + n \cdot 2\pi)$$

whenever the functions are defined.

EXAMPLE 2 Find (a) $\sin 45\pi$, and (b) $\cos 765°$.

Solution (a) By (15),

$$\sin 45\pi = \sin(\pi + 22 \cdot 2\pi) = \sin \pi = 0$$

(b) Since $360° = 2\pi$ rad, the equation corresponding to (15) for the cosine of an angle in degrees is $\cos \theta = \cos(\theta + n \cdot 360°)$. Hence,

$$\cos 765° = \cos(45° + 2 \cdot 360°) = \cos 45° = \frac{1}{\sqrt{2}} \qquad ■$$

Using periodicity and equations like those in Example 2a, we can easily form the following table of the circular functions of t whenever t is a multiple of $\frac{\pi}{2}$:

	$\sin t$	$\cos t$	$\tan t$	$\csc t$	$\sec t$	$\cot t$
$0 + n \cdot 2\pi$	0	1	0	–	1	–
$\frac{\pi}{2} + n \cdot 2\pi$	1	0	–	1	–	0
$\pi + n \cdot 2\pi$	0	−1	0	–	−1	–
$\frac{3\pi}{2} + n \cdot 2\pi$	−1	0	–	−1	–	0

Note that a dash is used to indicate the function is undefined.

EXERCISES

In Exercises 1–20, use identities (8)–(13) to find the values of the trigonometric functions of t.

1. $\sin t = \frac{5}{13}$

2. $\cos t = -\frac{7}{25}$

3. $\tan t = -\frac{4}{3}$

4. $\cot t = -\frac{12}{5}$

5. $\csc t = -\frac{25}{24}$

6. $\sec t = \frac{13}{5}$

7. $\sin t = \frac{1}{3}$, $W(t)$ in Q_2

8. $\cos t = -\frac{1}{5}$, $W(t)$ in Q_3

9. $\sec t = -2$, $W(t)$ in Q_2

10. $\cot t = \sqrt{3}$, $W(t)$ in Q_3

11. $\tan t = -1$, $W(t)$ in Q_4

12. $\csc t = 3$, $W(t)$ in Q_1

13. $\sin t = \frac{2}{3}$, $\tan t$ negative

14. $\cos t = -\frac{1}{4}$, $\sin(-t)$ negative

15. $\tan t = -\sqrt{3}$, $\sin(-t)$ positive

16. $\sin t = \frac{3}{5}$, $\cos(-t)$ positive

17. $\cot t = \frac{1}{2}$, $\cot(-t)$ negative

18. $\sec t = -\sqrt{5}$, $\tan t$ positive

19. $\cot t = \sqrt{3}$, $\cos t$ negative

20. $\csc t = 2$, $\sec t$ negative

In Exercises 21–40, use (15) to evaluate the given functions.

21. $\cos \dfrac{75\pi}{2}$

22. $\cot \dfrac{41\pi}{2}$

23. $\tan(-125\pi)$

24. $\csc \dfrac{123\pi}{2}$

25. $\sin 1710°$

26. $\cos 2340°$

27. $\sec(-1080°)$

28. $\cot(-3060°)$

29. $\sec \dfrac{65\pi}{4}$

30. $\sin \dfrac{87\pi}{4}$

31. $\tan\left(-\dfrac{97\pi}{4}\right)$

32. $\csc\left(-\dfrac{13\pi}{4}\right)$

33. $\cot \dfrac{29\pi}{3}$

34. $\cos \dfrac{94\pi}{3}$

35. $\sin\left(-\dfrac{41\pi}{3}\right)$

36. $\sec\left(-\dfrac{109\pi}{6}\right)$

37. $\csc 4725°$

38. $\tan 7770°$

39. $\cos(-2760°)$

40. $\cot(-11190°)$

2.4 Reference Angles and Reference Numbers

We now turn our attention to reference angles and reference numbers. The general concept involved is one of the most useful concepts in trigonometry, and it provides a vital connection between general trigonometry and right triangle trigonometry.

> **(16) Definition** If θ is an angle in standard position, the **reference angle** of θ is the smallest positive angle ϕ between the terminal side of θ and the x-axis.

If θ is a quadrantal angle, then the reference angle of θ is $\phi = 0 = 0°$ if the terminal side is on the x-axis or $\phi = \frac{\pi}{2} = 90°$ if the terminal side is on the y-axis. Otherwise Fig. 16 demonstrates where ϕ is, depending on the quadrant where the terminal side of θ lies. In each case, one of the many possible angles θ with the given terminal side is shown. For that terminal side, ϕ is unique.

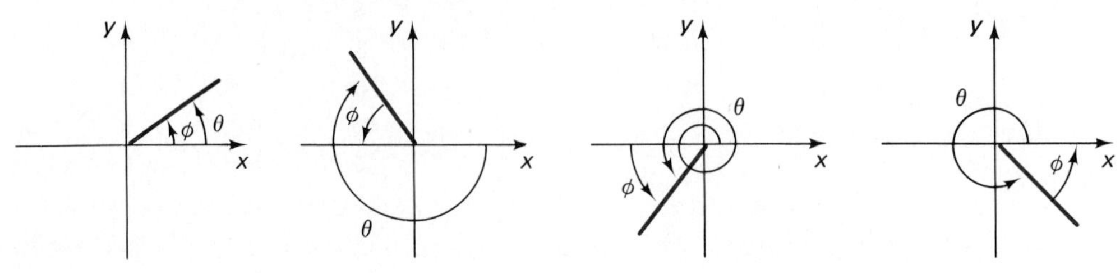

FIG. 16

EXAMPLE 1 Find the reference angle ϕ if θ is (a) 35°; (b) 114°; (c) 4.08; (d) 5.1963

Solution (a) $\phi = \theta = 35°$. See Fig. 17(a).
(b) $\phi = 180° - \theta = 180° - 114° = 66°$. See Fig. 17(b).
(c) $\phi = \theta - \pi = 4.08 - \pi \approx 0.938407$. See Fig. 17(c).
(d) $\phi = 2\pi - \theta = 2\pi - 5.1963 \approx 1.08689$. See Fig. 17(d).

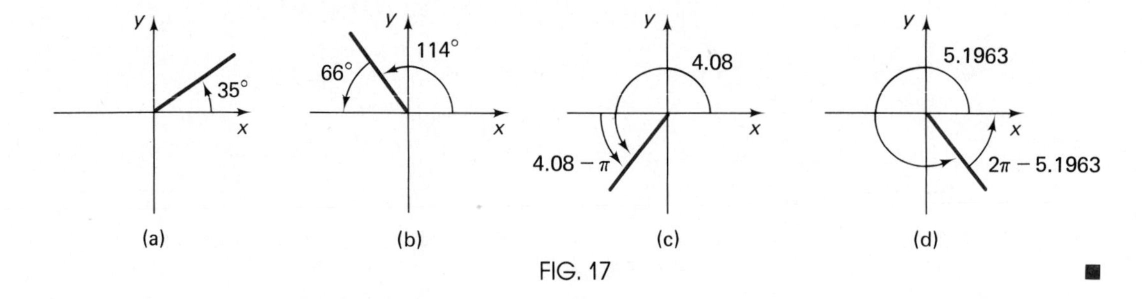

FIG. 17 ■

In general when you are given an angle θ, you first determine the quadrant where the terminal side of θ lies and then proceed as in Example 1 to find the reference angle. In doing the latter, you can add or subtract multiples of $2\pi = 360°$ without changing the reference angle, since complete rotations do not affect where the terminal side is located.

EXAMPLE 2 If $\theta = 7789°$, find the quadrant in which the terminal side lies, and then determine the reference angle ϕ.

Solution We divide 7789° by 360°, obtaining 21.6361. Thus 7789° is 21 complete rotations plus 0.6361 of a complete rotation. Since the last is between one half and three quarters of a complete rotation [it has measure $7789° - 21(360°) = 229°$], the terminal side is in Q_3. The reference angle is $\phi = 229° - 180° = 49°$. See Fig. 18. ■

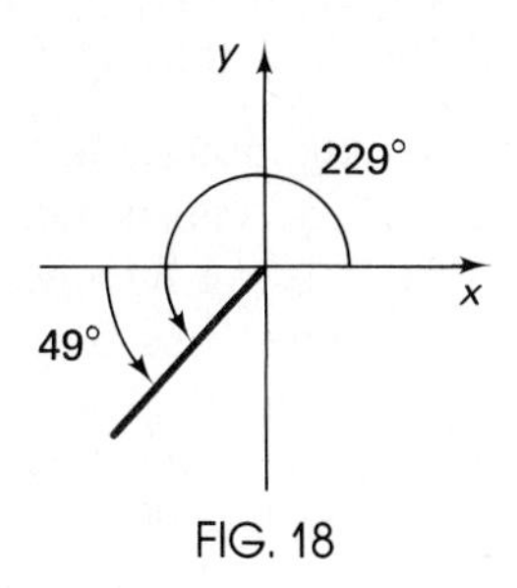

FIG. 18

Actually, we have already used reference angles in a few special cases, although we did not use the term at the time. If you refer back to Example 3 in Sec. 2 of this chapter, where we were finding the coordinates of $W(t)$ for t a multiple of $\frac{\pi}{6}$ or $\frac{\pi}{4}$, you will see that the angle ϕ we found there was the reference angle for an angle (in standard position) of measure t radians. At that time, we were illustrating the relationship between a real number and the associated angle. Given that relationship, it is not surprising that we use the term *reference number* when dealing with real numbers in a way analogous to the use of reference angle when dealing with angles.

> **(17) Definition** If t is a real number, its **reference number** is the shortest arc length s along the unit circle between $W(t)$ and the x-axis.

Observe that if t is a real number and θ is the associated angle with measure t radians, and if the reference number for t is s and the reference angle for θ is ϕ, then ϕ has measure s rad. See Fig. 19.

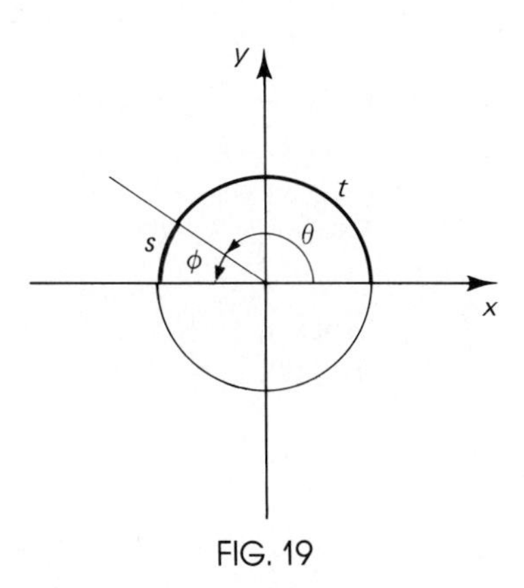

FIG. 19

EXAMPLE 3 Find the reference number of 5.1963.

Solution $W(5.1963)$ is in the fourth quadrant. Therefore the reference number is $2\pi - 5.1963 \approx 1.08689$. Compare this with (d) of Example 1. ■

Warning. If you are computing either reference numbers or reference angles in radians by hand, then unless you are using a fairly accurate approximation of π, t does not have to be very large before s is so inaccurate as to make it essentially useless for further computations. Even if you are using a calculator, it only stores 12 digits or so of any number (including π), which means that if t is large, say 10^8 or 10^{10}, the computation of s can be very inaccurate.

USING REFERENCE ANGLES

Now let us see how reference angles relate general trigonometry and right angle trigonometry. As we have seen, if θ is any angle in standard position, the trigonometric functions depend only on the terminal side, not on how many times or in what direction the initial side was rotated to get there. Assume for a moment that θ is not a quadrantal (between quadrant) angle. Then the terminal side of θ is in one of the quadrants. Now let P be any point (except the origin) on the terminal side, and drop a perpendicular from P to the x-axis. The possible cases are shown in Fig. 20. In each case, a natural right triangle

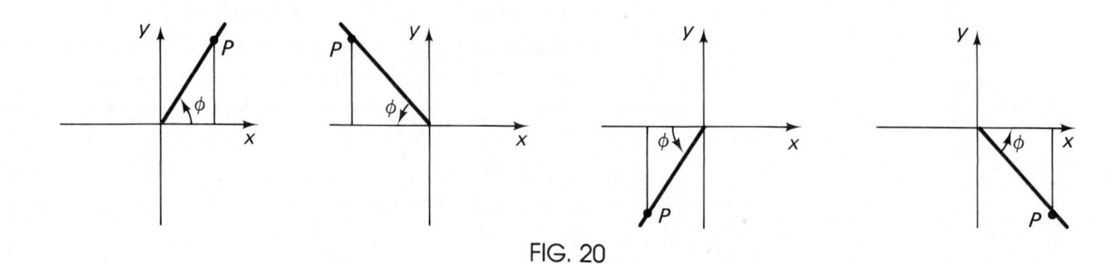

FIG. 20

is formed, and we have used ϕ to denote the angle between the leg along the x-axis and the hypotenuse. Notice that ϕ is also the reference angle of θ in each case. This construction should seem familiar; we used it with $P = W(t)$ in Example 3 of Sec. 2 of this chapter. If P has coordinates (x, y) and $r = \sqrt{x^2 + y^2}$, then all four right triangles have dimensions as shown in Fig. 21. Since $P = (x, y)$ is a point on the terminal side of θ, we can compute the trigonometric functions of θ and compare them with those of ϕ:

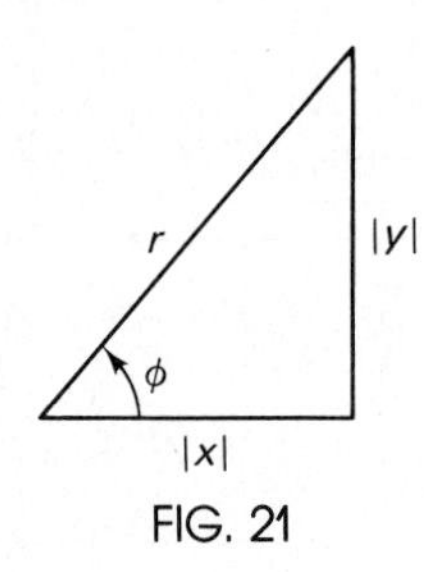

FIG. 21

$$\sin\theta = \frac{y}{r} \qquad \sin\phi = \frac{|y|}{r}$$

$$\cos\theta = \frac{x}{r} \qquad \cos\phi = \frac{|x|}{r}$$

$$\tan\theta = \frac{y}{x} \qquad \tan\phi = \frac{|y|}{|x|}$$

$$\csc\theta = \frac{r}{y} \qquad \csc\phi = \frac{r}{|y|}$$

$$\sec\theta = \frac{r}{x} \qquad \sec\phi = \frac{r}{|x|}$$

$$\cot\theta = \frac{x}{y} \qquad \cot\phi = \frac{|x|}{|y|}$$

The only possible differences are the signs.

> **(18)** If ϕ is the reference angle of an angle θ, then the value of a trigonometric function at θ and the value of that same function at ϕ agree, except possibly for the sign.

[Although quadrantal angles were excluded from the discussion because no right triangle is formed, result (18) holds for them as well.]

EXAMPLE 4 Suppose $(-3, 4)$ is on the terminal side of θ. Compare the trigonometric functions of θ and its reference angle ϕ.

Solution Since $(-3, 4)$ is on the terminal side and $(-3, 4)$ is in Q_2, we have the graph in Fig. 22 and the corresponding right triangle. Thus

$$\begin{array}{llll}
\sin\theta = \frac{4}{5}, & \sin\phi = \frac{4}{5} & \csc\theta = \frac{5}{4}, & \csc\phi = \frac{5}{4} \\
\cos\theta = -\frac{3}{5} & \cos\phi = \frac{3}{5} & \sec\theta = -\frac{5}{3}, & \sec\phi = \frac{5}{3} \\
\tan\theta = -\frac{4}{3}, & \tan\phi = \frac{4}{3} & \cot\theta = -\frac{3}{4}, & \cot\phi = \frac{3}{4}
\end{array}$$

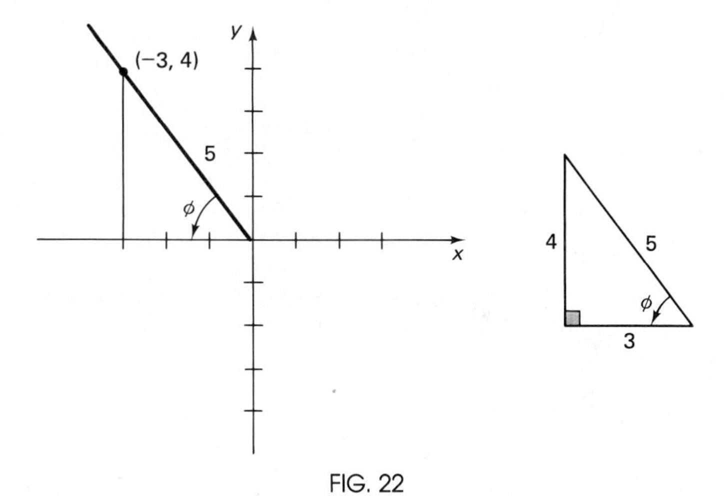

FIG. 22

We see that corresponding functions agree, except for some of the signs. ■

USING REFERENCE NUMBERS

From the relationship between reference angles (in radians) and reference numbers, it follows immediately that the result corresponding to (18) holds for real numbers.

> **(19)** If s is the reference number of the real number t, then the value of a circular function at t and the value of that same function at s agree, except possibly for the sign.

EXAMPLE 5 Find the trigonometric functions of $\frac{5\pi}{3}$ (without a calculator or tables).

Solution $W\left(\frac{5\pi}{3}\right)$ is in Q_4, and the reference number is $\frac{\pi}{3}$. We know that

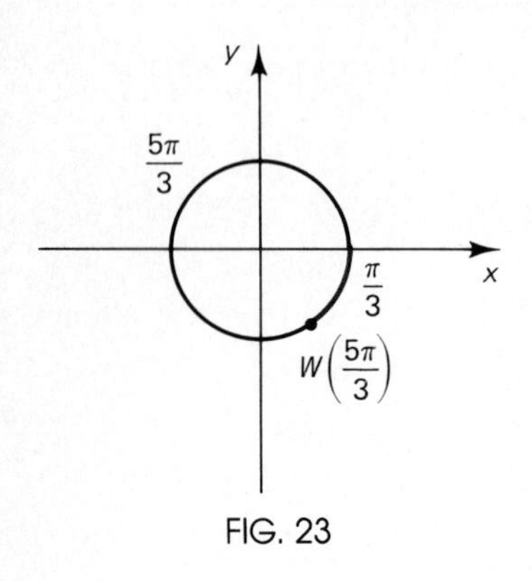

FIG. 23

$$\sin\frac{\pi}{3} = \frac{\sqrt{3}}{2}, \qquad \cos\frac{\pi}{3} = \frac{1}{2}, \qquad \tan\frac{\pi}{3} = \sqrt{3}$$

$$\csc\frac{\pi}{3} = \frac{2}{\sqrt{3}}, \qquad \sec\frac{\pi}{3} = 2, \qquad \cot\frac{\pi}{3} = \frac{1}{\sqrt{3}}$$

In Q_4, cosine and secant are positive, and the rest negative. Hence

$$\sin\frac{5\pi}{3} = -\frac{\sqrt{3}}{2}, \qquad \cos\frac{5\pi}{3} = \frac{1}{2}, \qquad \tan\frac{5\pi}{3} = -\sqrt{3}$$

$$\csc\frac{5\pi}{3} = -\frac{2}{\sqrt{3}}, \qquad \sec\frac{5\pi}{3} = 2, \qquad \cot\frac{5\pi}{3} = -\frac{1}{\sqrt{3}} \qquad ■$$

SPECIAL ANGLES

A calculation similar to the one in Example 5 would enable us to find the trigonometric functions of a number or angle whenever the reference number or angle is $\frac{\pi}{6}$, $\frac{\pi}{4}$, or $\frac{\pi}{3}$ (30°, 45°, or 60°). Because of this, we now use the term **special angle** to denote any angle with one of these three angles as reference angle. These special angles arise often, both in examples and in actual applications. Consequently, it is important to be able to compute the trigonometric functions of special angles. We strongly recommend you do Exercise 25, perhaps even several times, so that you can find these values with ease.

EXERCISES

In Exercises 1–16, the angle is in standard position. Find in which quadrant the terminal side lies, and determine the reference angle.

1. 237° **2.** −38° **3.** 103° **4.** 409°
5. −716° **6.** 903° **7.** −331°15′ **8.** 497.23°
9. $\frac{7\pi}{9}$ **10.** $\frac{38\pi}{3}$ **11.** $-\frac{24\pi}{5}$ **12.** $\frac{11\pi}{6}$
13. 2 **14.** 5 **15.** 15 **16.** −11

In Exercises 17–24, determine the reference number of the given real number.

17. $\frac{7\pi}{9}$ **18.** $\frac{38\pi}{3}$ **19.** $-\frac{12\pi}{5}$ **20.** $-\frac{65\pi}{7}$
21. 2 **22.** −11 **23.** 6 **24.** 16.257

25. Fill out the following table (without using a calculator or tables):

	$\frac{\pi}{6}$	$\frac{\pi}{4}$	$\frac{\pi}{3}$	$\frac{2\pi}{3}$	$\frac{3\pi}{4}$	$\frac{5\pi}{6}$	$\frac{7\pi}{6}$	$\frac{5\pi}{4}$	$\frac{4\pi}{3}$	$\frac{5\pi}{3}$	$\frac{7\pi}{4}$	$\frac{11\pi}{6}$
sin t												
cos t												
tan t												

	$-\frac{\pi}{6}$	$-\frac{\pi}{4}$	$-\frac{\pi}{3}$	$-\frac{2\pi}{3}$	$-\frac{3\pi}{4}$	$-\frac{5\pi}{6}$	$-\frac{7\pi}{6}$	$-\frac{5\pi}{4}$	$-\frac{4\pi}{3}$	$-\frac{5\pi}{3}$	$-\frac{7\pi}{4}$	$-\frac{11\pi}{6}$
sin t												
cos t												
tan t												

In Exercises 26–33, verify the equations without using a calculator or tables.

26. $\sin^2 \frac{5\pi}{6} + \cos^2 \frac{5\pi}{6} = 1$

27. $\cos\left(\frac{3\pi}{2} + \frac{\pi}{6}\right) = \sin \frac{\pi}{6}$

28. $\sin \frac{\pi}{3} \cos \frac{\pi}{6} + \cos \frac{\pi}{3} \sin \frac{\pi}{6} = 1$

29. $\cos 1230° + \cos(-390°) = 0$

30. $\sin\left(\pi + \frac{\pi}{4}\right) = -\sin \frac{\pi}{4}$

31. $\sin\left(\frac{3\pi}{2} + \frac{\pi}{6}\right) = -\cos \frac{\pi}{6}$

32. $\cos \frac{\pi}{3} \cos \frac{\pi}{6} + \sin \frac{\pi}{3} \sin \frac{\pi}{6} = \cos \frac{\pi}{6}$

33. $\sin 270° \cos 990° + \sin 540° \cos 720° = 0$

2.5 Evaluating Trigonometric Functions

EVALUATION USING A CALCULATOR

This is a place where using a claculator is significantly easier than using tables. The procedure for evaluating a trigonometric function at any angle is *exactly the same* as evaluating at an acute angle. Remember, ***make sure the calculator is in the proper mode.***

EXAMPLE 1 Find (a) $\sin 192°$, (b) $\cos 2$, (c) $\tan 90°$, and (d) $\cos \frac{5\pi}{6}$.

Solution (a) With the calculator in degree mode, press [192] [sin], obtaining -0.207912.

(b) The 2 could represent either a real number or an angle with measure 2 rad. In either case, with the calculator in radian mode, press [2] [cos], obtaining -0.416147.

(c) With the calculator in degree mode, press [90] [tan]. You should obtain the error indication on your calculator since $\tan 90°$ is undefined.

(d) With the calculator in radian mode, you could first compute $\frac{5\pi}{6}$, then press [cos], obtaining -0.866025. However, since the reference number (angle) is $\frac{\pi}{6}$, we already saw in the previous section that we could compute this without a calculator, obtaining $-\frac{\sqrt{3}}{2}$. This form of the answer is often preferable if the result is to be used in subsequent calculations. ■

The fundamental identities

$$\csc\theta = \frac{1}{\sin\theta}, \qquad \sec\theta = \frac{1}{\cos\theta}, \qquad \text{and} \quad \cot\theta = \frac{1}{\tan\theta}$$

are used when evaluating cosecant, secant, or cotangent on a calculator.

EXAMPLE 2 Find $\csc(-218.2°)$.

Solution With the calculator in degree mode, use $\csc(-218.2°) = \frac{1}{\sin(-218.2°)}$, and press [218.2] [+/−] [sin] [1/x], obtaining 1.61705. ■

Warning: It may be wise to use periodicity for large values of θ even if the calculator can handle them, because machine error begins to play a significant role for the larger values. For instance, one calculator computes $\sin\left(10^{10}\pi + \frac{1}{6}\pi\right)$ as 0 instead of $\sin\frac{1}{6}\pi = 0.5$. Moreover, if θ is measured in radians or represents a real number and $|\theta|$ is larger than say 10^7 or 10^8, then even using periodicity can lead to massive round-off error since the calculator uses only an 8–12-digit approximation of π. If θ is in either degrees or radians and is large, roughly $|\theta| \geq 10^{10}$, then there is almost no chance that the calculator will compute the trigonometric functions of that angle correctly.

Recall there was a similar warning about round-off errors in the previous section when we were discussing reference angles and reference numbers.

EXAMPLE 3 Suppose the calculator computes the trigonometric functions only for $0 \leq \theta \leq 2\pi$. Find (a) $\sin 478.2$ and (b) $\tan(-8712.41°)$.

Solution (a) Divide 478.2 by 2π, obtaining 76.107894. Thus

$$\begin{aligned} 478.2 &\approx 2\pi(76.107894) \\ &\approx 2\pi(76 + 0.107894) \\ &\approx 2\pi(76) + 2\pi(0.107894) \\ &\approx 2\pi(76) + 0.6779167 \end{aligned}$$

Consequently,

$$\begin{aligned} \sin 478.2 &= \sin[478.2 - 76(2\pi)] \\ &\approx \sin 0.6779167 \approx 0.627172 \end{aligned}$$

(b) Divide $-8712.41°$ by $360°$, obtaining -24.201139.

Thus $\quad -8712.41° \approx 360°(-24.201139)$

$$\approx -24(360°) - 0.201139(360°)$$

$$= -24(360°) - 72.41°$$

Consequently,

$$\tan(-8712.41°) = \tan[-8712.41° + 24(360°)]$$

$$= \tan(-72.41°).$$

However, $-74.21°$ is still not between 0° and 360°. But

$$\tan(-74.21°) = \tan(-72.41° + 360°)$$

$$= \tan 287.59° \approx -3.15431 \quad ■$$

EVALUATION USING TABLES

We use Table D1 or Table D2. The only angles appearing in these tables are between 0° and 90° (0 rad and $\frac{\pi}{2}$ rad). When the angle lies in this range, the tables are used as described in the previous chapter. When the angle θ is outside this range, we can still find the trigonometric functions of θ from these tables; however, we must use reference angles and determine the signs of the trigonometric functions. The procedure is as follows:

1. Determine in which quadrant the terminal side of θ lies.
2. Find the reference angle ϕ of θ.
3. Find the trigonometric function of ϕ.
4. From step 1, determine the sign of the trigonometric function of θ.
5. Attach the sign found in step 4 to the value found in step 3.

EXAMPLE 4 Find (a) sin 192°, (b) cos 2, and (c) csc(−218.2°).

Solution (a) 192° is in Q_3, and the reference angle is 12°. From Table D1, $\sin 12° \approx 0.2079$. Since sine is negative in Q_3, $\sin 192° \approx -0.2079$.

(b) The 2 may represent either a real number or an angle of measure 2 rad. In either case, we use Table D2. Since 2 is in Q_2, the reference number (angle) is $\pi - 2 \approx 3.14 - 2 = 1.14$. From Table D2, $\cos 1.14 \approx 0.4176$. Since cosine is negative in Q_2, $\cos 2 \approx -0.4176$.

Note that we have used the approximation 3.14 for π since Table D2 only contains arguments to two decimal places. This is not a very good approximation, so the answer is not very accurate (see Example 1).

(c) $-218.2°$ is in Q_2, and the reference angle is $38.2° = 38°12'$. Using Table D1 and interpolating, $\csc 38.2° \approx 1.617$. Since cosecant is positive in Q_2, $\csc(-218.2°) \approx 1.617$. ■

APPLICATIONS

In Sec. 2, we found that the wrapping function mapped the real number t to $W(t) = (x, y)$, and by definition, $x = \cos t$ and $y = \sin t$. Now that we know how to compute $\cos t$ and $\sin t$ for any real number t, we can find the coordinates $(\cos t, \sin t)$ of $W(t)$ for any real number t.

EXAMPLE 5 Find the coordinates of $W(-7.25)$.

Solution We know $x = \cos(-7.25)$ and $y = \sin(-7.25)$. Using a calculator,

$$x \approx 0.567924 \text{ and } y \approx -0.823081$$

Using tables, -7.25 is in Q_4 with reference number $7.25 - 2\pi \approx 7.25 - 6.28 = 0.97$. By Table D2, $\cos 0.97 \approx 0.5653$ and $\sin 0.97 \approx 0.8249$. Since cosine is positive and sine is negative in Q_4,

$$x \approx 0.5653 \text{ and } y \approx -0.8249 \quad \blacksquare$$

We have just seen how to find the coordinates (x, y) of $W(t)$, a point of distance 1 from the origin on the terminal side of θ. In a similar fashion, we can find the coordinates (x, y) of a point of distance r from the origin on the terminal side of θ. From Definition (1), we know that

$$\cos \theta = \frac{x}{r} \quad \text{and} \quad \sin \theta = \frac{y}{r}$$

Therefore

$$\textbf{(20)} \qquad x = r \cos \theta \quad \text{and} \quad y = r \sin \theta$$

EXAMPLE 6 Find the coordinates of the point on the terminal side of $\theta = 327°$ which is 5.8 units from the origin.

Solution By (20), $x = 5.8 \cos 327°$ and $y = 5.8 \sin 327°$. Using a calculator, $x \approx 4.86429$ and $y \approx -3.15891$. When using tables, we determine that the reference angle is 33° and θ is in Q_4. Thus $\cos 327° = \cos 33° \approx 0.8387$ and $\sin 327° = -\sin 33° \approx -0.5446$. Therefore $x \approx 5.8(0.8387) \approx 4.864$ and $y \approx 5.8(-0.5446) \approx -3.159$. $\blacksquare$

From (20), we can find formulas relating x, y, and θ:

$$r = \frac{x}{\cos \theta} \quad \text{and} \quad r = \frac{y}{\sin \theta}, \quad \text{or} \quad \frac{x}{\cos \theta} = \frac{y}{\sin \theta}$$

Thus

$$\textbf{(21)} \quad x = y\frac{\cos \theta}{\sin \theta} = y \cot \theta \quad \text{and} \quad y = x\frac{\sin \theta}{\cos \theta} = x \tan \theta$$

The formulas in (21) need not be memorized; it is better to remember how to derive them. They are useful if you are given θ and one of x and y and want to find the other one.

EXAMPLE 7 Find the coordinates of the point P on the terminal side of $\theta = 108°$ if $x = 8$.

Solution We already know $x = 8$. From (21), $y = x \tan \theta = 8 \tan 108°$. Using a calculator, we obtain $y \approx -24.6215$. When using tables, we observe θ is in Q_2 and tangent is negative there. The reference angle is $72°$, so $y \approx -8 \tan 72° \approx -8(3.078) \approx -24.62$. ■

In many applications, there are functions of a real variable which involve circular functions. Consequently, evaluating such a function requires evaluating the circular functions.

EXAMPLE 8 Evaluate $f(t) = 4.15 + t \sin t$ when $t = 2.91$.

Solution $f(2.91) = 4.15 + 2.91 \sin 2.91$. Using a calculator, $f(2.91) \approx 4.81793$. When using tables $W(2.91)$ is in Q_2, so $\sin 2.91$ is positive, and the reference number is $\pi - 2.91 \approx 0.23$.

Thus
$$\begin{aligned} f(2.91) &\approx 4.15 + 2.91 \sin 0.23 \\ &\approx 4.15 + 2.91(0.2280) = 4.81348 \\ &\approx 4.813 \end{aligned}$$

Note that we have rounded to four significant figures since this is the accuracy of the tables we used. ■

EXERCISES

In Exercises 1–6, find the values of the six trigonometric functions of θ.

1. $\theta = 146°$
2. $\theta = 661.2°$
3. $\theta = -489°$
4. $\theta = 2.6$
5. $\theta = -2.6$
6. $\theta = 9.21$

In Exercises 7–14, find the coordinates of the point P on the terminal side of the given θ satisfying the given condition.

7. $\theta = 167°$, $r = 16$
8. $\theta = -29°$, $x = 14$
9. $\theta = 4.89$, $y = -5$
10. $\theta = 10.2$, $r = 1200$
11. $\theta = -1.89$, $x = -2$
12. $\theta = -11.3$, $y = 680$
13. $\theta = 684°$, $y = -3$
14. $\theta = -219°$, $x = -10$

In Exercises 15–18, find the the coordinates of $W(t)$ for the given t.

15. $t = 0.62$
16. $t = 2.5$
17. $t = -4.14$
18. $t = -13.2$

In Exercises 19–28, evaluate the given expression, using a calculator or tables only when necessary.

19. $\sin 90° \csc 630° + \tan 720°$

20. $\cos \frac{5\pi}{2} + \cos 3\pi + \cos \frac{7\pi}{2} + \cos 4\pi$

21. $8 \sin 142° + 4 \tan 206°$

22. $\frac{1}{\cos (3.2 + \frac{1}{5})}$

23. $\sin 540° + \tan 990°$

24. $\sec(-900°) \sec 900° + \tan 900° \sin 405°$

25. $\tan 4.62 + 1.97 \sin 7.4$

26. $\sin^2(-42°) + \cos^2(-42°)$

ⓒ**27.** $\sqrt{\csc(-507.2°) + 2.0328}$

ⓒ**28.** $6.2^{4.7} + \cos^3 9.25$

In Exercises 29–38, evaluate the given function at the given real number.

29. $f(t) = 2 \tan t,\ t = 3$

30. $f(x) = 5 \cot x,\ x = 5.8$

31. $f(u) = \frac{1}{6} \cot^2 u,\ u = 0.96$

32. $f(t) = 5 \sin t \cos t,\ t = 1.4$

ⓒ**33.** $g(x) = 6 \sin^4 x - \tan^2 x,\ x = 2.19$

ⓒ**34.** $g(u) = \frac{1}{1 + \cos u} + 4 \sec u,\ u = 1.89$

ⓒ**35.** $f(t) = \tan(\sin t),\ t = 5.3$

ⓒ**36.** $g(t) = \sin[\cos(t - 1.93)],\ t = 2.65$

ⓒ**37.** $g(t) = t^3 + 3t \sin t,\ t = e^{2.1}$

ⓒ**38.** $g(x) = 5x^{1/3} - 6 \cot(x^3),\ x = -2^{\sqrt{\pi}}$

In Exercises 39–44, compute the value of the given expression under the assumption that your calculator computes the trigonometric functions only for $0 \le \theta \le 2\pi$ or $0° \le \theta \le 360°$.

ⓒ**39.** $\cos 1436.7°$

ⓒ**40.** $\tan 64{,}291°$

ⓒ**41.** $\sin(-42{,}385.6°)$

ⓒ**42.** $\cos(-81{,}277{,}033°)$

ⓒ**43.** $\sin 48{,}963.4$

ⓒ**44.** $\tan(-74{,}306)$

2.6 Graphs of the Trigonometric Functions

In this section we shall determine the graphs of the trigonometric functions. From our discussion of the periodicity of these functions, we know that if f is any trigonometric function, then $f(t + 2\pi) = f(t)$. Hence we need only find the graphs of each function on an interval of length 2π, say $0 \le t \le 2\pi$ or $-\frac{\pi}{2} \le t \le \frac{3\pi}{2}$, and then use $f(t + 2\pi) = f(t)$ to get the graph everywhere.

We begin with the sine and cosine functions. In determining their graphs, the presentation is simplified if we view these functions as circular functions.

We first consider the sine function. We shall use the interval $0 \le t \le 2\pi$. Recall from the definition of the wrapping function that if $W(t) = (a, b)$, then $\sin t = b$. See Fig. 24.

Now consider what happens to $\sin t$ as t increases from 0 to 2π. When $t = 0$, $\sin t = 0$.

As t increases from 0 to $\frac{\pi}{2}$, $b = \sin t$ increases from 0 to 1. See Fig. 25(a). As t increases from $\frac{\pi}{2}$ to π, $b = \sin t$ decreases from 1 to 0. See Fig. 25(b). As t increases from π to $\frac{3\pi}{2}$, $b = \sin t$ decreases from 0 to -1.

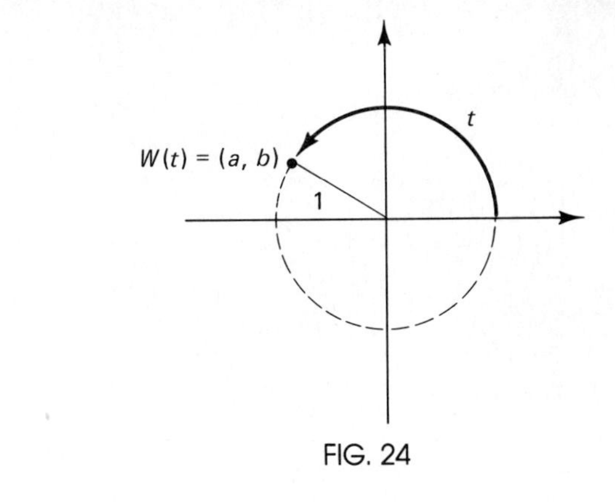

FIG. 24

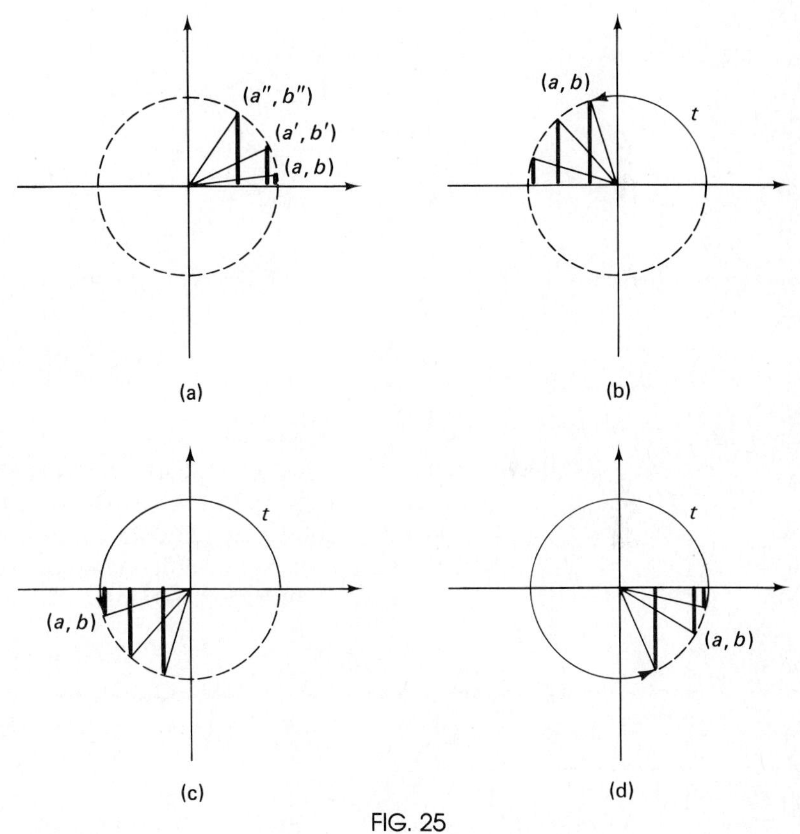

FIG. 25

See Fig. 25(c). As t increases from $\frac{3\pi}{2}$ to 2π, $b = \sin t$ increases from -1 to 0. See Fig. 25(d).

From this discussion, we now know the general nature of $y = \sin t$, $0 \leq t \leq 2\pi$. If we keep this in mind and plot several points, we can obtain a fairly accurate graph. We do this as in Fig. 26(a), using the table of special values

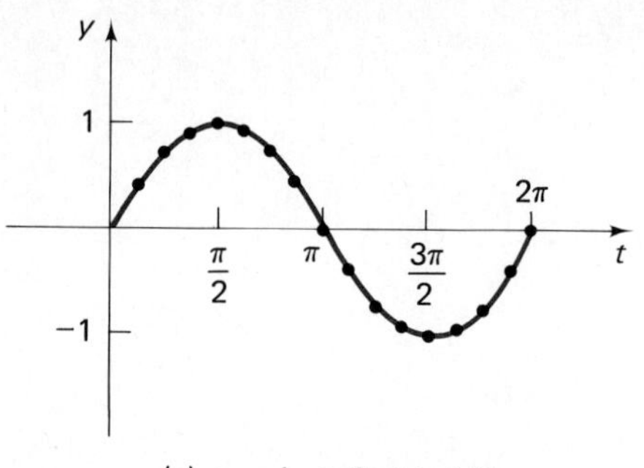

(a) $y = \sin t, 0 \leq t \leq 2\pi$

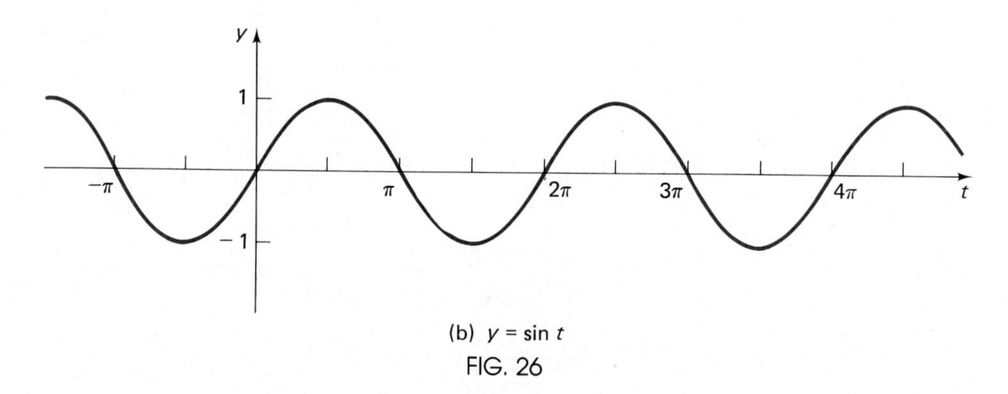

(b) $y = \sin t$

FIG. 26

(estimating $\frac{\sqrt{2}}{2} \approx 0.71$ and $\frac{\sqrt{3}}{2} \approx 0.87$) given as Table 1. Of course you could use a calculator or tables to obtain points instead of using Table 1. From the fact that $\sin (t + 2\pi) = \sin t$ for all t (i.e., periodicity), if we repeat the graph in Fig. 26(a) every 2π units, we obtain the graph of $y = \sin t$ for all real t. See Fig. 26(b). Note that the scale on the t-axis is different from the scale on the y-axis in order to show more of the graph.

TABLE 1

t	0	$\frac{\pi}{6}$	$\frac{\pi}{4}$	$\frac{\pi}{3}$	$\frac{\pi}{2}$	$\frac{2\pi}{3}$	$\frac{3\pi}{4}$	$\frac{5\pi}{6}$	π	$\frac{7\pi}{6}$	$\frac{5\pi}{4}$	$\frac{4\pi}{3}$	$\frac{3\pi}{2}$	$\frac{5\pi}{3}$	$\frac{7\pi}{4}$	$\frac{11\pi}{6}$
$\sin t$	0	$\frac{1}{2}$	$\frac{\sqrt{2}}{2}$	$\frac{\sqrt{3}}{2}$	1	$\frac{\sqrt{3}}{2}$	$\frac{\sqrt{2}}{2}$	$\frac{1}{2}$	0	$-\frac{1}{2}$	$-\frac{\sqrt{2}}{2}$	$-\frac{\sqrt{3}}{2}$	-1	$-\frac{\sqrt{3}}{2}$	$-\frac{\sqrt{2}}{2}$	$-\frac{1}{2}$

We next consider the cosine function. We again use the interval $0 \leq t \leq 2\pi$ and Fig. 24. Recall that if $W(t) = (a, b)$, then $\cos t = a$. We now follow what happens to $\cos t$ as t increases from 0 to 2π. When $t = 0$, $\cos t = 1$.

As t increases from 0 to $\frac{\pi}{2}$, $a = \cos t$ decreases from 1 to 0. As t increases from $\frac{\pi}{2}$ to π, $a = \cos t$ decreases from 0 to -1. As t increases from π to $\frac{3\pi}{2}$, $a = \cos t$ increases from -1 to 0. As t increases from $\frac{3\pi}{2}$ to 2π, $a = \cos t$ increases from 0 to 1. See Fig. 27.

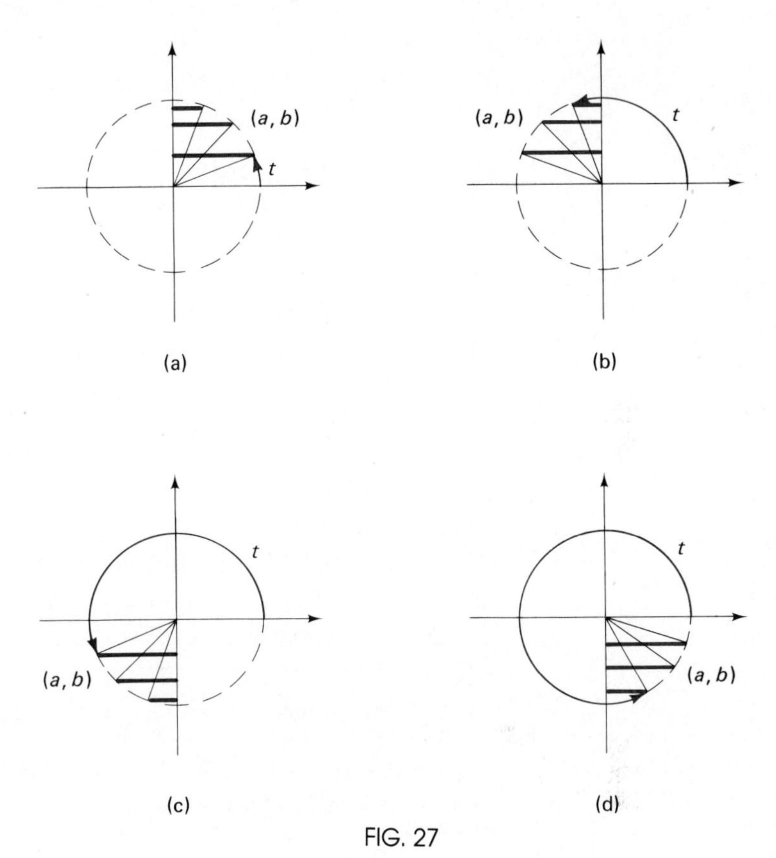

FIG. 27

From this discussion, we know the general nature of the graph of $y = \cos t$, $0 \leq t \leq 2\pi$. If we use this and plot some points from a table like Table 1, we obtain a fairly accurate graph, which is given in Fig. 28(a). From the fact that $\cos(t + 2\pi) = \cos t$ for all t (i.e., periodicity), if we repeat the graph in Fig. 28(a) every 2π units, we obtain the full graph of $y = \cos t$ for all real t. See Fig. 28(b).

From the graphs of $y = \sin t$ and $y = \cos t$, it is clear that there is no positive number $p < 2\pi$ for which $\sin(t + p) = \sin t$ for all t or $\cos(t + p) = \cos t$ for all t. Therefore, since we do know that $\sin(t + 2\pi) = \sin t$ and $\cos(t + 2\pi) = \cos t$ for all t, we have the following:

(22) The sine and cosine functions are periodic of period 2π.

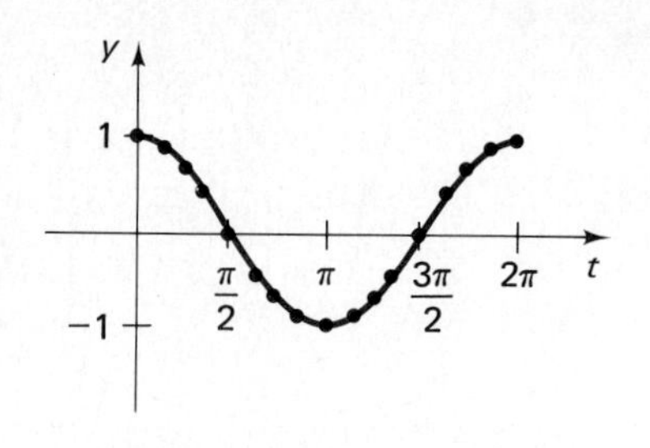

(a) $y = \cos t, 0 \leq t \leq 2\pi$

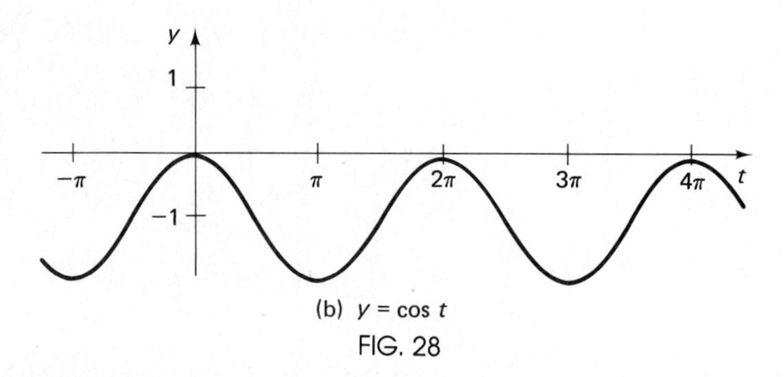

(b) $y = \cos t$

FIG. 28

By taking reciprocals, we obtain the following:

(23) The cosecant and secant functions are periodic of period 2π.

Indeed, the graphs of $y = \csc t$ and $y = \sec t$ can also be obtained by taking reciprocals, either directly from the graphs of $y = \sin t$ and $y = \cos t$ as indicated in Fig. 29(a) and (b), or from a table of values, such as Table 1. Note the vertical asymptotes to the two graphs.

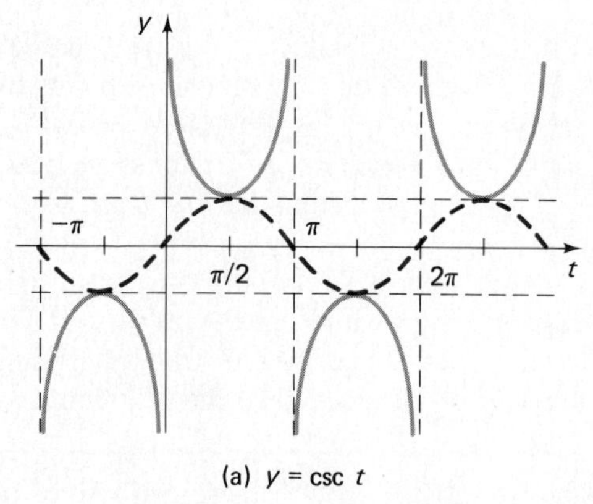

(a) $y = \csc t$

(b) $y = \sec t$

FIG. 29

TANGENT AND COTANGENT

We now consider the tangent function. Instead of viewing it as a circular function, it is easier to understand the behavior of the tangent function if we view it as a function of a general angle t measured in radians.

We shall use the interval $-\frac{\pi}{2} \leq t \leq \frac{3\pi}{2}$. When the terminal side of t is on the y-axis, i.e., for $t = -\frac{\pi}{2}, \frac{\pi}{2}$ or $\frac{3\pi}{2}$, $\tan t$ is undefined. When the terminal side of t is not on the y-axis, let P be the point where the terminal side of t meets the vertical line through $(1, 0)$ or through $(-1, 0)$. Thus, $P = (1, b)$ if $-\frac{\pi}{2} < t < \frac{\pi}{2}$, and $P = (-1, b)$ if $\frac{\pi}{2} < t < \frac{3\pi}{2}$. By the definition of $\tan t$

$$\tan t = \frac{b}{1} = b \qquad \text{if} \qquad -\frac{\pi}{2} < t < \frac{\pi}{2}$$

$$\tan t = \frac{b}{-1} = -b \qquad \text{if} \qquad \frac{\pi}{2} < t < \frac{3\pi}{2}$$

Now consider what happens to $\tan t$ as t increases from near $-\frac{\pi}{2}$ to near $\frac{3\pi}{2}$. When t is near $-\frac{\pi}{2}$ (but $t > -\frac{\pi}{2}$), $\tan t = b$ is large and negative. As t increases from near $-\frac{\pi}{2}$, through 0, and then approaches $\frac{\pi}{2}$, the numbers $b = \tan t$ increase from very large (in absolute value) negative numbers, pass through zero, and become very large positive numbers. See Fig. 30(a). As $t > \frac{\pi}{2}$ increases from near $\frac{\pi}{2}$, through π, and then approaches $\frac{3\pi}{2}$, the numbers b

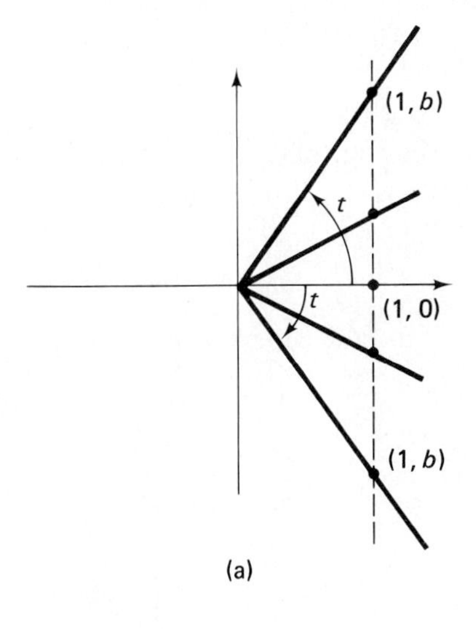

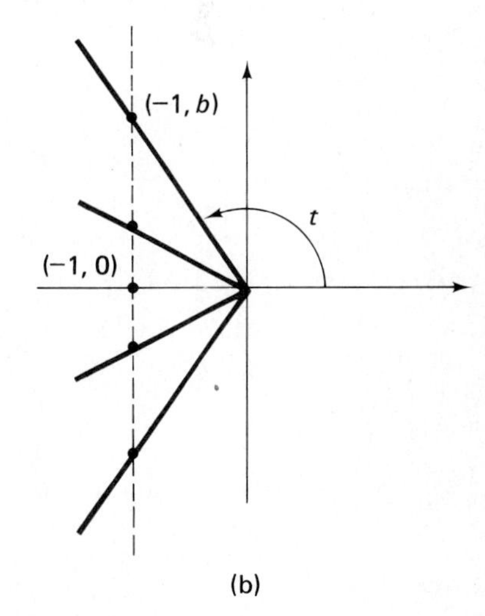

FIG. 30

decrease from very large positive numbers, pass through zero, and become very large (in absolute value) negative numbers. Hence the numbers $-b = \tan t$ increase from very large (in absolute value) negative numbers, go through zero, and become very large positive numbers. See Fig. 30(b).

From this discussion, we see that the general nature of the graph of $y = \tan t$, $-\frac{\pi}{2} < t < \frac{3\pi}{2}$, $t \neq \frac{\pi}{2}$, is like that sketched in Fig. 31. Of course, we could use a table of values or a calculator to plot points. Note in particular that the lines $t = -\frac{\pi}{2}, \frac{\pi}{2}, \frac{3\pi}{2}$ are vertical asymptotes.

From Fig. 31, it appears that $\tan t$ is periodic of period π. It can be shown that this is the case.

(24) The tangent function is periodic of period π.

From Fig. 31, we now use periodicity (observing that $\tan t$ is not defined for

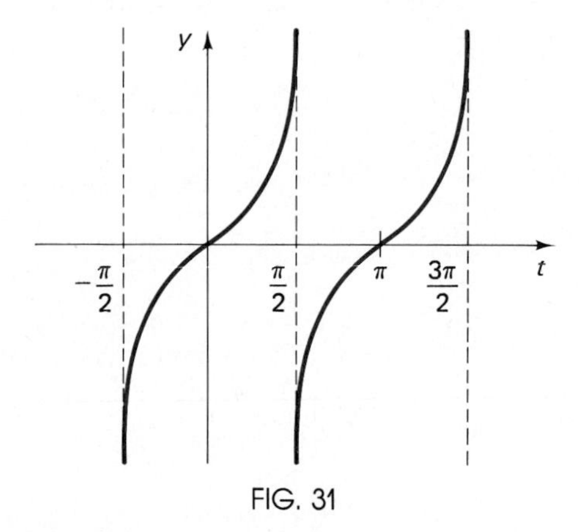

FIG. 31

$t = \frac{1}{2}\pi + k\pi$, k an integer) and obtain the full graph of $y = \tan t$, all $t \neq \frac{1}{2}\pi + k\pi$; see Fig. 32(a).

The graph of $y = \cot t$ given in Fig. 32(b) is obtained in a manner very similar to that of $y = \tan t$. The important things to remember are that $\cot t$ is positive for $0 < t < \frac{\pi}{2}$, $\cot t$ is undefined when t is a multiple of π, and the lines $t = k\pi$ (k an integer) are vertical asymptotes.

Since $\cot t = \frac{1}{\tan t}$, it follows from (24) that

(25) The cotangent function is periodic of period π.

To this point we have been considering the trigonometric functions as

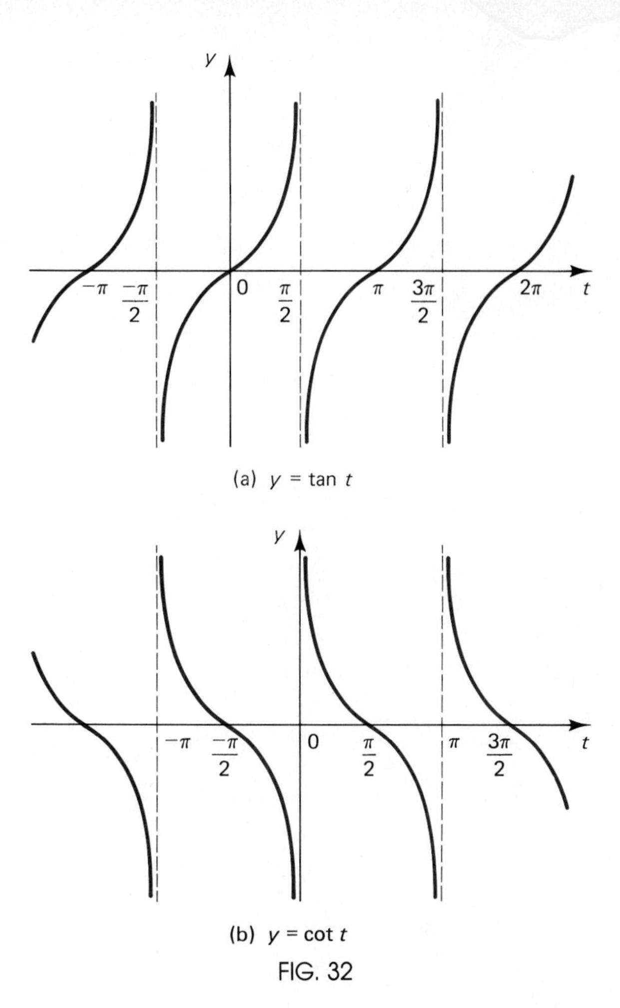

(a) $y = \tan t$

(b) $y = \cot t$

FIG. 32

functions of a real number t. Thus the graphs have the horizontal axis labeled t (and the vertical axis labeled y). We have been using t rather than the traditional x, since otherwise confusion might arise when reference is made to the coordinates of the wrapping function. From here on in this text, however, we shall seldom need the wrapping function. Consequently, henceforth we shall use the traditional x when dealing with the trigonometric functions. Of course, corresponding graphs will have horizontal axis labeled x.

EXAMPLE 1 Graph $\quad y = 2 + \cos x.$

Solution Note that for any value of x, the corresponding y-coordinate on the graph of $y = 2 + \cos x$ is just 2 more than the y-coordinate on the graph of $y = \cos x$. Thus the graph of $y = 2 + \cos x$ can be obtained by merely moving all the points on the graph of $y = \cos x$ up 2 units, i.e., by *translating* the graph up 2 units. See Fig. 33.

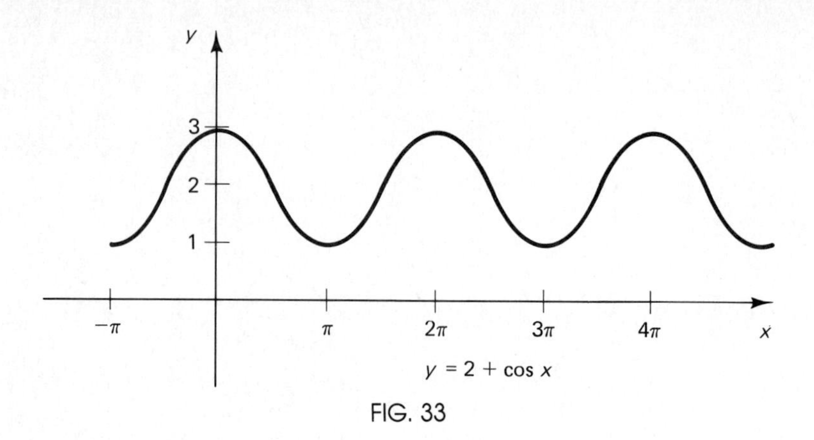

FIG. 33

■

EXERCISES

In Exercises 1–6, graph each function on the interval $0 \leq x \leq 2\pi$ by using a calculator or tables to find the approximate values of the functions for a whole series of x's and then plotting the points.

1. $y = \sin x$
2. $y = \cos x$
3. $y = \tan x$
4. $y = \csc x$
5. $y = \sec x$
6. $y = \cot x$

In Exercises 7–18, sketch the graph.

7. $y = |\sin x|$
8. $y = |\cos x|$
9. $y = |\tan x|$
10. $y = |\csc x|$
11. $y = |\sec x|$
12. $y = |\cot x|$
13. $y = \sin|x|$
14. $y = \cos|x|$
15. $y = \tan|x|$
16. $y = \csc|x|$
17. $y = \sec|x|$
18. $y = \cot|x|$

In Exercises 19–30, sketch the graph,

19. $y = 3 + \sin x$
20. $y = -2 + \sin x$
21. $y = -1 + \cos x$
22. $y = 5 + \cos x$
23. $y = 1 + \csc x$
24. $y = -2 + \csc x$
25. $y = -3 + \sec x$
26. $y = 1.5 + \sec x$
27. $y = 4 + \tan x$
28. $y = -1 + \tan x$
29. $y = -2 + \cot x$
30. $y = \pi + \cot x$

In Exercises 31–36, use the graphs in this section to determine how many numbers x in the interval $[-2, 6]$ satisfy the equation.

31. $\sin x = 0.2$
32. $\cos x = 0.2$
33. $\tan x = 0.2$
34. $\csc x = 5$
35. $\sec x = 5$
36. $\cot x = 5$

2.7 The Graphs of $y = a \sin(bx - c)$ and $y = a \cos(bx - c)$

In this section we see how to use the graphs of $y = \sin x$ and $y = \cos x$ to obtain the graphs of $y = a \sin(bx - c)$ and $y = a \cos(bx - c)$. We shall do this in stages, first finding the significance of a, then b, and finally c. This is done first without, then with graphing techniques (expansions, contractions, and translations). Thus knowledge of these techniques, although helpful, is not required.

From the graphs of $y = \sin x$ and $y = \cos x$, we know that both of these graphs regularly oscillate. (In fact, because of the shape of their graphs, these functions are sometimes called wave functions.) For any function whose graph is regularly oscillating, the **amplitude** is defined to be one half of the difference between the maximum value and the minimum value of the function. For a function such as $\sin x$ or $\cos x$ which regularly oscillates an equal amount above and below the x-axis, the amplitude is just the maximum value of the function. Thus it is immediate that $\sin x$ and $\cos x$ have amplitude 1.

Let us now consider the function $y = a \sin x$ (the function $y = a \cos x$ is analogous). The graph of $y = a \sin x$ can be obtained from the graph of $y = \sin x$ by multiplying the y-coordinate of every point on the graph of $y = \sin x$ by a. This can be seen by examining Example 1.

For those who know graphing techniques, we can be a little more precise: If $a > 0$, the graph of $y = a \sin x$ is obtained from the graph of $y = \sin x$ by expanding (if $a > 1$) or contracting (if $0 < a < 1$) in the y-direction, while if $a < 0$, we first expand or contract the graph of $y = \sin x$ in the y-direction and then reflect the result in the x-axis.

EXAMPLE 1 The graphs of $y = \sin x$, $y = 2 \sin x$, $y = \frac{1}{2} \sin x$, and $y = -2 \sin x$, all drawn on the same axes, are shown in Fig. 34.

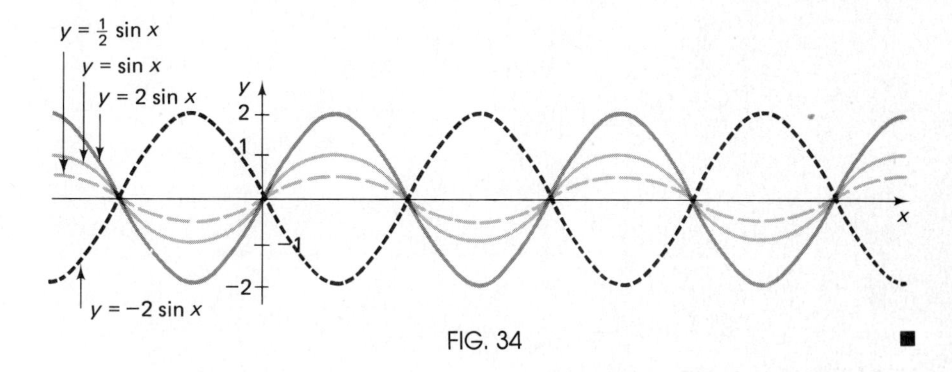

FIG. 34 ■

From our discussion it should be clear that $y = a \sin x$ (and similarly $y = a \cos x$) has amplitude $|a|$. Moreover, by the previous section, $y = \sin x$ and $y = \cos x$ are periodic of period 2π. Since multiplying the y-coordinate by a does not affect the period (see, for example, Fig. 34), $y = a \sin x$ and $y = a \cos x$ have period 2π.

(26) $y = a \sin x$ and $y = a \cos x$ have amplitude $|a|$, period 2π.

Next we consider the functions $y = \sin bx$ and $y = \cos bx$. First recall equation (13):

$$\sin(-x) = -\sin x \quad \text{and} \quad \cos(-x) = \cos x$$

By using this, equations like $y = \sin(-3x)$ and $y = \cos(-3x)$ can be rewritten as $y = -\sin 3x$ and $y = \cos 3x$. Since we already know how to handle the minus sign in front of the sine function (i.e., $a = -1$), *whenever graphing $y = \sin bx$ or $y = \cos bx$, we shall assume $b > 0$.* This simplifies some of the algebra.

Consider, then, the function $y = \sin bx$, $b > 0$. Suppose for a moment we put $\theta = bx$. Then the function becomes $y = \sin \theta$. This function has amplitude 1, and we can graph one period by graphing the function values for all values of θ, $0 \le \theta \le 2\pi$. But if we replace θ by bx, we have $0 \le bx \le 2\pi$. Dividing each term by b yields $0 \le x \le \frac{2\pi}{b}$ (the direction of the inequalities does not change since $b > 0$). Therefore, to graph one period of $y = \sin bx$, $b > 0$, we graph the function values as x goes from 0 to $\frac{2\pi}{b}$; i.e., the period is $\frac{2\pi}{b}$. Otherwise it is an ordinary sine function.

On the other hand, if you are using graphing techniques, recall that the effect on the graph of $y = \sin x$ of replacing x by bx is to expand (if $0 < b < 1$) or contact (if $b > 1$) the graph of $y = \sin x$ in the x-direction. Therefore, $y = \sin bx$, $b > 0$, is still periodic, but the period is now $\frac{2\pi}{b}$ (the amplitude is still 1).

(27) $y = \sin bx$ and $y = \cos bx$, $b > 0$,

have period $\frac{2\pi}{b}$, amplitude 1.

EXAMPLE 2 $y = \sin 2x$ has period $\frac{2\pi}{2} = \pi$, while $y = \sin \frac{1}{2}x$ has period $\frac{2\pi}{\frac{1}{2}} = 4\pi$. See Fig. 35 for the graphs, drawn on the same axes as $y = \sin x$.

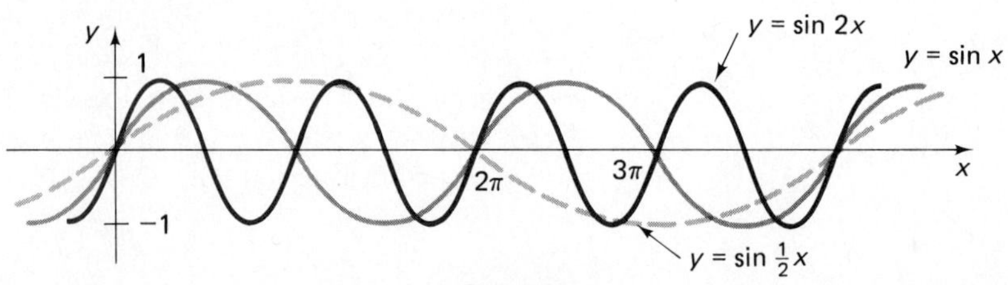

FIG. 35 ■

Now let us consider $y = \sin(bx - c)$, $b > 0$. First we write this in the form $y = \sin b\left(x - \frac{c}{b}\right)$. Then we proceed as before and set $\theta = b\left(x - \frac{c}{b}\right)$, and we think of the function as $y = \sin \theta$. Thus we still have a sine function, its amplitude is 1, and it goes through one period as θ goes from 0 to 2π; i.e., $0 \leq \theta \leq 2\pi$. Replacing θ by $b\left(x - \frac{c}{b}\right)$ yields $0 \leq b\left(x - \frac{c}{b}\right) \leq 2\pi$. We wish to determine for what values of x this holds, so we divide each term by b (a positive number), obtaining $0 \leq x - \frac{c}{b} \leq \frac{2\pi}{b}$, and then add $\frac{c}{b}$ to each term, obtaining $\frac{c}{b} \leq x \leq \frac{c}{b} + \frac{2\pi}{b}$. Therefore, to graph one period of $y = \sin b\left(x - \frac{c}{b}\right)$, $b > 0$, we graph the function values as x starts from $\frac{c}{b}$ and goes to $\frac{c}{b} + \frac{2\pi}{b}$, a distance of $\frac{2\pi}{b}$. Thus the period is still $\frac{2\pi}{b}$, but the starting point has been shifted from $x = 0$ to $x = \frac{c}{b}$. The number $\frac{c}{b}$ is called the **phase shift.**

For those who prefer to use graphing techniques, also write $y = \sin(bx - c)$, $b > 0$, in the form $y = \sin b\left(x - \frac{c}{b}\right)$. Then the graph can easily be obtained as follows: Start with the graph of $y = \sin x$. Replace x by bx, and obtain the graph of $y = \sin bx$ as we have just discussed. Then replace x by $x - \frac{c}{b}$. This causes a translation of the graph of $y = \sin bx$. The translation is $\left|\frac{c}{b}\right|$ units, to the right if $c > 0$ and to the left if $c < 0$. This number $\frac{c}{b}$ is what we just defined as the phase shift.

(28)

$$y = \sin(bx - c) = \sin b\left(x - \frac{c}{b}\right) \text{ and } y = \cos(bx - c) = \cos b\left(x - \frac{c}{b}\right),$$

$$b > 0, \text{ have period } \frac{2\pi}{b}, \text{ phase shift } \frac{c}{b}, \text{ and amplitude } 1$$

EXAMPLE 3 Graph $y = \sin(x + \frac{1}{4}\pi)$.

Solution The function is of the form $y = \sin(bx - c) = \sin b\left(x - \frac{c}{b}\right)$ with $b = 1$, $c = -\frac{1}{4}\pi$. Therefore its period is 2π, and its phase shift is $-\frac{1}{4}\pi$. We draw the graph by first drawing a normal period of length 2π, starting at $x = -\frac{1}{4}\pi$ (the heavy part of the graph), and then complete the graph using periodicity.

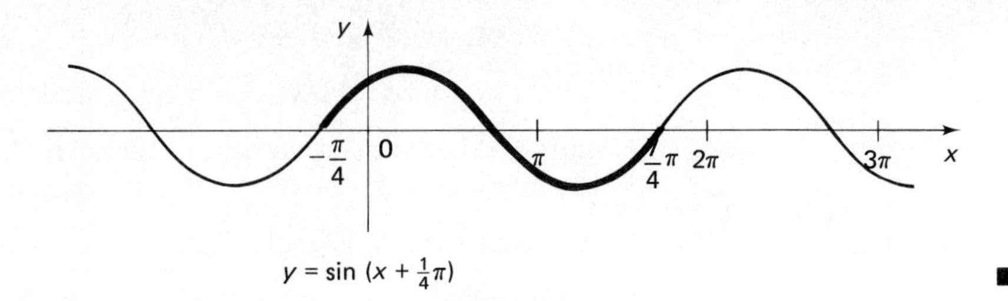

$y = \sin\,(x + \frac{1}{4}\pi)$ ■

EXAMPLE 4 Graph $y = \cos(\frac{2}{3}x - \frac{1}{3}\pi) = \cos\frac{2}{3}(x - \frac{1}{2}\pi)$.

Solution The function is of the form $y = \cos(bx - c) = \cos b\left(x - \dfrac{c}{b}\right)$ with $b = \dfrac{2}{3}$, $c = \dfrac{\pi}{3}$. Thus the period is $\dfrac{2\pi}{b} = \dfrac{2\pi}{\frac{2}{3}} = 3\pi$, and the phase shift is $\dfrac{c}{b} = \dfrac{\pi/3}{2/3} = \dfrac{\pi}{2}$. Some people have difficulty remembering the formula for these numbers. If so, just remember how we derived them: One period is $0 \le \dfrac{2}{3}\left(x - \dfrac{\pi}{2}\right) \le 2\pi$, or $0 \le x - \dfrac{\pi}{2} \le 3\pi$ (dividing by $\frac{2}{3}$), and finally $\dfrac{\pi}{2} \le x \le \dfrac{\pi}{2} + 3\pi$ $\left(\text{adding } \dfrac{\pi}{2}\right)$. Hence one period starts at $\dfrac{\pi}{2}$ (the phase shift) and goes for 3π units (the period). This is the heavy part of the graph; the graph is completed using periodicity.

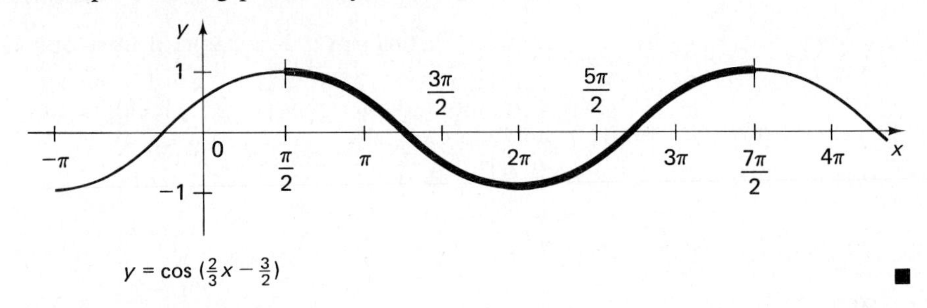

$y = \cos\,(\frac{2}{3}x - \frac{3}{2})$ ■

Finally we can now consider the functions $y = a\sin(bx - c)$ and $y = a\cos(bx - c)$. First, if b is negative, use $\sin(-\theta) = -\sin\theta$ or $\cos(-\theta) = \cos\theta$ to write it in the form where $b > 0$. For instance, if $y = 3\sin(-2x + 4)$, we write $y = 3\sin[-(2x - 4)] = 3[-\sin(2x - 4)]$, or $y = -3\sin(2x - 4)$, and graph this. Similarly, if $y = -7\cos(-\pi x - 5)$, we write $y = -7\cos[-(\pi x + 5)] = -7\cos(\pi x + 5)$ and graph this. Other than that, we know that the effect of multiplying by a is to change the amplitude to $|a|$, and if $a < 0$, to reflect the graph through the x-axis.

> **(29)** $$y = a\sin(bx - c) = a\sin b\left(x - \frac{c}{b}\right) \quad \text{and}$$
> $$y = a\cos(bx - c) = a\cos b\left(x - \frac{c}{b}\right),\ b > 0,$$
> have amplitude $|a|$, period $\dfrac{2\pi}{b}$, and phase shift $\dfrac{c}{b}$.

EXAMPLE 5 Graph $y = -3\sin(\frac{1}{2}x + 1)$.

Solution Here $a = -3$, $b = \frac{1}{2}$, $c = -1$, and $-3\sin(\frac{1}{2}x + 1) = -3\sin\frac{1}{2}[x - (-2)]$. Thus the amplitude is 3, the period is $\dfrac{2\pi}{1/2} = 4\pi$, and the phase shift is $\dfrac{-1}{1/2} = -2$. Since a is negative, we can draw the graph of $3\sin(\frac{1}{2}x + 1)$ and reflect it through the x-axis. See Fig. 36.

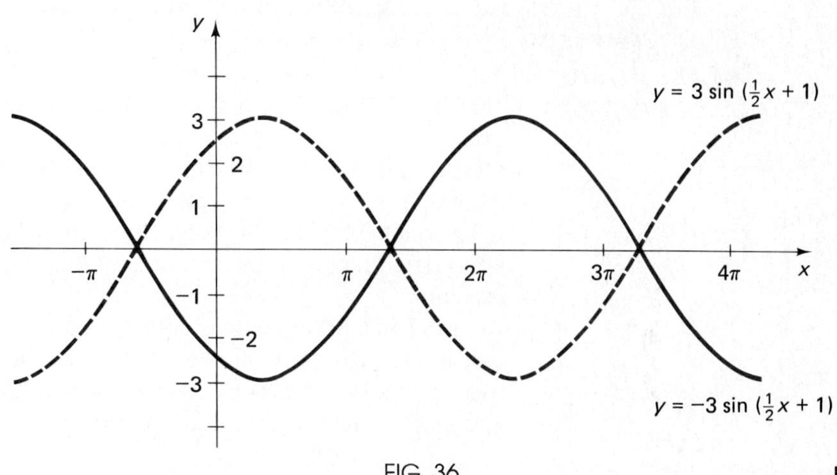

FIG. 36 ■

Occasionally you may run across equations like $y = -3\sin(\frac{1}{2}x + 1) + 4$. To graph this, first graph $y = -3\sin(\frac{1}{2}x + 1)$ as in Fig. 36, and then translate that graph four units upward.

EXERCISES

In Exercises 1–25, state the amplitude, period, and phase shift (if nonzero), and then sketch the graph.

1. $y = 2\cos x$
2. $y = \frac{1}{2}\cos x$
3. $y = -\frac{1}{2}\cos x$
4. $y = -\frac{1}{2}\sin x$
5. $y = 3\sin x$
6. $y = \cos 2x$
7. $y = \cos\frac{1}{2}x$
8. $y = \sin 3x$
9. $y = \cos(-3x)$
10. $y = \sin(-\pi x)$
11. $y = \sin(x - \frac{1}{3}\pi)$
12. $y = \cos(x + \frac{5}{6}\pi)$
13. $y = \sin(x + 1)$
14. $y = \cos(x - 3)$
15. $y = \cos(\pi - x)$
16. $y = \sin(2x - \frac{1}{2}\pi)$
17. $y = \cos(3x - \pi)$
18. $y = \sin(\frac{1}{2}x + \frac{1}{3}\pi)$
19. $y = \sin(-3x + 6)$
20. $y = \cos(-\pi x + \pi)$
21. $y = 3\sin(\pi x + \frac{1}{2}\pi)$
22. $y = -2\cos(3x - 4\pi)$
23. $y = 4\cos(-3x + 3)$
24. $y = \frac{1}{2}\sin(-2x + \pi)$
25. $y = -\frac{1}{3}\cos(2\pi - x)$

In Exercises 26–31, sketch the graph.

26. $y = 2\cos x + 3$
27. $y = \sin 2x - 1$
28. $y = \cos\frac{1}{2}x - \frac{1}{2}$
29. $y = \sin(x - \frac{1}{3}\pi) + \frac{1}{3}\pi$
30. $y = 2\sin(\frac{1}{2}x + \pi) + 2$
31. $y = \frac{1}{2}\cos(\pi x - \pi) + 3$

In Exercises 32–40, sketch the graph, using the techniques demonstrated in this section.

32. $y = \tan(x - \frac{1}{2}\pi)$
33. $y = \tan \frac{1}{2}x$
34. $y = \csc(x + \frac{1}{2}\pi)$
35. $y = \csc 2x$
36. $y = \sec(\frac{3}{2}x - \frac{1}{2}\pi)$
37. $y = \sec(\pi x + \frac{1}{4}\pi)$
38. $y = \tan 3x - 1$
39. $y = \csc \pi x + \pi$
40. $y = 2 \sec 3x - 4$

2.8 Inverse Trigonometric Functions

In the previous chapter we defined the six trigonometric functions of acute angles, and then we saw that each of these functions has an inverse function. Now that we have generalized the definitions of the six trigonometric functions, it is natural to ask if each of the functions still has an inverse. Recall that a function has an inverse function precisely when no horizontal line intersects the graph more than once, i.e., the function passes the **horizontal line test.** It is apparent from their graphs, Figs. 26(b), 28(b), 29, and 32, that each trigonometric function fails the horizontal line test, so it does not have an inverse function.

When a function does not have an inverse and it would be useful to have one, what we usually do is restrict the domain in such a way that the function, *with its restricted domain,* has an inverse. Accordingly, for each of the trigonometric functions, we shall restrict the domain subject to the following:

1. Each value in the range is taken on only once (i.e., the graph passes the horizontal line test).
2. The range of the function with the restricted domain is the same as the range of the original function.
3. The domain includes the most commonly used numbers (or angles), $0 < x < \dfrac{\pi}{2}$.
4. The graph is connected (if possible).

For the function $y = \sin x$, conditions 1–4 are satisfied if we restrict the domain to $-\dfrac{\pi}{2} \leq x \leq \dfrac{\pi}{2}$:

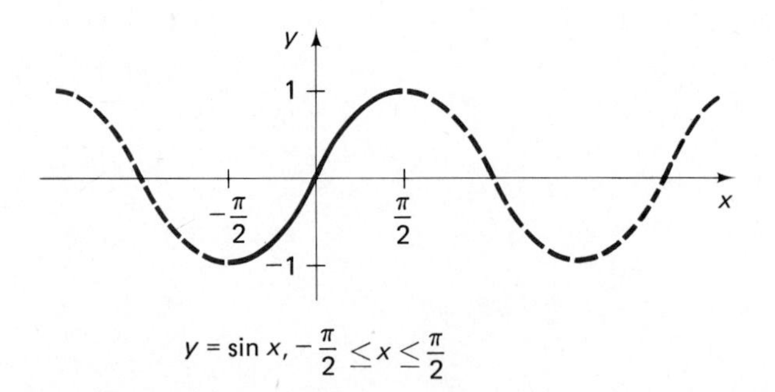

$y = \sin x, -\frac{\pi}{2} \leq x \leq \frac{\pi}{2}$

(30) Definition Suppose $-1 \leq x \leq 1$. Then $y = \sin^{-1} x$ (or $y = \arcsin x$) if and only if $x = \sin y$ and $-\dfrac{\pi}{2} \leq y \leq \dfrac{\pi}{2}$.

In other words, if $-1 \leq x \leq 1$, then the equation $y = \sin^{-1} x$ is saying "y is the number (angle) between $-\dfrac{\pi}{2}$ and $\dfrac{\pi}{2}$ whose sine is x."

Recall that the graph of the inverse f^{-1} can be obtained by reflecting the graph of f through the line $y = x$. Accordingly, the graph of $y = \sin^{-1} x$, $-1 \leq x \leq 1$, is obtained by reflecting the graph of $y = \sin x$, $-\dfrac{\pi}{2} \leq x \leq \dfrac{\pi}{2}$, through the line $y = x$. See Fig. 37.

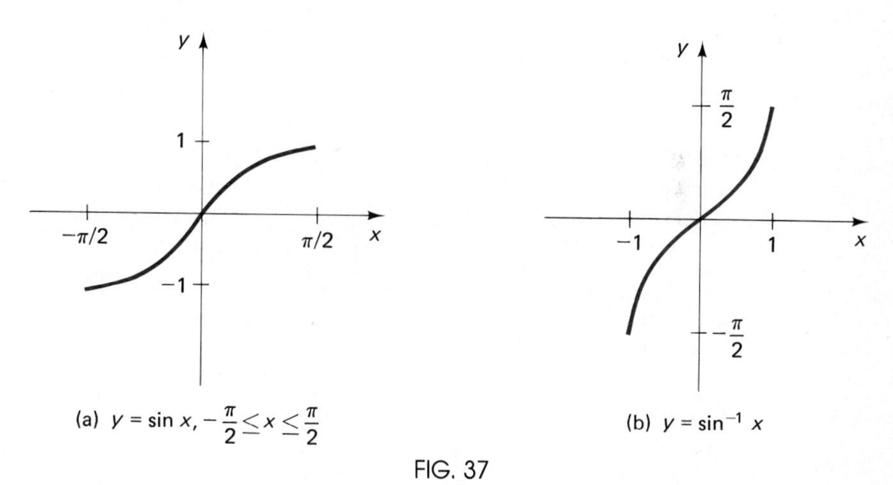

(a) $y = \sin x, -\frac{\pi}{2} \leq x \leq \frac{\pi}{2}$ (b) $y = \sin^{-1} x$

FIG. 37

For the function $y = \cos x$, conditions 1–4 are satisfied if we restrict the domain to $0 \leq x \leq \pi$. See Fig. 38(a).

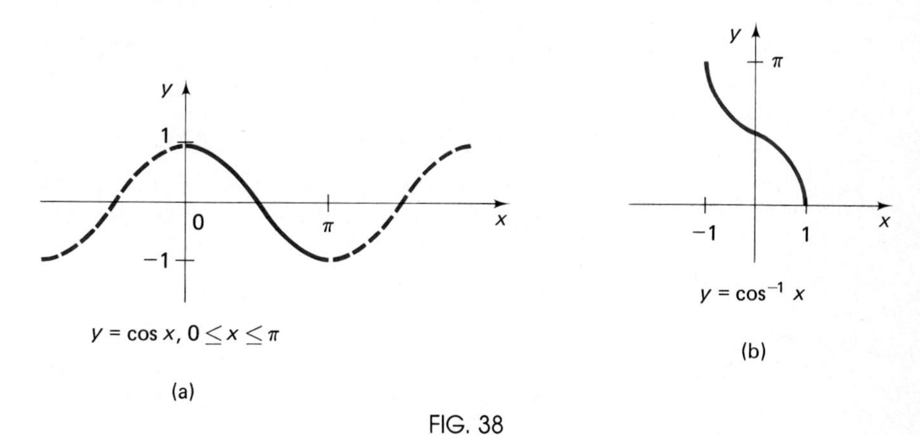

$y = \cos x, 0 \leq x \leq \pi$

(a)

$y = \cos^{-1} x$

(b)

FIG. 38

(31) Definition Suppose $-1 \le x \le 1$. Then $y = \cos^{-1} x$ (or $y = \arccos x$) if and only if $x = \cos y$ and $0 \le y \le \pi$.

In other words, if $-1 \le x \le 1$ then the equation $y = \cos^{-1} x$ is saying "y is the number (angle) between 0 and π whose cosine is x." The graph of $y = \cos^{-1} x$, Fig. 38(b), is obtained from Fig. 38(a) by reflection through the line $y = x$.

For the function $y = \tan x$, conditions 1–4 are satisfied if we restrict the domain to $-\frac{\pi}{2} < x < \frac{\pi}{2}$. See Fig. 39(a).

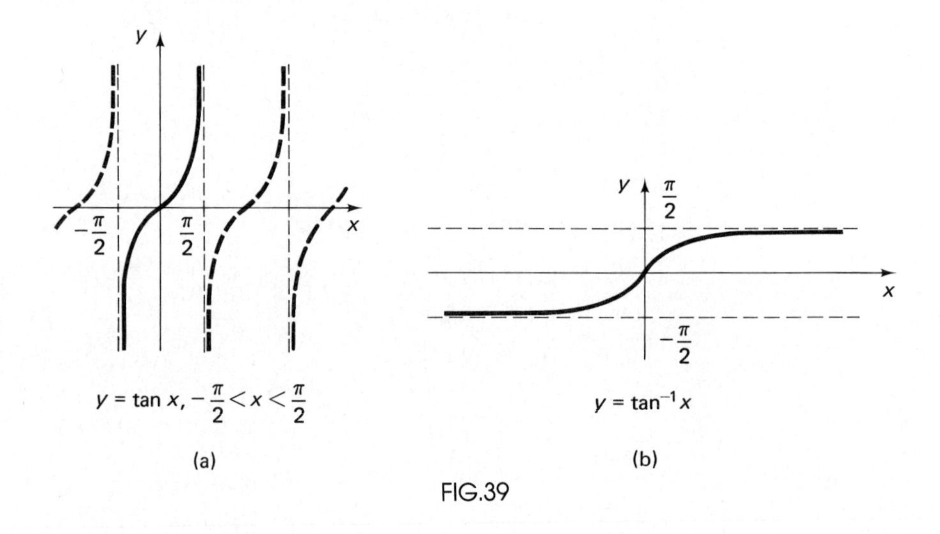

FIG.39

(32) Definition Suppose x is any real number. Then $y = \tan^{-1} x$ (or $y = \arctan x$) if and only if $x = \tan y$ and $-\frac{\pi}{2} < y < \frac{\pi}{2}$.

In other words, if x is any real number, the equation $y = \tan^{-1} x$ is saying "y is the number (angle) strictly between $-\frac{\pi}{2}$ and $\frac{\pi}{2}$ whose tangent is x." The graph of $y = \tan^{-1} x$ is shown in Fig. 39(b).

The remaining trigonometric functions also have inverses if their domains are appropriately restricted.

(33) Definition Suppose $|x| \ge 1$. Then $y = \csc^{-1} x$ (or $y = \operatorname{arccsc} x$) if and only if $x = \csc y$ and $-\frac{\pi}{2} \le y \le \frac{\pi}{2}$, $y \ne 0$.

(34) Definition Suppose $|x| \geq 1$. Then $y = \sec^{-1} x$ (or $y =$ arcsec x) if and only if $x = \sec y$ and $0 \leq y \leq \pi, y \neq \dfrac{\pi}{2}$.

(35) Definition Suppose x is any real number. Then $y = \cot^{-1} x$ (or $y =$ arccot x) if and only if $x = \cot y$ and $0 < y < \pi$.

In each of the definitions of the inverse trigonometric functions, y may be viewed as an angle measured in degrees. Just change the restriction on y to the corresponding inequality in degrees, such as $-90° \leq y \leq 90°$ in Definition (30).

EVALUATION USING A CALCULATOR

The process of finding inverse sine, inverse cosine, or inverse tangent on a calculator is exactly the same as was described in the previous chapter. First, ***make sure your calculator is in the proper mode,*** and remember that if you are computing with trigonometric functions or their inverses and you want all values to be real numbers, the calculator must be in radian mode. Then on either an RPN or an algebraic calculator, press (1) the number; (2) the button [ARC], [INV], [F], or the equivalent, whichever is on your calculator; and finally (3) the appropriate button of [sin], [cos], or [tan].

EXAMPLE 1 Find $\sin^{-1}(-0.5)$.

Solution In radian mode, press [.5] [+/−] [INV] [sin], obtaining -0.523599. In degree mode pressing the same buttons yields $-30°$. ■

EXAMPLE 2 Find arctan 56.47.

Solution Press [56.47] [INV] [tan]. In radian mode you would obtain 1.55309, while in degree mode you would obtain 88.9855°. ■

EXAMPLE 3 Find $\cos^{-1} 2.3$.

Solution If you press [2.3] [INV] [cos], you will get an error indication (in either mode), because $\cos^{-1}x$ is only defined for $-1 \leq x \leq 1$. ■

Even though your calculator does not have buttons for cosecant, secant, or cotangent, their inverse functions can still be evaluated on a calculator. For instance, suppose you are given x, $|x| \geq 1$, and you wish to find $\csc^{-1}x$. Set $y = \csc^{-1}x$. Then, using the notation $\Leftrightarrow$ for "if and only if,"

$$y = \csc^{-1}x \Leftrightarrow x = \csc y \quad \text{and } -\frac{\pi}{2} \leq y \leq \frac{\pi}{2}, y \neq 0 \text{ [by Definition (33)]}$$

$$\Leftrightarrow x = \frac{1}{\sin y} \quad \text{and } -\frac{\pi}{2} \le y \le \frac{\pi}{2}, y \ne 0 \left(\text{since } \csc y = \frac{1}{\sin y}\right)$$

$$\Leftrightarrow \frac{1}{x} = \sin y \quad \text{and } -\frac{\pi}{2} \le y \le \frac{\pi}{2}, y \ne 0 \text{ (taking reciprocals)}$$

$$\Leftrightarrow y = \sin^{-1}\left(\frac{1}{x}\right) \qquad \text{[by Definition (30)].}$$

Consequently, we have shown the following:

> **(36)** If $|x| \ge 1, \quad \csc^{-1}x = \sin^{-1}\left(\frac{1}{x}\right).$

Using formula (36), $\csc^{-1}x$ can easily be found on a calculator.

EXAMPLE 4 Find arccsc 5.3.

Solution By formula (36), arccsc 5.3 = arcsin $\left(\frac{1}{5.3}\right)$. Thus we press [5.3] [1/x] [INV] [sin], obtaining 0.189817 (in radian mode) or 10.8757° (in degree mode). ■

There is a formula similar to (36) for $\sec^{-1}x$ (see Exercise 81). By different methods, a formula for $\cot^{-1}x$ is

$$\cot^{-1}x = \tfrac{1}{2}\pi - \tan^{-1}x, \text{ for all } x$$

EVALUATION USING TABLES

In the previous chapter we showed how to evaluate inverse trigonometric functions when we were dealing with acute angles. In the more general situation, if we want to evaluate some inverse trigonometric function at x when x is positive, this is precisely the case we dealt with in the previous chapter. Consequently, we give only a simple example of this case here. For other examples, including interpolation, refer to Appendix D.

EXAMPLE 5 Find $\cos^{-1} 0.3256$.

Solution Recall we look in the body of Table D1 in the column labeled cos x (either at the top or bottom) for 0.3256. If the 0.3256 is in the column with cos x at the top, we look to the left for x, while if the cos x is at the bottom, we look to the right. In this case the latter holds, and we see $x = 1.2392$ or 71°. ■

The problem of evaluating an inverse trigonometric function at x when x is negative is more complicated. Refer to Appendix D for details. Here we just summarize the process.

1. Look up the same inverse function of $|x|$ to get the reference number (angle) u.
2. If $y = \sin^{-1}x$, $\csc^{-1}x$, or $\tan^{-1}x$, then x negative means y is between $-\frac{\pi}{2}$ and 0. In this case, $y = -u$.
3. If $y = \cos^{-1}x$, $\sec^{-1}x$, or $\cot^{-1}x$, then x negative means y is between $\frac{\pi}{2}$ and π. In this case, $y = \pi - u$.

EXAMPLE 6 Find $\sin^{-1}(-0.5)$ in degrees.

Solution From the table, we first find $\sin^{-1} 0.5 = 30°$. Since $y = \sin^{-1}x$ has $-90° \le y \le 90°$ and x is negative, $y = -30°$. ■

EVALUATION WITHOUT USING A CALCULATOR OR TABLES

There are many cases when you can evaluate an inverse trigonometric function without using a calculator or tables. When doing so, you gain much more understanding of the functions. $\sin^{-1}(-0.5)$, which we have evaluated above using both calculators and tables, is such a case.

EXAMPLE 7 Find $\sin^{-1}(-0.5)$ directly from the definition.

Solution Set $\theta = \sin^{-1}(-0.5)$. Then by Definition (30), $\theta = \sin^{-1}(-0.5)$ if and only if $\sin\theta = -0.5$ and $-\frac{\pi}{2} \le \theta \le \frac{\pi}{2}$. Since $\sin\theta$ is negative and $-\frac{\pi}{2} \le \theta \le \frac{\pi}{2}$, we have $-\frac{\pi}{2} \le \theta \le 0$ (i.e., θ is in Q_4). If ϕ is the reference angle of θ, then $\sin\phi = 0.5$. Therefore from our knowledge of special angles, $\phi = \frac{\pi}{6}$. But the only angle θ with $-\frac{\pi}{2} \le \theta \le 0$ and reference angle $\frac{\pi}{6}$ is $\theta = -\frac{\pi}{6}$. Of course, in degrees, $\theta = -30°$. ■

The following problem cannot be done using a calculator or tables, since there are no numbers involved. However, it can be done by techniques similar to the preceding.

EXAMPLE 8 Find $\sin(\cos^{-1}u)$ as an algebraic expression in u.

Solution Set $\theta = \cos^{-1}u$. We want $\sin\theta$ in terms of u. From Definition (31), $\theta = \cos^{-1}u$ if and only if $\cos\theta = u$ and $0 \le \theta \le \pi$. Since we have $\cos\theta = u$ and want $\sin\theta$ in terms of u, we use the fundamental identity relating $\sin\theta$ and $\cos\theta$, namely $\sin^2\theta + \cos^2\theta = 1$. We obtain

$$\sin^2\theta = 1 - \cos^2\theta = 1 - u^2 \quad \text{or} \quad \sin\theta = \pm\sqrt{1 - u^2}$$

The sign can be determined in this case. Since $0 \le \theta \le \pi$ and $\sin\theta$ is positive in Q_1 and Q_2, the sign must be $+$. Thus

$$\sin\theta = \sin(\cos^{-1}u) = \sqrt{1-u^2} \quad \blacksquare$$

From the definition of inverse functions, we know that $f(f^{-1}(x)) = x$ for all x in the domain of f^{-1} and that $f^{-1}(f(x)) = x$ for all x in the domain of f. Therefore these relations hold for the *restricted* trigonometric functions and their inverses.

EXAMPLE 9 (a) $\sin[\sin^{-1}(-0.17)] = -0.17$. (b) $\arctan\left(\tan\frac{\pi}{5}\right) = \frac{\pi}{5}$. (c) $\cos^{-1}(\cos 171°) = 171°$. ■

If the inverse function is on the inside, this always works. However, if the trigonometric function is on the inside, this works only if the angle is within the restricted domain of that function.

EXAMPLE 10 Find $\arcsin(\sin 589°)$.

Solution The answer is not 589°, because 589° is not in the restricted domain of the sine function, $-90° \le \theta \le 90°$. If we set $\theta = \arcsin(\sin 589°)$, then from Definition (30),

$$\theta = \arcsin(\sin 589°) \text{ if and only if}$$

$$\sin\theta = \sin 589° \text{ and } -90° \le \theta \le 90°$$

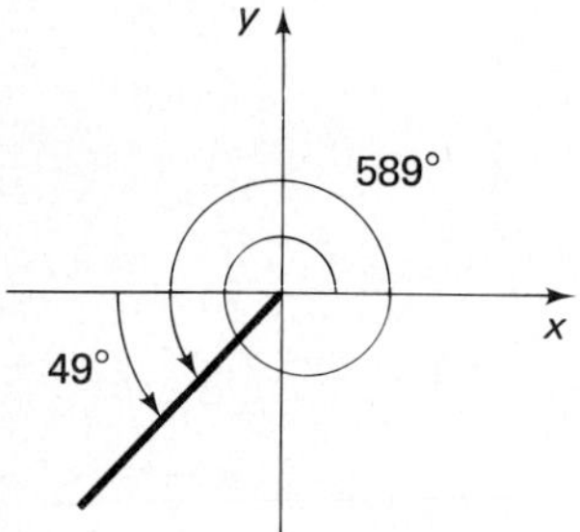

FIG. 40

The angle 589° is in Q_3 with reference angle 49°. See Fig. 40. Since sine is negative in Q_3, θ must satisfy the following: (1) $-90° \le \theta \le 90°$, (2) the reference angle of θ is 49°, and (3) $\sin\theta$ is negative. Clearly, $\theta = -49°$ is the desired angle. Thus $\arcsin(\sin 589°) = -49°$. Of course this problem could have been done on a calculator by pressing (in degree mode) [589] [sin] [INV] [sin]. ■

EXERCISES

In Exercises 1–8, evaluate, assuming the expression is (a) a real number or an angle measured in radians, and (b) an angle measured in degrees.

1. $\sin^{-1}0.829$
2. $\arcsin(-0.3502)$
3. $\arccos(-0.1276)$
4. $\cos^{-1}0.9781$
5. $\tan^{-1}(-0.4877)$
6. $\arctan 12.25$
7. $\operatorname{arccsc} 2.184$
8. $\csc^{-1}(-1.26)$

In Exercises 9–20, evaluate, assuming all numbers involved are real numbers.

9. $\cos 5^{-1}$
10. $\cos^{-1}5^{-1}$
11. $(\cos^{-1}5^{-1})^{-1}$
12. $[(\cos^{-1}5^{-1})^{-1}]^{-1}$
13. $\sin(\cos^{-1}0.788)$
14. $\cos(\arcsin 0.788)$
15. $\sec(\arctan 4.331)$
16. $4\sin^{-1}0.3448 + 9\cos 0.6283$
17. $\cos[\frac{1}{2}\tan^{-1}(-0.805)] - 36$

18. $\dfrac{\cos^{-1}(-0.645)}{\sin 0.5469}$

19. $\arcsin\left(\dfrac{4}{3.6 + 12.4} - \operatorname{arccsc} 4\right)$

ⓒ**20.** $[\tan^{-1}(-128)]^5 + 16 \arccos 3^{-5}$

In Exercises 21–36, evaluate the inverse functions directly from the definition (without using a calculator or tables).

21. $\sin^{-1}(\frac{1}{2})$

22. $\arcsin(-\frac{1}{2})$

23. $\arccos(-\frac{1}{2})$

24. $\cos^{-1}(\frac{1}{2})$

25. $\tan^{-1}(-1)$

26. $\arctan 1$

27. $\operatorname{arccot} 1$

28. $\cot^{-1}(-1)$

29. $\sin^{-1}\dfrac{1}{\sqrt{2}}$

30. $\operatorname{arccsc}(-\sqrt{2})$

31. $\operatorname{arcsec}\sqrt{2}$

32. $\cos^{-1}\left(-\dfrac{1}{\sqrt{2}}\right)$

33. $\sec^{-1}\left(-\dfrac{2}{\sqrt{3}}\right)$

34. $\operatorname{arccsc}\left(\dfrac{2}{\sqrt{3}}\right)$

35. $\operatorname{arccot}(-\sqrt{3})$

36. $\tan^{-1}\sqrt{3}$

In Exercises 37–48, state which one of (A), (B), and (C) holds without evaluating the problem: (A) the expression is undefined, (B) the expression could be evaluated without using a calculator or tables or (C) a calculator or tables should be used in the evaluation.

37. $\sin 0.5$

38. $\sin^{-1} 0.5$

39. $\cos 1.5$

40. $\arccos 1.5$

41. $\tan\left(-\dfrac{\pi}{2}\right)$

42. $\tan^{-1}\left(-\dfrac{\pi}{2}\right)$

43. $\csc 0$

44. $\csc^{-1} 0$

45. $\cot\sqrt{3}$

46. $\operatorname{arccot}\sqrt{3}$

47. $\sec(-0.5)$

48. $\sec^{-1}(-0.5)$

In Exercises 49–64, evaluate the expression without using a calculator or tables.

49. $\sin(\arcsin 0.2)$

50. $\cos(\arccos 0.643)$

51. $\tan^{-1}\left(\tan\dfrac{11\pi}{3}\right)$

52. $\tan(\tan^{-1} 12.9)$

53. $\sin^{-1}(\sin 38°)$

54. $\arccos[\cos(-16°)]$

55. $\cos^{-1}\left(\cos\dfrac{5\pi}{8}\right)$

56. $\tan^{-1}\left(\tan\dfrac{\pi}{3}\right)$

57. $\sin^{-1}(\sin 642°)$

58. $\csc(\operatorname{arccsc} 6.89)$

59. $\arctan[\tan(-843°)]$

60. $\arcsin[\sin(-11.7352)]$

61. $\arccos(\cos 6514°)$

62. $\sin^{-1}[\sin(-11.7352°)]$

63. $\tan[\tan^{-1}(-479)]$

64. $\tan^{-1}[\tan(-479°)]$

65. Suppose $\dfrac{\pi}{2} < x \leq \pi$. Then $\sin^{-1}(\sin x) = x$ is not true. Using reference angles, find m and b such that, for $\dfrac{\pi}{2} < x \leq \pi$, $\sin^{-1}(\sin x) = mx + b$.

In Exercises 66–70, rewrite the given expression as an algebraic expression in u.

66. $\cos(\sin^{-1} u)$

67. $\sec(\sin^{-1} u)$

68. $\sin(\sec^{-1} u)$

69. $\tan(\sec^{-1} u)$

70. $\cot(\operatorname{arccsc} u)$

In Exercises 71–80, sketch the graphs by applying your knowledge of graphing to the graphs of the inverse trigonometric functions.

71. $y = 2\sin^{-1}x$
72. $y = \frac{1}{2}\arcsin x$
73. $y = \tan^{-1}(x + \pi)$
74. $y = \arctan(x - 2)$
75. $y = 2\arccos x$
76. $y = \arcsin(x - 5)$
77. $y = \dfrac{5}{\pi}\tan^{-1}x$
78. $y = 2\sin^{-1}(3x - 4)$
79. $y = \frac{1}{3}\arccos(\frac{1}{2}x + 2)$
80. $y = \dfrac{1}{\pi}\arctan(x + 3)$

81. Develop a formula for $\sec^{-1}x$ by using similar steps to those used to develop formula (36).

Ⓒ**82.** Use the formula developed in Exercise 81 to evaluate $\sec^{-1}(-21.73)$ and arcsec 1.489.

83. Graph $y = \dfrac{\pi}{2} - \tan^{-1}x$ and compare it with the graph of $y = \cot^{-1}x$.

2.9 Applications

Anyone who has sat on a beach watching the waves roll in has at least an intuitive idea as to what wave motion is. The surface of the water is shaped a little like a sine curve (if you take a cross section in the right direction), and the crests and troughs of the waves move along through the water. If you get into the water and float for a while, you find yourself bobbing up and down and not moving with the waves. Hence you can realize that the water itself is not flowing along in the waves; it is energy moving through the water that causes the waves. (For example, a boat moving through the water temporarily pushes the water aside, and the energy from this push travels through the water in waves.) The water molecules themselves just move around in one region, transmitting this energy along.

Water waves provide one example of many in which energy moves through our environment in *waves*. Others include sound, light, and even earthquake waves. Trigonometry is the basic mathematical tool that is used to analyze wave motion. Our purpose in this section is to discuss briefly a few familiar physical phenomena and give an indication as to how trigonometry is used to describe them. Unfortunately, we are forced to keep the mathematics fairly simple, for otherwise the mathematics and physics become too deep for this brief survey.

Wave motion can be quite complicated, so we shall consider only the simplest wave motion.

(37) Definition A phenomenon is said to be **simple harmonic** or have **simple harmonic motion** if it can be described by

$$y = a\sin(bx - c)$$

where $b > 0$ and x is distance. If so, it has amplitude $|a|$, wavelength $\lambda = \frac{2\pi}{b}$, and phase shift $\frac{c}{b}$. See Fig. 41.

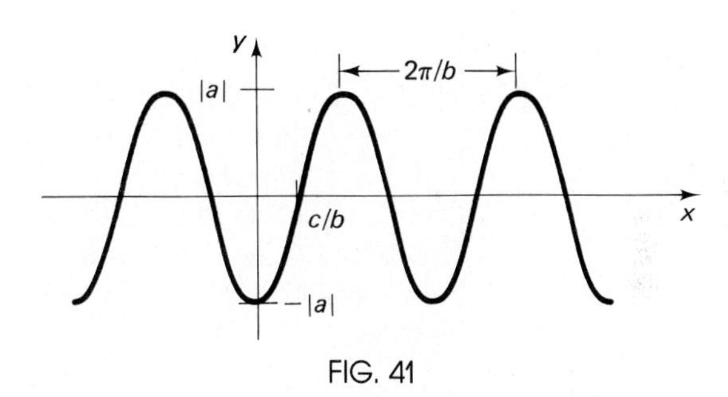

FIG. 41

In the situations we describe, the waves will be moving as time changes. Different kinds of waves move at different speeds. The time it takes for one complete wave to pass a fixed point is called the **period** of the motion; the number of waves that pass a fixed point in one unit of time is called the **frequency** of the motion. These two quantities are related by

(38) $$\text{Frequency} = \frac{1}{\text{period}}$$

In addition, it is not hard to see that there is a relationship between frequency, wavelength, and the speed that the waves are traveling:

(39) $$\text{Frequency} \times \text{wavelength} = \text{speed}$$

We now discuss a few natural phenomena.

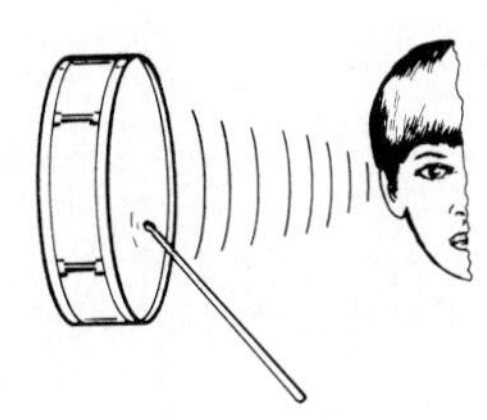

SOUND

The term *sound* has two distinct uses. A physiologist uses the word sound in connection with the sense of hearing and the effects on a human ear produced by certain vibrations in the air. However, we shall follow the usage of the physicists, who consider sound to be those air vibrations themselves.* The air vibrations are caused by something (such as a drum head or a violin string) moving back and forth. The vibrating object pushes the air aside, causing sound waves in somewhat the same way that a boat pushes water aside, causing water waves.

When made to vibrate, a tuning fork will produce a **pure tone**; this means that only one frequency is produced. A tuning fork that produces a middle C vibrates at approximately 260 cycles per second (cps) or 260 Hz (1 Hz is 1 hertz, which is 1 cps). Now sound moves through air at approximately 1100 feet per second. From relationship (39), frequency $\times$ wavelength = speed, we see that the wavelength λ of middle C is approximately

$$\lambda \approx \frac{1100}{260} \approx 4.23 \text{ ft}$$

This is of great interest to anyone who wants to build an open organ pipe to produce a middle C. An open pipe must be exactly one-half wavelength long to produce that frequency of sound (as a fundamental frequency). A closed pipe (for example, a cola bottle) need only be one-fourth wavelength (because the closed end reflects the wave and hence doubles the effect of the pipe). Try blowing across a bottle filled to different levels with fluid, and hear what happens.

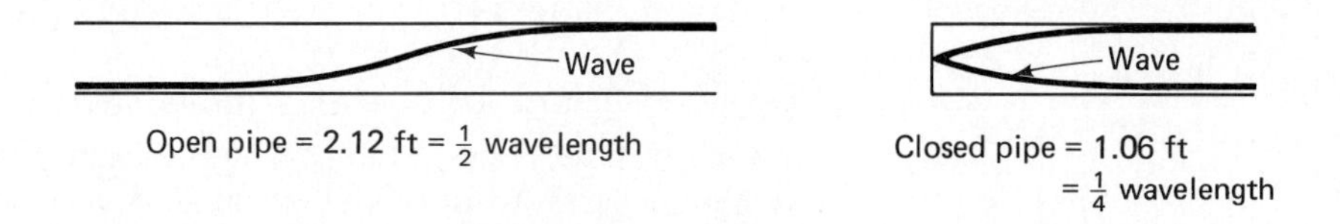

Length of pipe to produce middle C

If you have ever tuned a stringed instrument, you know that the tighter the string, the higher the sound which it produces. This occurs because a tighter string vibrates more quickly, and the higher the frequency (and correspondingly the shorter the wavelength), the higher the pitch of the sound.

If the stringed instrument is like a violin or a guitar, you also know that by pressing the string against the neck, a higher sound will be produced when the string is vibrated. Again, a shorter string (of the same tension) vibrates more quickly and hence produces a higher sound.

Usually when an object such as a drum, a bell, or a string is vibrating, more than one frequency is involved. The lowest frequency is called the **fundamental frequency.** The other frequencies that are present are multiples of the fundamental frequency and are called **overtones.** If the multiples are integers, the overtones are called **harmonics.**

Harmonics can be seen quite vividly on a stringed instrument, such as a

* If a tree falls on a deserted island, does it produce a sound? It depends on which sense of the word *sound* you use.

violin. Suppose a violin string is tuned to produce middle C, so that it is vibrating at approximately 260 cps. See Fig. 42(a). But this is just the fundamental frequency; at the same time the string is vibrating twice as quickly, 520 cps, as indicated in Fig. 42(b), producing the C above middle C. This is not all, for at the same time the string is also vibrating at three times the fundamental frequency (i.e., 780 cps), etc. In fact, the string is vibrating at all integral multiples of the fundamental frequency, producing all harmonics. Indeed, you can physically hear the harmonics. If you lightly touch a vibrating string in the middle, this will cancel all odd multiples of the fundamental frequency (including the fundamental frequency itself), and the sound will immediately rise an octave (to the first harmonic). Any reasonably experienced violinist knows how to produce a few harmonics this way (though he or she may not know why it works).

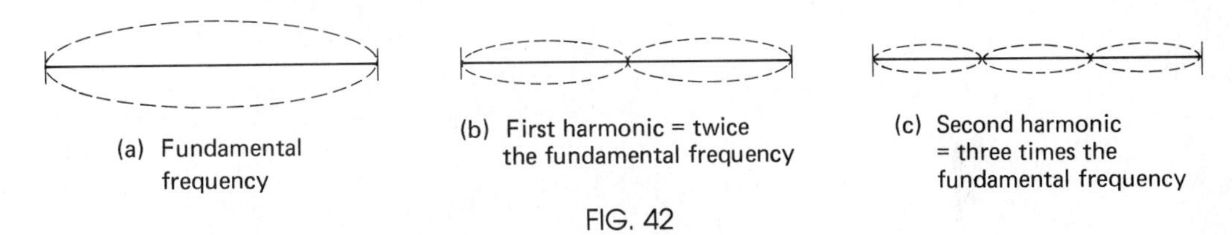

(a) Fundamental frequency

(b) First harmonic = twice the fundamental frequency

(c) Second harmonic = three times the fundamental frequency

FIG. 42

RADIO WAVES AND OTHER ELECTROMAGNETIC WAVES

The signal sent from the broadcast tower of a radio station to the antenna of your radio is an electromagnetic wave. There is a main wave present which is called a carrier wave, and this wave is basically a sine wave. If the station is an AM station, then the amplitude of the carrier wave is changed to "carry" sound waves, as indicated in Fig. 43(a). AM stands for **amplitude modulation.** If the station is an FM station, then the frequency of the carrier wave is changed to "carry" sound waves, as indicated in Fig. 43(b). FM stands for **frequency modulation.**

To obtain a rough idea as to dimensions, the middle of the AM dial is approximately 1200 kilohertz (kHz), which stands for 1200 kilocycles per second or $1{,}200{,}000 = 1.2 \times 10^6$ cps. Radio waves travel at the speed of light, which is approximately 3×10^8 meters per second. From relation (39), frequency $\times$ wavelength = speed, we see that the wavelength of this radio band is approximately $\dfrac{3 \times 10^8}{1.2 \times 10^6}$ meters $\approx$ 200 meters long. (Using 1 meter $\approx$ 1 yard, we see that this is about two football fields long.) For a comparison, CB radios broadcast at wavelengths of about 11 meters.

There are many other electromagnetic waves besides radio waves. These include television waves, microwaves, cosmic rays, and many others. In a famous work, Maxwell* showed that light could be explained as electro-

* James Clerk Maxwell (1831–1879), a Scottish theoretical physicist who became famous for his mathematical theory of electricity, magnetism, and light.

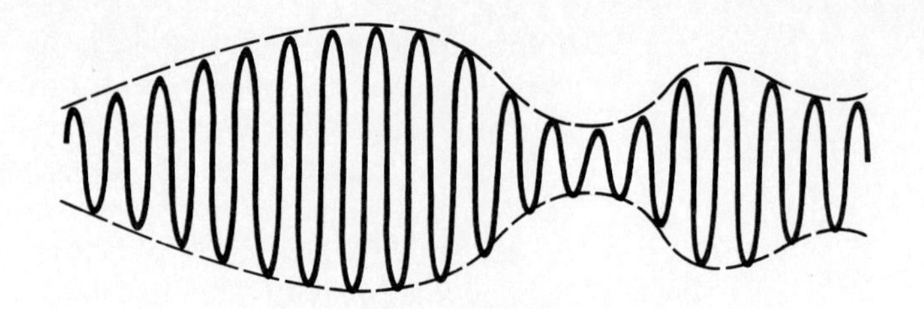

(a) AM: Carrier wave *amplitude* varied by sound source. Frequency is constant.

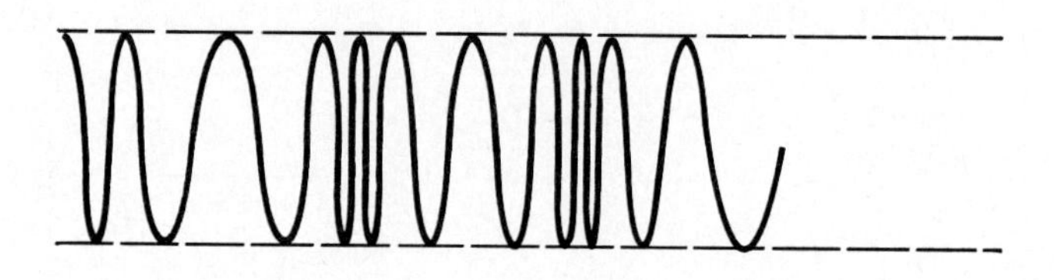

(b) FM: Carrier wave *frequency* varied by sound source. Amplitude is constant.

FIG. 43

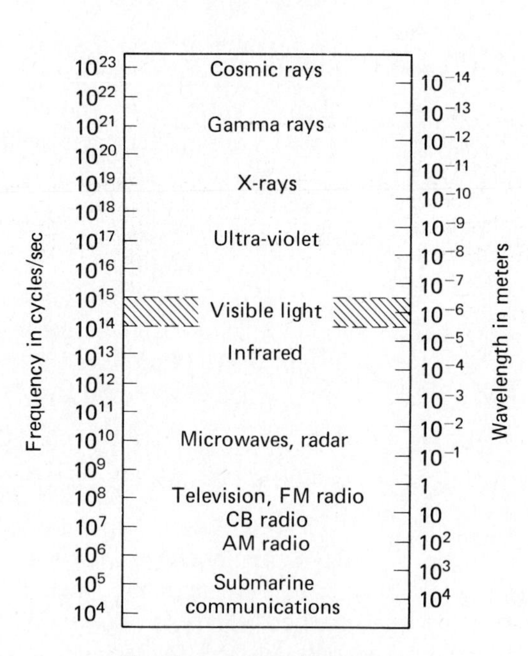

The Electromagnetic Spectrum
(Freq. $\times$ Wavelength $\approx 3 \times 10^8$ m/sec)

FIG. 44

magnetic waves. There is a whole spectrum of electromagnetic waves as Fig. 44 illustrates.

ELECTRICITY

A good analogy to electricity flowing through a wire is water flowing through a pipe. The amount of water flowing corresponds to the amount of electricity flowing, which is called **current.** Current is usually denoted by I (and is measured in **amperes** or **amps**). The amount of force pushing the water through the pipe corresponds to the amount of force pushing the electricity through the wire, which is called **electromotive force** (*EMF*). *EMF* is usually denoted by E (and is measured in **volts**). The electricity in your house (if it is in the United States) is being pushed at about 110 volts. Something that resists water flowing (like a water wheel) corresponds to something that resists electricity flowing (like a light bulb). This is called **resistance.** Resistance is usually denoted by R (and is measured in **ohms**). The basic relationship between these quantities is **Ohm's Law:***

(40) $$E = IR$$

(It is curious to note $\frac{1}{R}$ is called **conductance** and is measured in mhos, so named by some punster.)

Normally, water flows through a pipe in one direction. When electricity flows in one direction, this is called **direct current** and abbreviated *DC*. For example, batteries produce *DC* or direct current.

There are many instances in which **alternating current** (abbreviated *AC*) is produced. The analogy to keep in mind is a pipe with a person at each end controlling the water flow. First one person pushes some water through the pipe and then stops, then the other person pushes some water back the other way and then stops. This keeps going back and forth. Alternating current goes back and forth through a wire in the same way. (The reason alternating current is used is that generators naturally produce alternating current.) In the U.S., public power companies provide electricity which is alternating at 60 cycles per second.

Suppose you have a simple circuit with an *AC* power source and a resistance. This is diagrammed in Fig. 45. Then E is a simple periodic function of time,

(41) $$E = a \sin(\omega t)$$

For example, the power source could be a wall socket and the resistance a light bulb. Then $a = 110\sqrt{2}$ (volts) if t is in seconds, and $\omega = 120\pi$ (to get 60

*Discovered experimentally in 1826 by the German physicist George Simon Ohm (1787–1854).

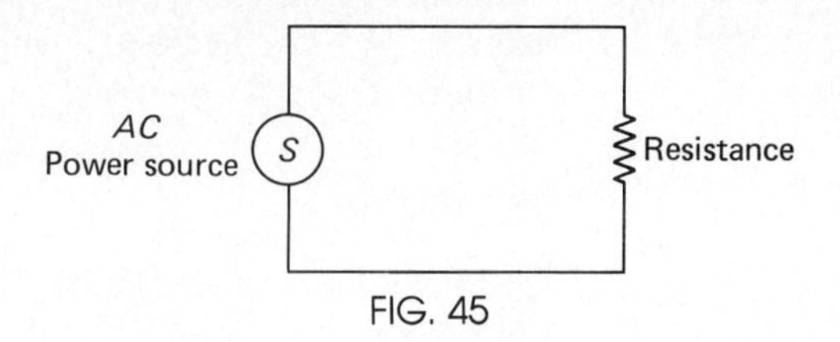

FIG. 45

cycles per second). It will then also happen that the current, I, is exactly "in phase" with E, i.e.,

(42) $$I = \frac{a}{R}\sin(\omega t)$$

The functions E and I are graphed in Fig. 46.

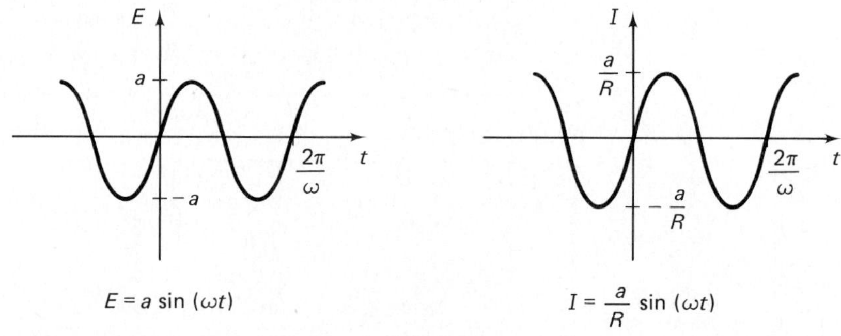

FIG. 46

Ohm's Law, $E = IR$, still applies. For example, suppose you had a simple circuit as above with an AC power source and a light bulb. If you doubled the voltage E (say you started up another generator), then R would stay the same, so from $E = IR$, I would have to double. Thus the bulb would burn twice as brightly (if it did not burn out). If instead you put a second identical bulb in the original circuit, R would double but E would stay the same. From $E = IR$, the current would be halved, so each bulb would be half as bright as the one alone.

Changes in the flow of electricity produce some very strange but very interesting and useful effects. (The fact that AC current is changing all the time is very important. The following effects cannot happen in a DC circuit.) There is an electronic mechanism known as a condenser. We will not describe how condensers are made or why they work, but for our purposes a condenser can be thought of as a magic electronic black box that leaves the voltage unchanged but *produces a phase shift of the current*. To illustrate what this means, suppose you had a simple circuit with an AC power source and a condenser. (This is diagrammed in Fig. 47.) Then the current is speeded up or thrown "ahead" exactly 90° or $\frac{\pi}{2}$ of the *EMF!* That is, if E is given by

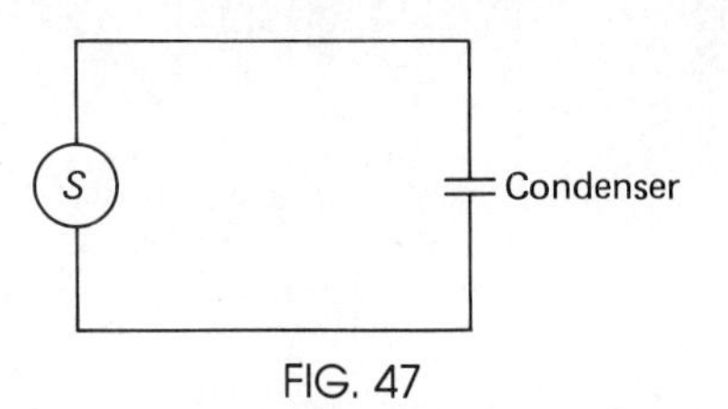

FIG. 47

(43) $$E = a\sin(\omega t)$$

then I is given by

(44) $$I = a\omega C\sin\left(\omega t + \frac{\pi}{2}\right)$$

where C is called **capacitance.** Graphically E and I are given in Fig. 48.

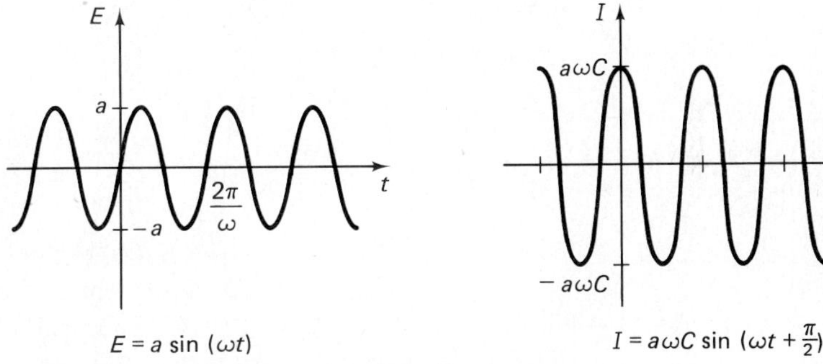

FIG. 48

Ohm's Law for this circuit is essentially $E = I\left(\frac{1}{\omega C}\right)$, except for the phase shift.

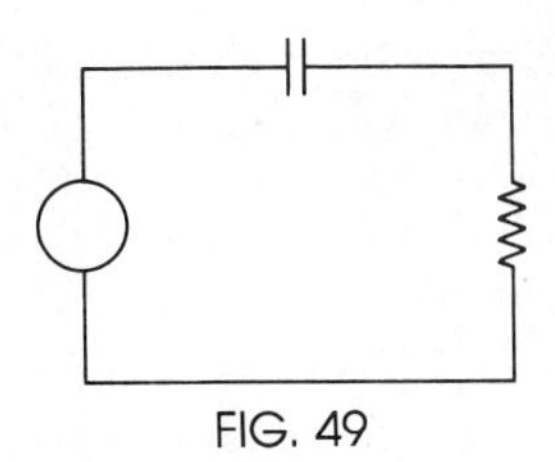
FIG. 49

The situation becomes even more interesting when you have a simple circuit with both a condenser and a resistor, as is diagrammed in Fig. 49. Then there is a phase shift ϕ of the current I forward, where ϕ is between 0 and $\frac{\pi}{2}$. In fact, $\phi = \tan^{-1}\left(\frac{1}{\omega CR}\right)$, so ϕ is a nice function of R and C. For this situation, if E is given by

(45) $$E = a\sin(\omega t)$$

Then I is given by

(46) $$I = \frac{a}{Z}\sin(\omega t + \phi)$$

where $Z = \sqrt{R^2 + \frac{1}{(\omega C)^2}}$ (Z is called **impedance**). Ohm's Law for this circuit is essentially $E = IZ$ except for the phase shift. The graphs of E and I are given in Fig. 50. For example, suppose you have a circuit as diagrammed in

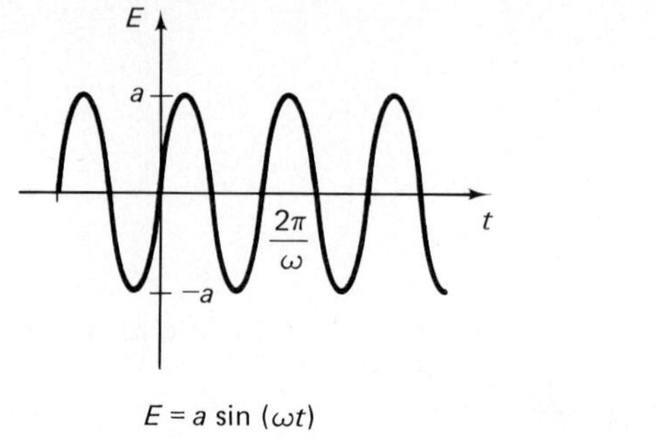

$E = a \sin(\omega t)$ $\qquad$ $I = \dfrac{a}{Z} \sin(\omega t + \phi)$

FIG. 50

Figure 49, where the voltage is 250 volts (so $a = 250\sqrt{2}$), $\omega = 120\pi$, the resistance R is 75 (ohms), and the condenser gives $\omega C = 0.01$. Then $E = 250\sqrt{2}\sin(120\pi t)$ and $Z = \sqrt{75^2 + \dfrac{1}{0.01^2}} = \sqrt{75^2 + 100^2} = 125$ (ohms). The phase shift of I is

$$\phi = \tan^{-1}\left(\frac{1}{0.01(75)}\right) = \tan^{-1}\left(\frac{100}{75}\right) = \tan^{-1}\left(\frac{4}{3}\right) \approx 53.13^\circ$$

$$\approx 0.927295$$

so that $I = 2\sqrt{2}\sin(120\pi t + 0.927295)$.

As a final note, you might notice the above formulas depend upon not only C and R, but also the frequency ω. Consequently, the same circuit will have different effects for different frequencies. This is an indication as to why instruments such as radios and television sets, which receive many different frequencies, can have one circuit "tune in" the frequencies you want and "tune out" the rest.

As circuits become more complicated, complex numbers are used to help describe and explain the various phenomena. This will be indicated in the chapter on complex numbers.

EXERCISES

In Exercises 1–8, assume that middle C has a frequency of approximately 260 Hz and that the A above middle C is 440 Hz (1 Hz is 1 cps). Assume all data are correct to three significant figures.

1. What is the wavelength of C above middle C?

2. What is the wavelength of the A above middle C?

3. What is the frequency of A above the C above middle C?

4. What is the frequency of the C below middle C?

5. If the speed of sound in water is 4800 feet per second, what is the wavelength of middle C in water?

6. The speed of sound in steel is about 16,000 feet per second. Suppose that an A above middle C tuning fork is vibrating at one end of a 20-foot steel beam and that the sound is coming out of the other end. How many wavelengths are there from one end of the beam to the other?

7. How long should you make an open organ pipe to produce an A above middle C? How long should a closed pipe be to produce that A?
8. Suppose Neptune wanted to build a water organ to play under the sea. How long should he make an open pipe to produce an A above middle C? A closed pipe? (See Exercise 5.)
9. Red light has a wavelength of about 7×10^{-7} meter. What is its frequency?
10. A certain radar set, used by police to check for speeders, uses microwaves of length 0.1 meter. What is the frequency?
11. The ideal length of a radio transmitting antenna is any multiple of one-half wavelength. What is the shortest antenna a radio station should use if it is broadcasting at 1400 kHz on the AM dial? Give the answer to the nearest meter.
12. What is the shortest antenna a radio station should use if it is broadcasting at 103 megahertz (MHz) on the FM dial? (See Exercise 11.)
13. A simple *AC* circuit has a generator and electric heater. If a smaller generator is attached so that the voltage in the circuit is increased by 20%, how much more heat does the heater give?
14. In the circuit in problem 13, if instead two more identical heaters are attached (in series), how much heat does the original heater give?
15. A simple *AC* circuit has a resistance and condenser as indicated in Fig. 49. Assume the voltage is 130 volts, $\omega = 120\pi$, the resistance is 120, and the condenser gives $\omega C = 0.02$. Find the phase shift and the current function I.
16. In an *AC* circuit as in Fig. 49, the voltage is 110, $\omega = 120\pi$, the resistance is 100, and $\omega C = 0.03$. Find the phase shift and the current function I.
17. In an *AC* circuit as in Fig. 49, the voltage is 75, $\omega = 120\pi$, the resistance is 21, and $\omega C = 0.125$. Find the phase shift and the current function I.

REVIEW EXERCISES

In Exercises 1–5, use the information given to find the values of the six trigonometric functions of θ.

1. $(-1, 4)$ is on the terminal side of θ.
2. $(-3, -2)$ is on the terminal side of θ.
3. For some point on the terminal side of θ, $r = 5$, $x = 2$.
4. For some point on the terminal side of θ, $r = 5$, $y = -1$.
5. For every point on the terminal side of θ, $y = -3x$.

In Exercises 6–13, find the values of the trigonometric functions of θ.

6. $\cos\theta = -\frac{5}{13}$
7. $\tan\theta = -\frac{12}{5}$
8. $\sin\theta = 0$
9. $\cos\theta = \frac{3}{5}$, $\sin(-\theta)$ negative
10. $\sin\theta = \frac{3}{5}$, $\cos(-\theta)$ negative
11. $\tan\theta = -5$, $\sin\theta$ positive
12. $\csc\theta = 3$, $\sec\theta$ positive
13. $\sec\theta = -4$, $\sin(-\theta)$ positive

In Exercises 14 and 15, find the coordinates of the point P on the terminal side of the given θ satisfying the given condition.

14. $\theta = -432°$, $r = 25$
15. $\theta = 8.63$, $x = -2$

In Exercises 16–21, find the coordinates of $W(t)$.

16. $W\left(\frac{7\pi}{2}\right)$
17. $W(9\pi)$

18. $W\left(\frac{11\pi}{6}\right)$

19. $W\left(-\frac{2\pi}{3}\right)$

20. $W(0.48)$

21. $W(2.95)$

In Exercises 22–32, evaluate the given expression, using a calculator or tables only when necessary.

22. $\sin^2 1465° + \cos^2 1465°$

23. $\tan 540° \sec 493° + \cos 450° \sin(-493°)$

24. $6 \tan 3.63 + 1.895$

25. $\frac{\sin 4.619}{\cos 4.619} - \tan 4.619 + \cot 4.619$

26. $\sin(-63°) + \cos(-63°)$

27. $\sin \frac{11\pi}{6} \cos 3\pi - \tan \frac{13\pi}{4} \sec \frac{4\pi}{3}$

28. $\arcsin(-0.309)$

29. $\tan 1127° \cos 1127° + \sin(-1127°)$

30. $\tan 1950° + \sin 1950° + \cos 1950°$

31. $3 \sin 1.3 + \cos^{-1} 0.6293$

32. $\sin(1 + \arctan 5.976)$

In Exercises 33–36, find the reference angle, and then evaluate the trigonometric functions of the reference angle and attach the appropriate sign.

33. $\sin 11.47$

34. $\tan(-4.198)$

35. $\cos 1265°$

36. $\sec(-15{,}688°)$

In Exercises 37–40, state the amplitude, period, and phase shift (if nonzero), and then sketch the graph.

37. $y = 3 \cos\left(x - \frac{\pi}{3}\right)$

38. $y = -\frac{1}{2} \sin\left(2x + \frac{\pi}{2}\right)$

39. $y = 2 \sin\left(\frac{1}{2}x - \frac{\pi}{4}\right) - 2$

40. $y = \cos(\pi - 2x)$

In Exercises 41 and 42, sketch the graph.

41. $y = 3 \sin^{-1} x$

42. $y = \cos^{-1}(\frac{1}{2}x)$

In Exercises 43–46, evaluate without a calculator or tables.

43. $\sin^{-1}(-\frac{1}{2})$

44. $\tan^{-1}\sqrt{3}$

45. $\sin^{-1}(\sin 658°)$

46. $\cos^{-1}\left[\cos\left(\frac{-17\pi}{7}\right)\right]$

CHAPTER 3

Analytic and Geometric Trigonometry

Many of the identities and formulas presented in this chapter were known to the Greeks 2000 years ago (in different but equivalent forms). The addition and half-angle formulas as well as the Law of Sines were probably known to Hipparchus (about 150 B.C.) and definitely to Ptolemy (about 150 A.D.). The Law of Cosines was known to Euclid (about 300 B.C.).

The Greeks worked with the chord function (where crd $2\theta = 2 \sin \theta$), but the identity $\sin^2 \theta + \cos^2 \theta = 1$ became obvious soon after the Hindus invented the sine function (around 300–400 A.D.).

The Arabs discovered many identities. For example, identities like

$$2 \cos x \cos y = \cos(x + y) + \cos(x - y)$$

were discovered by ibn-Yunus (around 1000 A.D.). Such identities were used by astronomers in the sixteenth century to convert large products into more manageable sums, before logarithms were invented by Napier for this purpose.

Vectors were invented much later in Europe by physicists and did not become a topic of serious mathematical study until the nineteenth century.

3.1 Fundamental Trigonometric Identities

An equation involving a variable is called an **identity** if equality holds for every value of the variable for which all terms in the equation are defined. For instance, $x + 2 = \dfrac{(x + 2)(x - 1)}{x - 1}$ is an identity; the right-hand side is undefined at $x = 1$, but for every other value of x, equality holds. On the other hand, $x^2 - 4 = 0$ is not an identity; the terms are defined for all real values of x, but equality holds only when $x = 2$ or $x = -2$. An equation of this type is called a **conditional equation**. In this section, we shall begin the study of **trigonometric identities**, that is, identities that involve trigonometric functions, while in the next section we shall study conditional trigonometric equations.

We have already seen a number of trigonometric identities. We restate the fundamental ones here.

Fundamental Trigonometric Identities

(1) $\csc \theta = \dfrac{1}{\sin \theta}$, $\sec \theta = \dfrac{1}{\cos \theta}$, $\cot \theta = \dfrac{1}{\tan \theta}$

(2) $\tan \theta = \dfrac{\sin \theta}{\cos \theta}$, $\cot \theta = \dfrac{\cos \theta}{\sin \theta}$

(3) $\sin^2 \theta + \cos^2 \theta = 1$, $\tan^2 \theta + 1 = \sec^2 \theta$, $1 + \cot^2 \theta = \csc^2 \theta$

In this section, we shall examine some of the uses of these identities. We begin with the most important use, namely simplifying expressions.

SIMPLIFYING EXPRESSIONS

In the next several examples, we illustrate how to use the fundamental identities to simplify trigonometric expressions. The techniques we develop will be used throughout this chapter.

EXAMPLE 1 Transform $(\sin x + \cos x)^2$ to $1 + 2 \sin x \cos x$ using the fundamental identities.

Solution We start by squaring the expression.

Note: We use the identity

$$\sin^2 x + \cos^2 x = 1$$

$$(\sin x + \cos x)^2 = \sin^2 x + 2 \sin x \cos x + \cos^2 x$$
$$= 1 + 2 \sin x \cos x \quad ■$$

EXAMPLE 2 Use the fundamental identities to rewrite the expression

$$\sec x - \sin x \tan x$$

in terms of sines and cosines and then simplify.

Solution

$$\sec x - \sin x \tan x = \frac{1}{\cos x} - \sin x \frac{\sin x}{\cos x} \qquad \text{[by (1) and (2)]}$$

$$= \frac{1 - \sin^2 x}{\cos x}$$

$$= \frac{\cos^2 x}{\cos x} \qquad \text{[by (3), } \sin^2 x + \cos^2 x = 1\text{]}$$

$$= \cos x \quad ■$$

Note: We first use

$$\sec x = \frac{1}{\cos x}, \ \tan x = \frac{\sin x}{\cos x}$$

EXAMPLE 3 Simplify the expression $\dfrac{\tan x \csc^2 x}{1 + \tan^2 x}$

Solution

$$\frac{\tan x \csc^2 x}{1 + \tan^2 x} = \frac{\tan x \csc^2 x}{\sec^2 x} \qquad \text{[by (3), } \tan^2 x + 1 = \sec^2 x\text{]}$$

$$= \frac{\dfrac{\sin x}{\cos x} \dfrac{1}{\sin^2 x}}{\dfrac{1}{\cos^2 x}} \qquad \text{[by (1) and (2)]}$$

$$= \frac{\cos x}{\sin x}$$

$$= \cot x \qquad \left[\text{by (2), } \cot x = \frac{\cos x}{\sin x}\right] \quad ■$$

Note: First apply any obvious simplifying identities, here

$$1 + \tan^2 x = \sec^2 x$$

Next, rewrite everything in terms of sines and cosines to simplify.

OTHER APPLICATIONS

We now turn to other uses of the fundamental identities. We begin by reviewing an application you have already seen, and then we extend it.

EXAMPLE 4 If $\tan \theta = -\frac{4}{3}$ and θ is in the second quadrant, find the values of the other trigonometric functions.

Solution First, $\cot \theta = \dfrac{1}{\tan \theta} = \dfrac{1}{-4/3} = -\dfrac{3}{4}$. Next, using $\tan^2 \theta + 1 = \sec^2 \theta$ and $1 + \cot^2 \theta = \csc \theta$, we see

$$\sec\theta = \pm\sqrt{1+\tan^2\theta} = \pm\sqrt{1+\left(-\frac{4}{3}\right)^2} = \pm\sqrt{1+\frac{16}{9}} = \pm\sqrt{\frac{25}{9}}$$

$$= \pm\frac{5}{3}$$

$$\csc\theta = \pm\sqrt{1+\cot^2\theta} = \pm\sqrt{1+\left(-\frac{3}{4}\right)^2} = \pm\sqrt{1+\frac{9}{16}} = \pm\sqrt{\frac{25}{16}}$$

$$= \pm\frac{5}{4}$$

Since θ is in the second quadrant, $\sec\theta$ is negative and $\csc\theta$ is positive. Thus, $\sec\theta = -\frac{5}{3}$ and $\csc\theta = \frac{5}{4}$.

Finally,

$$\sin\theta = \frac{1}{\csc\theta} = \frac{1}{5/4} = \frac{4}{5} \quad\text{and}\quad \cos\theta = \frac{1}{\sec\theta} = \frac{1}{-5/3} = -\frac{3}{5} \quad \blacksquare$$

The next example is similar.

EXAMPLE 5 If θ is in the fourth quadrant, express the other trigonometric functions in terms of $\sec\theta$.

Solution First, $\cos\theta = \frac{1}{\sec\theta}$. Next, using $\sin^2\theta + \cos^2\theta = 1$ and $\tan^2\theta + 1 = \sec^2\theta$, we see

$$\sin\theta = \pm\sqrt{1-\cos^2\theta} = \pm\sqrt{1-\frac{1}{\sec^2\theta}} = \pm\sqrt{\frac{\sec^2\theta - 1}{\sec^2\theta}}$$

$$\tan\theta = \pm\sqrt{\sec^2\theta - 1}$$

Since $\sin\theta$ and $\tan\theta$ are both negative in the fourth quadrant

$$\sin\theta = -\sqrt{\frac{\sec^2\theta - 1}{\sec^2\theta}} \quad\text{and}\quad \tan\theta = -\sqrt{\sec^2\theta - 1}$$

Finally,

$$\cot\theta = \frac{1}{\tan\theta} = -\frac{1}{\sqrt{\sec^2\theta - 1}}$$

$$\csc\theta = \frac{1}{\sin\theta} = -\sqrt{\frac{\sec^2\theta}{\sec^2\theta - 1}} \quad \blacksquare$$

The type of algebraic manipulations we did in Examples 4 and 5 are sometimes used to rewrite certain expressions.

EXAMPLE 6 Substitute $x = a\tan\theta$ into the expression $\sqrt{a^2 + x^2}$ and simplify.

Solution

$$\begin{aligned}\sqrt{a^2 + x^2} &= \sqrt{a^2 + a^2\tan^2\theta}\\ &= \sqrt{a^2(1 + \tan^2\theta)}\\ &= \sqrt{a^2\sec^2\theta}\\ &= |a \sec \theta|\end{aligned}$$

Until we know more about a and θ, we cannot drop the absolute value signs. ■

Note: In doing the exercises, you may wish to use the identities

$$\sin(-\theta) = -\sin \theta \qquad \text{and} \qquad \cos(-\theta) = \cos(\theta)$$

from the previous chapter.

EXERCISES

In Exercises 1–6, match the trigonometric expression with one of those from **(a)–(f)**.

(a) $-\sin x$ **(b)** $\sin^2x + \cos^2x$ **(c)** $\dfrac{\sin x}{\cos x}$

(d) $\cot x$ **(e)** $\dfrac{1}{\sin x}$ **(f)** $\tan^2x + 1$

1. $\tan x$ **2.** $\sin(-x)$ **3.** $\sec^2x$

4. 1 **5.** $\dfrac{\cos x}{\sin x}$ **6.** $\csc x$

In Exercises 7–12, match the trigonometric expression with one of those from **(a)–(f)**.

(a) $\sin(-x)$ **(b)** -1 **(c)** $\cot x$

(d) $\sec x$ **(e)** $\cos x$ **(f)** $\cos^2x$

7. $\dfrac{1}{\cos(-x)}$ **8.** $-\tan x \cos x$ **9.** $\dfrac{\csc x}{\sec x}$

10. $\dfrac{1}{\tan^2x + 1}$ **11.** $(\cos^2x - 1)\csc^2x$ **12.** $(1 + \sin x)(1 + \sin(-x))\sec x$

In Exercises 13–18, use the fundamental identities to transform the first expression into the second.

13. $\cot \theta \sin \theta$, $\cos \theta$ **14.** $\tan \theta \csc \theta$, $\sec \theta$

15. $\dfrac{\sec x}{\tan x}$, $\csc x$ **16.** $\dfrac{\csc y}{\sec y}$, $\cot y$

17. $\sin \alpha(\csc \alpha - \sin \alpha)$, $\cos^2\alpha$ **18.** $\dfrac{1 - \sin^2\beta}{\cos^2\beta}$, 1

In Exercises 19–28, simplify.

19. $\cos \theta + \tan \theta \sin \theta$ **20.** $\cot^2\phi(1 + \tan^2\phi)$

21. $\dfrac{\tan(-\theta)}{\sec\theta}$

22. $\cot(-\theta)\sin\theta$

23. $\dfrac{1+\cos x}{1+\sec x}$

24. $\dfrac{\tan^2 y+1}{\cot^2 y}$

25. $\cos\alpha(\sec\alpha+\tan(-\alpha))$

26. $(\sin\beta+\cot\beta\cos\beta)\tan\beta$

27. $\dfrac{\tan\theta\sin\theta}{\sec^2\theta-1}$

28. $\sin^4\phi-\cos^4\phi+\cos^2\phi$

In Exercises 29–32, find the values of the trigonometric functions.

29. $\cos\theta=-\frac{5}{13},\quad \theta \text{ in } Q_3$

30. $\cot\theta=\frac{7}{24},\quad \theta \text{ in } Q_3$

31. $\sin\theta=\frac{1}{2},\quad \cos\theta<0$

32. $\tan\theta=-2,\quad \cos\theta>0$

In Exercises 33–36, express each of the other trigonometric functions in terms of the given function.

33. $\sin\theta,\quad \theta \text{ in } Q_2$

34. $\cos\theta,\quad \theta \text{ in } Q_3$

35. $\csc\theta,\quad \theta \text{ in } Q_4$

36. $\cot\theta,\quad \theta \text{ in } Q_1$

In Exercises 37–40, substitute the trigonometric expression for x and simplify.

37. $\sqrt{4-x^2},\quad x=2\sin\theta$

38. $\sqrt{x^2-9},\quad x=3\sec\theta$

39. $\sqrt{9+49x^2},\quad x=\frac{3}{7}\tan\theta$

40. $\dfrac{\sqrt{x^2-a^2}}{x},\quad x=a\sec\theta,\ a>0.$

3.2 Conditional Trigonometric Equations

A conditional trigonometric equation is an equation involving trigonometric functions which is not an identity. In other words, there are values of the variables for which the terms in the equation are all defined but equality does not hold. A simple example is $\sin x=1$. The terms are defined for all x, but equality only holds when $x=\frac{1}{2}\pi+n\cdot 2\pi$, where n is an integer.

When we are given a conditional trigonometric equation, we usually are interested in finding its solutions. To do so, we may use any method normally employed in finding the solutions to an ordinary (conditional) equation, such as factoring, the quadratic formula, etc. In addition, we may have to use trigonometric identities and our knowledge of the trigonometric functions. We shall demonstrate some of these methods in the following examples.

EXAMPLE 1 Find all solutions to $2\cos x+\sqrt{2}=0.$

Solution It is clearly easy to solve the equation for $\cos x$, obtaining $\cos x=-\dfrac{\sqrt{2}}{2}$. One solution to this equation is $x=\cos^{-1}\left(-\dfrac{\sqrt{2}}{2}\right)=\dfrac{3\pi}{4}$. However, there are many more solutions. For any such solution x, the reference angle must be $\phi=\cos^{-1}\dfrac{\sqrt{2}}{2}=\dfrac{\pi}{4}$. Since $\cos x$ is negative $\left(\cos x=-\dfrac{\sqrt{2}}{2}\right)$, the terminal side of x must be in the second or third quadrant. The solutions are all those angles with terminal side in the second or third quadrant and reference angle $\frac{\pi}{4}$. That is, the solutions are all the numbers $x = \frac{3}{4}\pi + n \cdot 2\pi$ and $x = \frac{5}{4}\pi + n \cdot 2\pi$, where n is an integer. See Fig. 1.

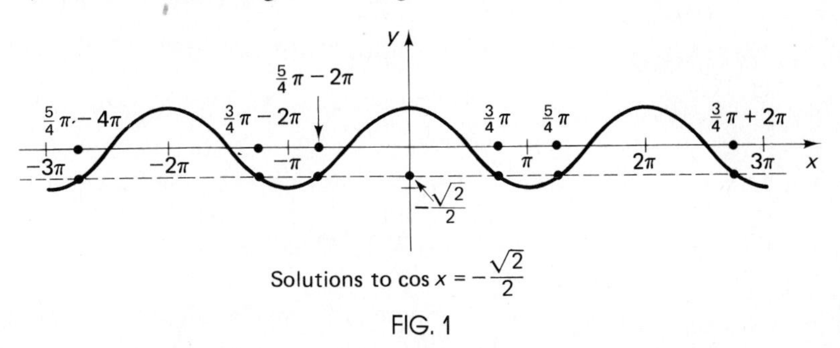

FIG. 1 ■

Remark: As in Example 1, the solutions to most conditional trigonometric equations are written in terms of n, where n is any integer. Hereafter, we shall leave the phrase "where n is any integer" as understood. Also, hereafter we shall write $n \cdot 2\pi$ as $2n\pi$.

Most conditional equations are more complicated than that in Example 1, but often the final steps are as in the preceding example. One of the techniques frequently used in solving conditional equations is factoring.

EXAMPLE 2 Find all solutions to $\cos x \cot x = \cos x$.

Solution It would be wrong to divide by $\cos x$, as you would miss all solutions where $\cos x = 0$. Instead, subtract $\cos x$ from both sides:

$$\cos x \cot x - \cos x = 0$$

Then factor out $\cos x$:

$$\cos x(\cot x - 1) = 0$$

The only way the product of two factors can be zero is if one of the factors is zero:

$$\cos x = 0 \qquad \text{or} \qquad \cot x - 1 = 0$$

Now $\cos x = 0$ has solutions $x = \frac{1}{2}\pi + n\pi$, while $\cot x = 1$ has solutions $x = \frac{1}{4}\pi + n\pi$. Thus, $\cos x \cot x = \cos x$ has solutions $x = \frac{1}{2}\pi + n\pi$ and $x = \frac{1}{4}\pi + n\pi$. ■

It may be necessary to use a trigonometric identity before factoring.

EXAMPLE 3 Find all solutions to $-3\cos^2 x - 7\sin x + 5 = 0$.

Solution If we use the identity $\cos^2 x = 1 - \sin^2 x$, the equation becomes a quadratic equation in $\sin x$:

$$-3(1 - \sin^2 x) - 7 \sin x + 5 = 0$$

$$3 \sin^2 x - 7 \sin x + 2 = 0$$

Thinking $u = \sin x$, this quadratic factors as

$$(3 \sin x - 1)(\sin x - 2) = 0$$

This equation is true if either

$$3 \sin x - 1 = 0 \qquad \text{or} \qquad \sin x - 2 = 0$$

If $\sin x - 2 = 0$, then $\sin x = 2$, but this has no solutions.

If $3 \sin x - 1 = 0$, then $\sin x = \frac{1}{3}$, so x is in Q_1 or Q_2 with reference angle $\sin^{-1}(\frac{1}{3}) \approx 0.339837$. Therefore $x \approx 0.339837 + 2n\pi$ or $x \approx (\pi - 0.339837) + 2n\pi \approx 2.80176 + 2n\pi$. ■

Some problems require solutions only in a finite interval. In addition, if the problem asks for the answer in degrees, we express the answers that way.

EXAMPLE 4 Find all solutions to $4 \sin^2 x \tan x = \tan x$ which are in the interval $[0°, 360°)$.

Solution As in Example 2, we must subtract $\tan x$ from both sides and factor.

$$4 \sin^2 x \tan x - \tan x = 0$$

$$\tan x(4 \sin^2 x - 1) = 0$$

$$\tan x(2 \sin x - 1)(2 \sin x + 1) = 0$$

This is true if either

$$\tan x = 0 \qquad \text{or} \qquad 2 \sin x - 1 = 0 \qquad \text{or} \qquad 2 \sin x + 1 = 0$$

We solve each of these in the interval $[0°, 360°)$.

- □ If $\tan x = 0$, then $x = 0°, 180°$.
- □ If $2 \sin x - 1 = 0$, then $\sin x = \frac{1}{2}$ or $x = 30°, 150°$.
- □ If $2 \sin x + 1 = 0$, then $\sin x = -\frac{1}{2}$ or $x = 210°, 330°$.

Altogether, the solutions are

$$0°, 30°, 150°, 180°, 210°, 330° \qquad ■$$

A technique often employed in solving equations involving radicals is squaring both sides. This may be employed here, but we must remember that extraneous solutions may be introduced, so all solutions should be checked in the original equation.

EXAMPLE 5 Find all solutions to $\sin x = \cos x - 1$.

Solution Since $\sin^2 x = 1 - \cos^2 x$, $\sin x = \pm\sqrt{1 - \cos^2 x}$. Thus the equation we wish to solve is

$$\pm\sqrt{1 - \cos^2 x} = \cos x - 1$$

Squaring, we obtain

$$1 - \cos^2 x = \cos^2 x - 2\cos x + 1$$

Thus

$$\begin{aligned} -\cos^2 x &= \cos^2 x - 2\cos x \\ 0 &= 2\cos^2 x - 2\cos x \\ 0 &= 2\cos x(\cos x - 1) \end{aligned}$$

Consequently $\cos x = 0$, in which case $x = \frac{1}{2}\pi + n\pi$, or $\cos x = 1$, in which case $x = 2n\pi$. We now check these in our original equation, $\sin x = \cos x - 1$. When $x = 2n\pi$, both sides are zero, so the solutions check. However, when $x = \frac{1}{2}\pi + n\pi$, the right-hand side is -1, but the left-hand side is -1 only when n is odd (the left-hand side is $+1$ when n is even). Thus the solutions are $x = \frac{1}{2}\pi + n\pi$, n odd, and $x = 2n\pi$. We sometimes write $x = \frac{1}{2}\pi + n\pi$, n odd, as $x = \frac{1}{2}\pi + (2k + 1)\pi$. ■

When an equation can be put into quadratic form but it is not easy to factor, we usually use the quadratic formula.

EXAMPLE 6 Find all solutions to $\sin^2 x - 3\sin x + 1 = 0$ in $[0, 2\pi)$.

Solution By the quadratic formula,

$$\sin x = \frac{3 \pm \sqrt{3^2 - 4 \cdot 1 \cdot 1}}{2 \cdot 1} = \frac{3 \pm \sqrt{5}}{2} \approx 2.61803,\ 0.381966$$

Now $\sin x = 2.61803$ has no solution, since $|\sin x| \leq 1$. If $\sin x = 0.381966$, then the reference angle is $\sin^{-1}(0.381966) \approx 0.391923$. Since $\sin x > 0$, x is in Q_1 or Q_2 with this reference angle. Thus the solutions in $[0, 2\pi)$ are

$$x \approx 0.391923 \quad \text{and} \quad x \approx \pi - 0.391923 \approx 2.74967 \quad ■$$

EXERCISES

Find all solutions for each of the following equations:

1. $\sqrt{2}\sin x - 1 = 0$

2. $\sqrt{3}\cot x - 1 = 0$

3. $\cot x + \sqrt{3} = 0$

4. $\sqrt{3}\tan x + 1 = 0$

5. $2\cos x - 1 = 0$

6. $2\cos x + \sqrt{3} = 0$

7. $\sec x - 2 = 0$

8. $4\sin x + 3 = 0$

9. $2\cos x - 3 = 0$

10. $3\cos x - 2 = 0$

11. $\sin x\cos x = 2\sin x$

12. $2\sin x\cos x = \sin x$

13. $\sec^2 x - \sec x - 2 = 0$

14. $4\sin^2 x - 3 = 0$

15. $6 \sin^2 x - \sin x = 1$

16. $\tan^2 x + 6 \tan x + 5 = 0$

17. $-\sin^2 x - 4 \cos x + 4 = 0$

18. $\cos^4 x - \sin^4 x = 0$

19. $11 + \cos^2 x + \sin x = 0$

20. $\sec^2 x + 3 \tan x - 1 = 0$

21. $\sin x = -\cos x$

22. $\sin x = 1 - \cos x$

In Exercises 23–38, find all solutions of the given equation in the interval $[0, 2\pi)$. *Also give the equivalent solution in degrees in the interval* $[0°, 360°)$.

23. $1 - 4 \sin^2 x = 0$

24. $\tan^2 x = \tan x$

25. $2 \cos^2 x = 1 - \cos x$

26. $2 \sin^2 x + 1 = 3 \sin x$

27. $\sec x \tan^2 x = \sec x$

28. $\sec x \csc x = 2 \sec x$

29. $2 \cos^2 x + \cos x = 6$

30. $\sin x - \cos x = 1$

31. $1 - \sin x = \sqrt{3} \cos x$

32. $\tan x + \sec x = 1$

33. $\sin x = 1 + \cos x$

34. $\csc x + \cot x = 1$

35. $\tan^5 x - 9 \tan x = 0$

36. $4 \sin x \cos x + 2 \sin x + 2 \cos x + 1 = 0$

37. $4 \sin^2 x + \sin x - 1 = 0$

38. $3 \cos^2 x + 5 \cos x + 1 = 0$

3.3 Trigonometric Identities

In this section we extend the techniques of the first section of this chapter to verify more complicated identities. These are explained in the examples as we go along, and then collected at the end. We begin by discussing the three general approaches to proving identities.

THE METHODS OF PROOF

Suppose we wish to verify an equation such as

$$\frac{\tan \theta + \cos \theta}{\sin \theta} = \sec \theta + \cot \theta$$

This equation has a left side, L, and a right side, R, and we must show $L = R$ (wherever both sides are defined). There are three general methods of proceeding.

Method 1. Work with the left side only, using algebra and known identities to transform it into the right side

$$\begin{aligned} L &= L_1 \\ &= L_2 \\ &\vdots \\ &= L_k \end{aligned}$$

so that $L_k = R$. Then $L = R$.

EXAMPLE 1 Prove $\sin \theta \tan \theta + \cos \theta = \sec \theta$ is an identity.

Solution As the left side is complicated and the right side is not, we use Method 1 and transform the left side into the right.

$$\sin\theta\tan\theta + \cos\theta = \sin\theta\frac{\sin\theta}{\cos\theta} + \cos\theta \qquad \left[\text{by (2), } \tan\theta = \frac{\sin\theta}{\cos\theta}\right]$$

$$= \frac{\sin^2\theta}{\cos\theta} + \frac{\cos^2\theta}{\cos\theta} \qquad (\cos\theta \text{ is the least common denominator})$$

$$= \frac{\sin^2\theta + \cos^2\theta}{\cos\theta} \qquad \text{(adding)}$$

$$= \frac{1}{\cos\theta} \qquad [\text{by (3), } \sin^2\theta + \cos^2\theta = 1]$$

$$= \sec\theta \qquad \left[\text{by (1), } \sec\theta = \frac{1}{\cos\theta}\right] \quad \blacksquare$$

Method 2. Work with the right side only, using algebra and known identities to transform it to the left side.

$$\begin{aligned} R &= R_1 \\ &= R_2 \\ &\vdots \\ &= R_m \end{aligned}$$

so that $R_m = L$. Then $R = L$.

EXAMPLE 2 Verify the identity $\csc\theta\sec\theta = \cot\theta + \tan\theta$.

Solution Here, the right side is more complicated than the left, so we use Method 2 and transform the right side into the left.

Hint: At some stage, you may not be sure what to do next, such as here. If you keep in mind the expression you want in the end, this usually dictates what to do.

$$\begin{aligned} \cot\theta + \tan\theta &= \frac{\cos\theta}{\sin\theta} + \frac{\sin\theta}{\cos\theta} \\ &= \frac{\cos^2\theta + \sin^2\theta}{\sin\theta\cos\theta} \\ &= \frac{1}{\sin\theta\cos\theta} \\ &= \frac{1}{\sin\theta}\frac{1}{\cos\theta} \\ &= \csc\theta\sec\theta \quad \blacksquare \end{aligned}$$

Method 3. Transform each side *independently* into the same expression.

$$
\begin{array}{l|r}
L = L_1 & R = R_1 \\
\quad = L_2 & = R_2 \\
\quad \vdots & \vdots \\
\quad = L_n & = R_p
\end{array}
$$

If $L_n = R_p$ then $L = R$.

Note: When using Method 3, make sure you work with each side *independently*. In particular, do not miss the warning at the end of this section.

EXAMPLE 3 Verify the identity $\dfrac{\tan\theta + \cos\theta}{\sin\theta} = \sec\theta + \cot\theta$.

Solution Here, we choose to work both sides *independently*, using Method 3, above, and transform everything into sines and cosines,

Hint: Sometimes it helps to first change everything into sines and cosines, using the fundamental identities, and then work with the more complicated side.

$$
\begin{aligned}
\frac{\tan\theta + \cos\theta}{\sin\theta} &= \frac{\dfrac{\sin\theta}{\cos\theta} + \cos\theta}{\sin\theta} \\
&= \frac{\sin\theta}{\sin\theta\cos\theta} + \frac{\cos\theta}{\sin\theta} \\
&= \frac{1}{\cos\theta} + \frac{\cos\theta}{\sin\theta}
\end{aligned}
$$

$$
\sec\theta + \cot\theta = \frac{1}{\cos\theta} + \frac{\cos\theta}{\sin\theta}
$$

These are identical, so the original equation is an identity. ■

FURTHER CONSIDERATIONS

We now examine four more examples of verifying identities. These illustrate further considerations which help solve identities and give more examples of the above methods.

EXAMPLE 4 Verify the identity

$$
\frac{1 + \sin x}{\cos x} = \frac{\cos x}{1 - \sin x}
$$

Solution There is no "more complicated" side, so we arbitrarily choose the left side. Looking at the right side, we see $1 - \sin x$ in the denominator. One way to get a term like that in the denominator of the left side is to multiply the left side by $1 = \dfrac{1 - \sin x}{1 - \sin x}$:

$$\frac{1 + \sin x}{\cos x} = \frac{1 + \sin x}{\cos x}\frac{1 - \sin x}{1 - \sin x}$$

Hint: Again, looking ahead told us what to try.

$$= \frac{1 - \sin^2 x}{\cos x(1 - \sin x)}$$

$$= \frac{\cos^2 x}{\cos x(1 - \sin x)}$$

$$= \frac{\cos x}{1 - \sin x} \quad ■$$

EXAMPLE 5 Verify the identity $\quad 2 \sin^2 t - 1 = \sin^4 t - \cos^4 t$

Solution Sometimes it pays to factor:

Hint: Look for ways to use a fundamental identity such as $\sin^2 \theta + \cos^2 \theta = 1$.

$$\begin{aligned}\sin^4 t - \cos^4 t &= (\sin^2 t - \cos^2 t)(\sin^2 t + \cos^2 t)\\ &= (\sin^2 t - \cos^2 t) \cdot 1\\ &= \sin^2 t - (1 - \sin^2 t)\\ &= 2 \sin^2 t - 1 \quad ■\end{aligned}$$

EXAMPLE 6 Verify the identity $\quad \sin 2t + \cos 2t \cot 2t = \csc 2t$

Solution Do not be confused by the $2t$. That is just an angle or number. You can think of it as θ if you like.

$$\begin{aligned}\sin 2t + \cos 2t \cot 2t &= \sin 2t + \cos 2t \frac{\cos 2t}{\sin 2t}\\ &= \frac{\sin^2 2t + \cos^2 2t}{\sin 2t}\\ &= \frac{1}{\sin 2t}\\ &= \csc 2t \quad ■\end{aligned}$$

If an identity involves trigonometric functions of different angles, we have to be very careful. For instance, $\sin^2 2t + \cos^2 t = 1$ is *not* an identity. We shall discuss equations like this in later sections.

EXAMPLE 7 Verify the identity

$$\frac{\cos^2 \theta + \cot \theta}{\cos^2 \theta - \cot \theta} = \frac{\cos^2 \theta \tan \theta + 1}{\cos^2 \theta \tan \theta - 1}$$

Solution Working with each side separately, we transform everything into sines and cosines, and then simplify.

Hint: Method 3 is particularly called for when both sides are somewhat complicated.

$$\frac{\cos^2\theta + \cot\theta}{\cos^2\theta - \cot\theta} = \frac{\cos^2\theta + \dfrac{\cos\theta}{\sin\theta}}{\cos^2\theta - \dfrac{\cos\theta}{\sin\theta}}$$

$$= \frac{\cos\theta\left(\cos\theta + \dfrac{1}{\sin\theta}\right)}{\cos\theta\left(\cos\theta - \dfrac{1}{\sin\theta}\right)}$$

$$= \frac{\cos\theta + \dfrac{1}{\sin\theta}}{\cos\theta - \dfrac{1}{\sin\theta}}$$

$$= \frac{\dfrac{\cos\theta\sin\theta + 1}{\sin\theta}}{\dfrac{\cos\theta\sin\theta - 1}{\sin\theta}}$$

$$= \frac{\cos\theta\sin\theta + 1}{\cos\theta\sin\theta - 1}$$

$$\frac{\cos^2\theta\tan\theta + 1}{\cos^2\theta\tan\theta - 1} = \frac{\cos^2\theta\dfrac{\sin\theta}{\cos\theta} + 1}{\cos^2\theta\dfrac{\sin\theta}{\cos\theta} - 1}$$

$$= \frac{\cos\theta\sin\theta + 1}{\cos\theta\sin\theta - 1}$$

These two are equal, so the original equation is an identity. ■

Unfortunately, there is no general method of proof of identities which works for all identities. However, here is a list of some hints that we have suggested.

1. Work with the more complicated side (if there is one).
2. If you do not see directly how to reduce one side to the other, use fundamental identities (1) and (2) to change everything into sines and cosines.
3. Look for ways to use identity (3), $\sin^2\theta + \cos^2\theta = 1$, or one of its other forms: $\cos^2\theta = 1 - \sin^2\theta = (1 - \sin\theta)(1 + \sin\theta)$, $\sin^2\theta = 1 - \cos^2\theta = (1 - \cos\theta)(1 + \cos\theta)$.
4. Above all, keep in mind the expression that you want in the end—this usually dictates what you should do.

Warning: When proving identities, it is important *not* to start with the equality you are trying to verify and "prove" it by ending up with a true statement. This is not a valid method of proof, as is illustrated by the following simple example.

Suppose $\quad -1 = +1.$

Then $\quad (-1)^2 = (+1)^2.$

Thus $\quad 1 = 1.$

Since $1 = 1$, can we conclude that $-1 = +1$? Of course not. You must remember that ending up with a true statement does not imply you began with one (a type of false argument some politicians try). You will not make such errors if you work with one side at a time.

EXERCISES

In Exercises 1–40, verify the identities.

1. $\sin\theta \cot\theta = \cos\theta$

2. $\sin\theta \csc\theta = 1$

3. $\dfrac{\sec\theta}{\csc\theta} = \tan\theta$

4. $\cot\theta \tan\theta = 1$

5. $\sin x + \cot x \cos x = \csc x$

6. $\sin\theta(\csc\theta - \sin\theta) = \cos^2\theta$

7. $1 - 2\sin^2 t = 2\cos^2 t - 1$

8. $\dfrac{\sin x}{1 + \cos x} + \dfrac{1 + \cos x}{\sin x} = 2\csc x$

9. $\dfrac{\csc^2 x}{1 + \tan^2 x} = \cot^2 x$

10. $\dfrac{1 + \tan u}{1 - \tan u} + \dfrac{1 + \cot u}{1 - \cot u} = 0$

11. $\dfrac{\sin\theta}{\sin\theta - \cos\theta} = \dfrac{1}{1 - \cot\theta}$

12. $(\csc t - \cot t)(\sec t + 1) = \tan t$

13. $(\sin 2y + \cos 2y)(\tan 2y + \cot 2y)$
$= \sec 2y + \csc 2y$

14. $\dfrac{\sin u + \cos u}{\sec u + \csc u} = \dfrac{\sin u}{\sec u}$

15. $\dfrac{\sec^2 3\theta}{\sec 3\theta + 1} = \dfrac{\sec 3\theta - 1}{\sin^2 3\theta}$

16. $\sec^2\theta \cot^2\theta - \cos^2\theta \csc^2\theta = 1$

17. $\dfrac{\sin^3\theta + \cos^3\theta}{\sin\theta + \cos\theta} = 1 - \sin\theta\cos\theta$

18. $\csc^4\theta - \cot^4\theta = \csc^2\theta + \cot^2\theta$

19. $\dfrac{\csc^2\theta - 1}{\cos\theta} = \cot\theta\csc\theta$

20. $\dfrac{\sin u}{1 - \cos u} = \dfrac{\tan u}{\sec u - 1}$

21. $\cos^4\theta - \sin^4\theta = 2\cos^2\theta - 1$

22. $\dfrac{\tan^2 u}{\sec u + 1} = \dfrac{1 - \cos u}{\cos u}$

23. $(\sec^2\theta - \tan^2\theta)^3 = 1$

24. $(\cot\theta - \csc\theta)^3(\cot\theta + \csc\theta)^3 = -1$

25. $\dfrac{\sin^2\theta + 2}{\sin^2\theta - 2} = \dfrac{1 - 3\sec^2\theta}{1 + \sec^2\theta}$

26. $\dfrac{\cot x}{\csc x - 1} = \dfrac{\csc x + 1}{\cot x}$

27. $\dfrac{\tan\theta - \cot\theta}{\cos\theta + \sin\theta} = \sec\theta - \csc\theta$

28. $\dfrac{\sin x + \cos x}{\tan^2 x - 1} = \dfrac{\cos^2 x}{\sin x - \cos x}$

29. $\dfrac{1 + \sin\theta}{1 - \sin\theta} = \dfrac{\csc\theta + 1}{\csc\theta - 1}$

30. $\tan^2\theta - \sin^2\theta = \tan^2\theta\sin^2\theta$

31. $\dfrac{1 + \cos^2\phi}{\sin^2\phi} = 2\csc^2\phi - 1$

32. $\dfrac{\sin\theta}{\sin\theta + \cos\theta} = \dfrac{\tan\theta}{1 + \tan\theta}$

33. $\sec\theta + \tan\theta = \dfrac{\cos\theta}{1 - \sin\theta}$

34. $\sin^4\theta + \cos^2\theta = \cos^4\theta + \sin^2\theta$

35. $\dfrac{\sin^2\theta}{\sin^2\theta + \cos\theta} = \dfrac{\tan\theta}{\tan\theta + \csc\theta}$

36. $\dfrac{\cot\phi - 1}{1 - \tan\phi} = \dfrac{1}{\tan\phi}$

37. $\left(\dfrac{\sec^3 x}{\tan^6 x}\right)^2\left(\dfrac{\cos^2 x}{\cot^4 x}\right)^3 = 1$

38. $\sec^4\theta(1 - \sin^4\theta) = \sec^2\theta + \dfrac{1}{\cot^2\theta}$

39. $\dfrac{\cos\alpha - \sin\alpha\tan\beta}{\sin\alpha + \cos\alpha\tan\beta} = \dfrac{\cot\alpha - \tan\beta}{1 + \tan\beta\cot\alpha}$

40. $\dfrac{\sin\alpha\cos\beta - \cos\alpha\sin\beta}{\cos\alpha\cos\beta + \sin\alpha\sin\beta} = \dfrac{\tan\alpha - \tan\beta}{1 + \tan\alpha\tan\beta}$

In Exercises 41–48, either verify the identity or find a value of θ for which equality does not hold. (If you use a calculator to find such a θ, make sure that inequality is not due just to calculator error).

41. $\sin^2\theta(1 + \tan^2\theta) = \tan\theta$

42. $\dfrac{\sin\theta\cot\theta + \cos\theta}{\cot\theta} = 2\cos\theta$

43. $\dfrac{\sin\theta}{1 + \cos\theta} = \dfrac{1 - \cos\theta}{\sin\theta}$

44. $\dfrac{\sec\theta}{\sin\theta} - \dfrac{\sin\theta}{\cos\theta} = \cot\theta$

45. $\dfrac{\sin^4\theta - \cos^4\theta}{\tan^4\theta - 1} = \cos^2\theta$

46. $\csc^2\theta\tan^2\theta - \sin^2\theta\sec^2\theta = \tan^4\theta$

47. $\sec^2\theta + \csc^2\theta = 1$

48. $\tan\theta + \cot\theta = \tan\theta\csc^2\theta$

3.4 The Addition Formulas

In this section, we shall derive the following identities, which are called **the addition formulas:**

(4) $\sin(u + v) = \sin u\cos v + \cos u\sin v$

(5) $\sin(u - v) = \sin u\cos v - \cos u\sin v$

(6) $\cos(u + v) = \cos u\cos v - \sin u\sin v$

(7) $\cos(u - v) = \cos u\cos v + \sin u\sin v$

(8) $\tan(u + v) = \dfrac{\tan u + \tan v}{1 - \tan u\tan v}$

(9) $\tan(u - v) = \dfrac{\tan u - \tan v}{1 + \tan u\tan v}$

As we can see from these, an addition formula for a trigonometric function is an identity that relates the trigonometric function of the sum or difference of two angles to trigonometric functions of the individual angles.

Now let us derive the addition formulas (4)–(9). We begin by deriving the addition formula for $\cos(u - v)$. To do this, let $P = (1, 0)$ and let u and v be any two angles in standard position. Let Q and R be the points on the terminal side of u and v, respectively, which are on the unit circle. See Fig. 2 for one possibility. Thus

$$Q = (\cos u, \sin u) \qquad \text{and} \qquad R = (\cos v, \sin v)$$

Locate S on the unit circle so that the angle POS equals the difference $u - v$, as indicated in Fig. 2. Thus

$$S = (\cos(u - v), \sin(u - v))$$

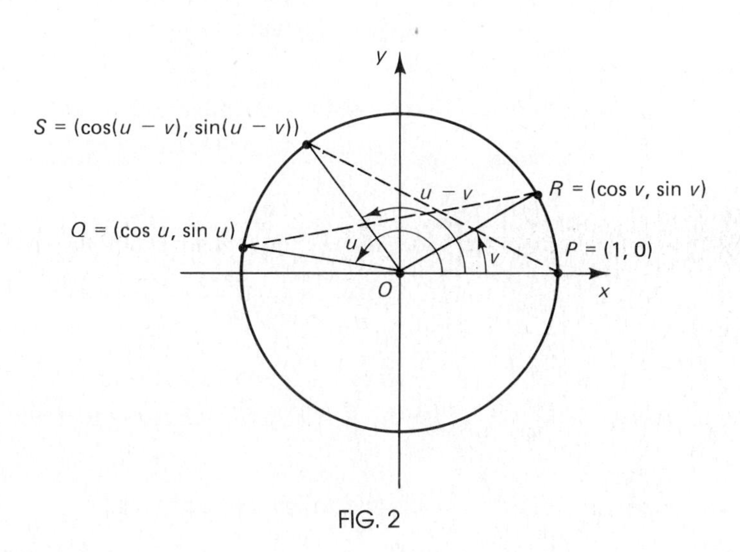

FIG. 2

Now the angle ROQ also equals $u - v$. Since the central angles POS and ROQ are equal, the distance between P and S is equal to the distance between R and Q,

$$d(P, S) = d(R, Q)$$

By the distance formula,

$$\sqrt{[\cos(u - v) - 1]^2 + \sin^2(u - v)}$$
$$= \sqrt{(\cos u - \cos v)^2 + (\sin u - \sin v)^2}$$

Squaring both sides and multiplying out, we have

$$\cos^2(u - v) - 2\cos(u - v) + 1 + \sin^2(u - v)$$
$$= \cos^2 u - 2\cos u \cos v + \cos^2 v + \sin^2 u - 2\sin u \sin v + \sin^2 v$$

We reorder this equation and use the identity $\sin^2 \theta + \cos^2 \theta = 1$ three times to obtain

$$2 - 2\cos(u - v) = 2 - 2\cos u \cos v - 2\sin u \sin v$$

Simplifying, we obtain Equation (7):

$$\text{(7)} \qquad \cos(u - v) = \cos u \cos v + \sin u \sin v$$

Thus we have derived our first addition formula. Fortunately, we do not have to go to so much work to derive the others, for they may be obtained from this one. Before we do that, let us look at an example. The main reason for doing problems like the following is to become familiar with the formulas and how they work so that you can readily use them when they arise in more complicated situations.

EXAMPLE 1 Use an addition formula to find the exact value of $\cos 15°$.

Solution We use $u = 45°$ and $v = 30°$ since we know the trigonometric functions of these angles and that their difference is $15°$:

$$\cos 15° = \cos(45° - 30°) = \cos 45° \cos 30° + \sin 45° \sin 30°$$

$$= \frac{\sqrt{2}}{2}\frac{\sqrt{3}}{2} + \frac{\sqrt{2}}{2}\frac{1}{2} = \frac{\sqrt{2}}{4}(\sqrt{3} + 1)$$

You might check on your calculator that the decimal approximation to this is the same as what your calculator gives for $\cos 15°$. ■

To simplify our derivation of the remaining addition formulas, we first verify the following identities:

$$\textbf{(10)} \qquad \cos\left(\frac{\pi}{2} - x\right) = \sin x$$

$$\textbf{(11)} \qquad \sin\left(\frac{\pi}{2} - x\right) = \cos x$$

These identities are obviously true if x is an acute angle in a right triangle. For the other acute angle is $\frac{1}{2}\pi - x$, and the side adjacent to $\frac{1}{2}\pi - x$ is the side opposite to x, and vice versa. See Fig. 3. Thus

$$\cos\left(\frac{\pi}{2} - x\right) = \frac{\text{side adj. to } (\frac{1}{2}\pi - x)}{\text{hypotenuse}}$$

$$= \frac{\text{side opp. to } x}{\text{hypotenuse}} = \sin x$$

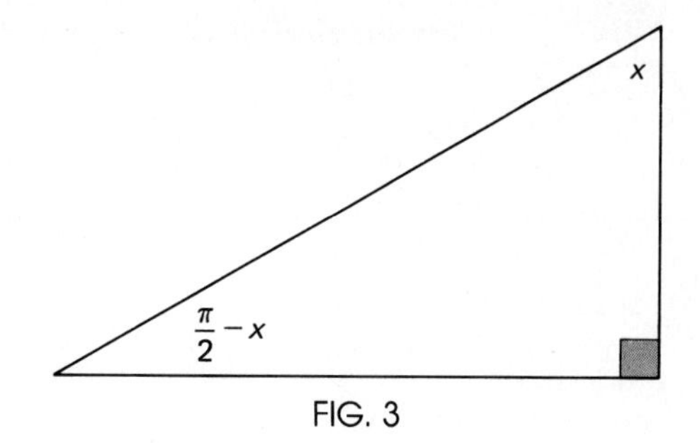

FIG. 3

and similarly, $\sin\left(\dfrac{\pi}{2} - x\right) = \cos x$. We now show that these identities are true for any angle x.

Verification of (10): $\cos\left(\dfrac{\pi}{2} - x\right) = \sin x$:

$$\begin{aligned} \cos\left(\frac{\pi}{2} - x\right) &= \cos\frac{\pi}{2}\cos x + \sin\frac{\pi}{2}\sin x \qquad \text{[by (7)]} \\ &= 0 \cdot \cos x + 1 \cdot \sin x \\ &= \sin x \quad \blacksquare \end{aligned}$$

Verification of (11): $\sin(\frac{1}{2}\pi - x) = \cos x$.
Let $y = \frac{1}{2}\pi - x$. By (10), $\sin y = \cos(\frac{1}{2}\pi - y)$. Replace y by $\frac{1}{2}\pi - x$:

$$\sin(\tfrac{1}{2}\pi - x) = \cos[\tfrac{1}{2}\pi - (\tfrac{1}{2}\pi - x)] = \cos x \quad \blacksquare$$

We also need the following two identities, which we derived in the previous chapter using the definitions of the trigonometric functions:

(12) $$\cos(-x) = \cos x$$

(13) $$\sin(-x) = -\sin x$$

Since $\tan(-x) = \dfrac{\sin(-x)}{\cos(-x)} = \dfrac{-\sin x}{\cos x} = -\tan x$, we also have

(14) $$\tan(-x) = -\tan x$$

Finally we are in position to derive the remaining addition formulas.

Verification of (6): $\cos(u + v) = \cos u \cos v - \sin u \sin v$:

$$\begin{aligned} \cos(u + v) &= \cos[u - (-v)] \\ &= \cos u \cos(-v) + \sin u \sin(-v) \qquad \text{[by (7)]} \\ &= \cos u \cos v + \sin u(-\sin v) \qquad \text{[by (12) and (13)]} \\ &= \cos u \cos v - \sin u \sin v \quad \blacksquare \end{aligned}$$

Verification of (4): $\sin(u + v) = \sin u \cos v + \cos u \sin v$:

$$\begin{aligned}
\sin(u+v) &= \cos\left[\frac{\pi}{2} - (u+v)\right] && \text{[by (10)]}\\
&= \cos\left[\left(\frac{\pi}{2} - u\right) - v\right]\\
&= \cos\left(\frac{\pi}{2} - u\right)\cos v + \sin\left(\frac{\pi}{2} - u\right)\sin v && \text{[by (7)]}\\
&= \sin u \cos v + \cos u \sin v && \text{[by (10) and (11)]}
\end{aligned}$$

■

Verification of (5): $\sin(u - v) = \sin u \cos v - \cos u \sin v$:

$$\begin{aligned}
\sin(u-v) &= \sin[u + (-v)]\\
&= \sin u \cos(-v) + \cos u \sin(-v) && \text{[by (4)]}\\
&= \sin u \cos v + \cos u(-\sin v) && \text{[by (12) and (13)]}\\
&= \sin u \cos v - \cos u \sin v \quad ■
\end{aligned}$$

Verification of (8): $\tan(u + v) = \dfrac{\tan u + \tan v}{1 - \tan u \tan v}$:

$$\begin{aligned}
\tan(u+v) &= \frac{\sin(u+v)}{\cos(u+v)}\\
&= \frac{\sin u \cos v + \cos u \sin v}{\cos u \cos v - \sin u \sin v} && \text{[by (4) and (6)]}\\
&= \frac{\dfrac{\sin u \cos v}{\cos u \cos v} + \dfrac{\cos u \sin v}{\cos u \cos v}}{\dfrac{\cos u \cos v}{\cos u \cos v} - \dfrac{\sin u \sin v}{\cos u \cos v}} && \text{[divide top and bottom by } \cos u \cos v \text{ (if nonzero)]}\\
&= \frac{\dfrac{\sin u}{\cos u} + \dfrac{\sin v}{\cos v}}{1 - \dfrac{\sin u}{\cos u}\dfrac{\sin v}{\cos v}}\\
&= \frac{\tan u + \tan v}{1 - \tan u \tan v} \quad ■
\end{aligned}$$

Note: In the third step we used $\cos u \cos v \neq 0$. If $\cos u \cos v = 0$, then either $\cos u = 0$ or $\cos v = 0$, which in turn means either $\tan u$ or $\tan v$ is not defined. Since an equation is an identity if equality holds whenever both sides are defined, equation (8) is still an identity.

Note: Formula (9) allows us to prove that the tangent function has period π. See Exercise 21.

Verification of (9): $\tan(u - v) = \dfrac{\tan u - \tan v}{1 + \tan u \tan v}$:

$$\begin{aligned} \tan(u - v) &= \tan[u + (-v)] \\ &= \frac{\tan u + \tan(-v)}{1 - \tan u \tan(-v)} \qquad \text{[by (8)]} \\ &= \frac{\tan u - \tan v}{1 + \tan u \tan v} \qquad \text{[by (14)]} \end{aligned}$$ ■

Now that we have derived the addition formulas, let us look at some examples.

EXAMPLE 2 Suppose $\sin u = \frac{1}{3}$, $\cos v = -\frac{3}{4}$, u is in Q_2, and v is in Q_2. Find $\sin(u + v)$, $\cos(u + v)$, and $\tan(u + v)$, and determine in which quadrant $u + v$ lies.

Solution From the given information,

$$\cos u = -\sqrt{1 - \sin^2 u} = -\tfrac{2}{3}\sqrt{2}, \quad \tan u = \tfrac{1}{3}/(-\tfrac{2}{3}\sqrt{2}) = -\tfrac{1}{4}\sqrt{2},$$
$$\sin v = \sqrt{1 - \cos^2 v} = \tfrac{1}{4}\sqrt{7}, \quad \tan v = \tfrac{1}{4}\sqrt{7}/(-\tfrac{3}{4}) = -\tfrac{1}{3}\sqrt{7}.$$

Thus

$$\begin{aligned} \sin(u + v) &= \sin u \cos v + \cos u \sin v = \frac{1}{3}\left(\frac{-3}{4}\right) + \left(\frac{-2\sqrt{2}}{3}\right)\frac{\sqrt{7}}{4} \\ &= -\frac{3 + 2\sqrt{14}}{12} \end{aligned}$$

$$\begin{aligned} \cos(u + v) &= \cos u \cos v - \sin u \sin v = \left(\frac{-2\sqrt{2}}{3}\right)\left(\frac{-3}{4}\right) - \frac{1}{3}\frac{\sqrt{7}}{4} \\ &= \frac{6\sqrt{2} - \sqrt{7}}{12} \end{aligned}$$

$$\begin{aligned} \tan(u + v) &= \frac{\tan u + \tan v}{1 - \tan u \tan v} = \frac{-\frac{1}{4}\sqrt{2} - \frac{1}{3}\sqrt{7}}{1 - (-\frac{1}{4}\sqrt{2})(-\frac{1}{3}\sqrt{7})} \\ &= -\frac{3\sqrt{2} + 4\sqrt{7}}{12 - \sqrt{14}} \end{aligned}$$

Since $\sin(u + v)$ is negative and $\cos(u + v)$ is positive, $u + v$ must be in Q_4. ■

EXAMPLE 3 Verify the identity $\quad \cos\left(x + \dfrac{\pi}{2}\right) = -\sin x.$

Solution

$$\cos\left(x + \frac{\pi}{2}\right) = \cos x \cos \frac{\pi}{2} - \sin x \sin \frac{\pi}{2}$$
$$= (\cos x) \cdot 0 - (\sin x) \cdot 1$$
$$= -\sin x \quad ■$$

AN IDENTITY FOR $a \sin cx + b \cos cx$

Suppose we start with the expression $A \sin(cx + d)$, where A, c, and d are constants. Then we can use addition formulas for sine to rewrite this as follows:

(15) $$A \sin(cx + d) = A[\sin cx \cos d + \cos cx \sin d]$$
$$= (A \cos d) \sin cx + (A \sin d) \cos cx$$

Since $A \cos d$ and $A \sin d$ are constants, we can set $a = A \cos d$, $b = A \sin d$ and obtain

(16) $$A \sin(cx + d) = a \sin cx + b \cos cx$$

EXAMPLE 4

$$3 \sin\left(2x + \frac{\pi}{3}\right) = 3 \cos \frac{\pi}{3} \sin 2x + 3 \sin \frac{\pi}{3} \cos 2x$$
$$= \tfrac{3}{2} \sin 2x + \tfrac{3}{2} \sqrt{3} \cos 2x \quad ■$$

Example 4 illustrates that it is quite straightforward to go from the left side of Eq. 16 to the right side. It is a more challenging problem to go the other way. In particular in Example 4, if we are given $\frac{3}{2} \sin 2x + \frac{3}{2}\sqrt{3} \cos 2x$, we would have to show this is $3 \sin\left(2x + \frac{\pi}{3}\right)$. This type of problem arises in a variety of situations—for example, in solving certain types of differential equations.

To see how to do this, suppose we are given $a \sin cx + b \cos cx$. We can see from (15) that we want to find constants A and d such that

(17) $$A \cos d = a \qquad \text{and} \qquad A \sin d = b$$

To solve this, we square the sides of both equations and add.

$$A^2 \cos^2 d = a^2 \qquad \text{and} \qquad A^2 \sin^2 d = b^2$$
$$A^2 \cos^2 d + A^2 \sin^2 d = a^2 + b^2$$
$$A^2(\cos^2 d + \sin^2 d) = a^2 + b^2$$

Taking A to be positive and recognizing $\cos^2 d + \sin^2 d = 1$, we obtain $A = \sqrt{a^2 + b^2}$. We substitute this value for A back into (17) and obtain

(18) $$\cos d = \frac{a}{\sqrt{a^2 + b^2}}, \qquad \sin d = \frac{b}{\sqrt{a^2 + b^2}}, \qquad A = \sqrt{a^2 + b^2}$$

If we restrict d to $0 \leq d < 2\pi$, this will give us a unique solution.

EXAMPLE 5 Rewrite $-\sqrt{3}\sin 5x + \cos 5x$ as $A\sin(5x + d)$.

Solution Here, $a = -\sqrt{3}$, $b = 1$, and we know

$$A = \sqrt{a^2 + b^2} = \sqrt{(-\sqrt{3})^2 + 1^2} = \sqrt{4} = 2$$

In addition, we know

$$\cos d = \frac{a}{A} = -\frac{\sqrt{3}}{2} \quad \text{and} \quad \sin d = \frac{b}{A} = \frac{1}{2}$$

If we take d in the interval $[0, 2\pi)$, we must have $d = \frac{5\pi}{6}$. Thus

$$-\sqrt{3}\sin 5x + \cos 5x = 2\sin\left(5x + \frac{5\pi}{6}\right) \quad \blacksquare$$

EXERCISES

In Exercises 1–8, use $15° = 45° - 30°$, $\frac{5\pi}{12} = \frac{\pi}{4} + \frac{\pi}{6}$, *etc., and the addition formulas to find the exact value of the functions.*

1. $\sin 15°$

2. $\tan 15°$

3. $\cos\frac{5\pi}{12}$

4. $\sin\frac{5\pi}{12}$

5. $\tan\frac{7\pi}{12}$

6. $\cos 195°$

7. $\csc\frac{11\pi}{12}$

8. $\cos 345°$

In Exercises 9–14, use an addition formula to find the exact value.

9. $\sin 12° \cos 33° + \cos 12° \sin 33°$

10. $\sin 14° \cos 44° - \cos 14° \sin 44°$

11. $\cos 15° \cos 75° + \sin 15° \sin 75°$

12. $\cos 15° \cos 75° - \sin 15° \sin 75°$

13. $\cos\frac{\pi}{12}\sin\frac{5\pi}{12} + \sin\frac{\pi}{12}\cos\frac{5\pi}{12}$

14. $\dfrac{\tan 15° - \tan 75°}{1 + \tan 15° \tan 75°}$

In Exercises 15–18, find $\sin(u + v)$, $\sin(u - v)$, $\cos(u + v)$, $\cos(u - v)$, $\tan(u + v)$, *and* $\tan(u - v)$ *from the given information.*

15. $\sin u = \frac{3}{5}$, $\cos v = -\frac{5}{13}$, u is in Q_2, v is in Q_3

16. $\sec u = 3$, $\sec v = \frac{3}{2}$, $\csc u$ is positive, $\csc v$ is negative

17. $\tan u = -\sqrt{3}$, $\cot v = -\sqrt{2}$, u is in Q_2, v is not in Q_2

18. $\csc u = -\frac{5}{4}$, $\tan u = \frac{4}{3}$, $\sec v = -\frac{25}{7}$, $\sin v = -\frac{24}{25}$

In Exercises 19–36, verify the identities.

19. $\sin(x + \pi) = -\sin x$

20. $\cos(x + \pi) = -\cos x$

21. $\tan(x + \pi) = \tan x$

22. $\sin\left(x + \frac{\pi}{2}\right) = \cos x$

23. $\cos\left(\frac{3\pi}{2} - x\right) = -\sin x$

24. $\csc\left(\frac{3\pi}{2} - x\right) = -\sec x$

25. $\sec(x + 2\pi) = \sec x$

26. $\sec\left(x - \frac{\pi}{2}\right) = \csc x$

27. $\sin\left(x + \frac{\pi}{6}\right) = \frac{\sqrt{3}}{2}\sin x + \frac{1}{2}\cos x$

28. $\cos\left(x + \frac{\pi}{4}\right) = \frac{\sqrt{2}}{2}(\cos x - \sin x)$

29. $\cos\left(\frac{\pi}{3} - x\right) = \frac{1}{2}\cos x + \frac{\sqrt{3}}{2}\sin x$

30. $\sin\left(\frac{\pi}{4} - x\right) = \frac{\sqrt{2}}{2}(\cos x - \sin x)$

31. $\tan\left(x - \frac{\pi}{4}\right) = \frac{\tan x - 1}{\tan x + 1}$

32. $\tan\left(x + \frac{\pi}{4}\right) = \frac{\tan x + 1}{1 - \tan x}$

33. $\sin\left(x + \frac{5\pi}{4}\right) = -\frac{\sqrt{2}}{2}(\sin x + \cos x)$

34. $\cos\left(x - \frac{3\pi}{4}\right) = -\frac{\sqrt{2}}{2}(\cos x - \sin x)$

35. $\sin(u + v)\sin(u - v) = \sin^2 u - \sin^2 v$

36. $\cos(u + v)\cos(u - v) = \cos^2 u - \sin^2 v$

In Exercises 37–40, verify the identities. (The identities are called the ***product formulas****. They allow you to replace the product on the right with the sum on the left.)*

37. $\sin(u + v) + \sin(u - v) = 2\sin u\cos v$

38. $\sin(u + v) - \sin(u - v) = 2\cos u\sin v$

39. $\cos(u + v) + \cos(u - v) = 2\cos u\cos v$

40. $\cos(u - v) - \cos(u + v) = 2\sin u\sin v$

In Exercises 41–44, verify the identities. Use Exercises 37–40 with $a = u + v$, $b = u - v$. (The identities are called the ***sum formulas****.)*

41. $\sin a + \sin b = 2\sin\frac{a+b}{2}\cos\frac{a-b}{2}$

42. $\sin a - \sin b = 2\cos\frac{a+b}{2}\sin\frac{a-b}{2}$

43. $\cos a + \cos b = 2\cos\frac{a+b}{2}\cos\frac{a-b}{2}$

44. $\cos b - \cos a = 2\sin\frac{a+b}{2}\sin\frac{a-b}{2}$

In Exercises 45–50, use Exercises 37–40 to express the product as a sum (or difference).

45. $2\sin 3\theta\cos\theta$

46. $6\cos 4\theta\cos 7\theta$

47. $-3\sin 5\theta\cos 2\theta$

48. $\cos(-5t)\cos 3t$

49. $\sin 12t\sin 6t$

50. $3\cos 5t\sin 8t$

In Exercises 51–56, use Exercises 41–44 to express the sum (or difference) as a product.

51. $\sin 5\theta + \sin 3\theta$

52. $\cos 2\theta + \cos 6\theta$

53. $5\cos 3\theta + 5\cos 4\theta$

54. $\sin(-3t) - \sin t$

55. $\sin 2x - \sin(-x)$

56. $\cos 7t + \cos 3t$

In Exercises 57–64, rewrite the expression as $A\sin(cx + d)$.

57. $\sin x + \cos x$

58. $3\sin 2x - 3\cos 2x$

59. $4\sqrt{3}\sin 7x - 4\cos 7x$

60. $-3\sin 3x - 3\sqrt{3}\cos 3x$

61. $2\sin\pi x + 2\sqrt{3}\cos\pi x$

62. $4\cos 3x$

[C] 63. $3.12\sin 5x + 4.92\cos 5x$

[C] 64. $0.013\sin\sqrt{2}x - 0.051\cos\sqrt{2}x$

65. Let l be a nonvertical straight line with equation $y = mx + b$. Let h be any horizontal line intersecting l, and let θ be the smallest positive angle with initial side h and terminal side l. See the accompanying figure. Show $\tan \theta = m$.

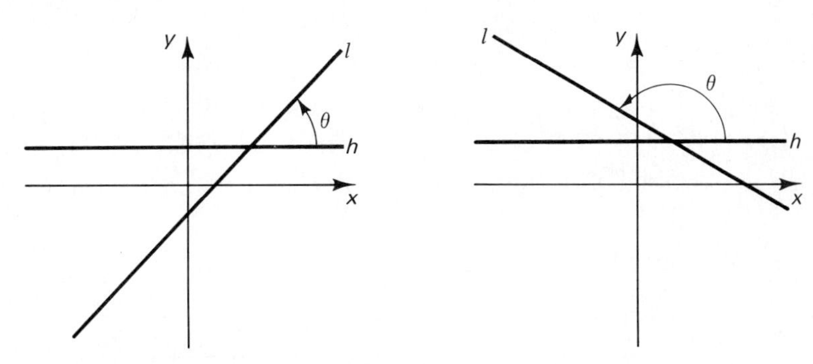

In Exercises 66–68, find the angle θ described in Exercise 65 for the given line.

66. $x + y = -2$ **67.** $\sqrt{3}x - y = 4$ **68.** $x + \sqrt{3}y = -5$

69. Let l_1 and l_2 be two nonvertical straight lines with equations $y = m_1x + b_1$ and $y = m_2x + b_2$, respectively. Let ϕ be the angle between them. See the accompanying figure. Show

$$\tan \phi = \pm \frac{m_1 - m_2}{1 + m_1m_2}$$

(*Hint:* $\phi = \theta_2 - \theta_1$. Use an addition formula.)

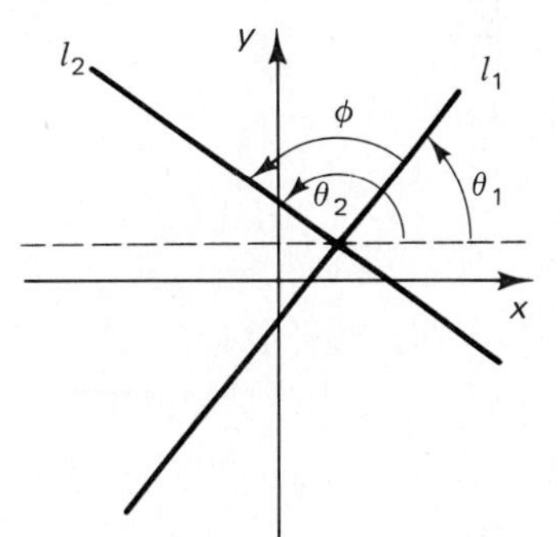

In Exercises 70–72, find the angle ϕ between the two given lines.

70. $\sqrt{3}x + y = 3$
$x + \sqrt{3}y = 2$

71. $y - \sqrt{3}x = -1$
$y + \sqrt{3}x = 1$

72. $2x + 3y = 1$
$x - 5y = 1$

73. Let l_1 and l_2 be two nonvertical straight lines with slopes m_1 and m_2, respectively. Use the results of Exercise 69 to show l_1 and l_2 are perpendicular if and only if $m_1m_2 = -1$. (*Hint:* If l_1 and l_2 are perpendicular, what is $\tan \phi$?)

3.5 The Multiple-Angle Formulas

In this section, we shall obtain formulas for the trigonometric functions of $2u$ (the **double-angle formulas**) and $\frac{1}{2}u$ (the **half-angle formulas**) in terms of the trigonometric functions of u. There are many aspects of integral calculus where these formulas play a key role.

If we start with the addition formula

$$\sin(u + v) = \sin u \cos v + \cos u \sin v$$

and let $u = v$, the formula becomes

$$\sin 2u = \sin(u + u) = \sin u \cos u + \cos u \sin u$$

Thus

$$\textbf{(19)} \qquad \sin 2u = 2 \sin u \cos u$$

Similarly, if we start with $\cos(u + v) = \cos u \cos v - \sin u \sin v$ and let $u = v$, we obtain

$$\cos 2u = \cos(u + u) = \cos u \cos u - \sin u \sin u$$

Thus

$$\textbf{(20)} \qquad \cos 2u = \cos^2 u - \sin^2 u$$

There are two other forms of the double-angle formula for cosine. Using $\cos^2 u = 1 - \sin^2 u$, we obtain $\cos 2u = (1 - \sin^2 u) - \sin^2 u$, or

$$\textbf{(21)} \qquad \cos 2u = 1 - 2 \sin^2 u$$

Using $\sin^2 u = 1 - \cos^2 u$, we obtain $\cos 2u = \cos^2 u - (1 - \cos^2 u)$, or

$$\textbf{(22)} \qquad \cos 2u = 2 \cos^2 u - 1$$

In the same manner as for $\sin 2u$ and $\cos 2u$, one can obtain the formula for $\tan 2u$:

$$\textbf{(23)} \qquad \tan 2u = \frac{2 \tan u}{1 - \tan^2 u}$$

EXAMPLE 1 Verify the identity $\sin 2u = \tan u(1 + \cos 2u)$.

Solution If we start with the right-hand side, we must choose which of (20), (21), or (22) to use for $\cos 2u$. The best choice is (22) since the 1's then drop out:

$$\begin{aligned} \tan u(1 + \cos 2u) &= \tan u[1 + (2 \cos^2 u - 1)] \\ &= \tan u(2 \cos^2 u) \\ &= 2 \frac{\sin u}{\cos u} \cos^2 u \\ &= 2 \sin u \cos u \\ &= \sin 2u \qquad \blacksquare \end{aligned}$$

Identities (21) and (22) can easily be transformed into two other identities called the *half-angle formulas*:

(24)	$$\sin\frac{u}{2} = \pm\sqrt{\frac{1 - \cos u}{2}}$$

(25)	$$\cos\frac{u}{2} = \pm\sqrt{\frac{1 + \cos u}{2}}$$

where the sign is determined by the quadrant in which $\frac{u}{2}$ lies.

Let us see how to obtain (24). From (21),

$$\cos 2v = 1 - 2\sin^2 v$$

If we solve this for $\sin v$, we obtain

$$2\sin^2 v = 1 - \cos 2v$$

$$\sin^2 v = \frac{1 - \cos 2v}{2}$$

$$\sin v = \pm\sqrt{\frac{1 - \cos 2v}{2}}$$

Now if we let $u = 2v$, then $v = \frac{u}{2}$, and the formula becomes

$$\sin\frac{u}{2} = \pm\sqrt{\frac{1 - \cos u}{2}}$$

The derivation of (25) is similar.

EXAMPLE 2 Find $\sin 22.5°$ and $\cos 22.5°$ by the half-angle formulas.

Solution

$$\sin 22.5° = \sqrt{\frac{1 - \cos 45°}{2}} = \sqrt{\frac{1 - \sqrt{2}/2}{2}} = \frac{\sqrt{2 - \sqrt{2}}}{2}$$

and

$$\cos 22.5° = \sqrt{\frac{1 + \cos 45°}{2}} = \sqrt{\frac{1 + \sqrt{2}/2}{2}} = \frac{\sqrt{2 + \sqrt{2}}}{2}$$

The signs are both + since 22.5° is in Q_1. ■

EXERCISES

In Exercises 1–6, use the given information and the double-angle formulas to find $\sin 2u$, $\cos 2u$, *and* $\tan 2u$.

1. $u = \frac{\pi}{3}$

2. $\sin u = \frac{4}{5}$, u in Q_2

3. $\cos u = -\frac{5}{13}$, $\tan u < 0$

4. $\csc u = 3$, u acute

5. $\cot u = -\frac{4}{3}$, $90° < u < 180°$

6. $\sec u = \frac{25}{7}$, $\csc u = -\frac{25}{24}$

In Exercises 7–14, use the half-angle formulas to find the exact value of the function.

7. $\sin 15°$

8. $\cos 67.5°$

9. $\cos\frac{5\pi}{8}$

10. $\sin\left(\frac{-5\pi}{8}\right)$

11. $\cos\frac{u}{2}$ if $\cos u = \frac{1}{8}$, $-90° < u < 0$

12. $\sin\frac{u}{2}$ if $\sin u = \frac{1}{8}$, $450° < u < 540°$

13. $\sin\frac{\theta}{2}$ if $\sec\theta = -7$, $\frac{\pi}{2} < \theta < \pi$

14. $\cos\frac{\theta}{2}$ if $\tan\theta = \frac{3}{4}$, $\pi < \theta < \frac{3\pi}{2}$

In Exercises 15–32, verify the identities.

15. $(\sin x + \cos x)^2 = 1 + \sin 2x$

16. $\tan x + \cot x = 2\csc 2x$

17. $\tan x = \dfrac{\sin 2x}{1 + \cos 2x}$

18. $\tan x = \dfrac{1 - \cos 2x}{\sin 2x}$

19. $\sin 4x = 2\sin 2x \cos 2x$

20. $\cos 3x = 4\cos^3 x - 3\cos x$

21. $\sin 3x = 3\sin x - 4\sin^3 x$

22. $\tan 3x = \dfrac{3\tan x - \tan^3 x}{1 - 3\tan^2 x}$

23. $\sin 4x = \cos x(4\sin x - 8\sin^3 x)$

24. $\cos 4x = 8\cos^4 x - 8\cos^2 x + 1$

25. $\cos^4 x - \sin^4 x = \cos 2x$

26. $\dfrac{1 + \sin 2x}{\cos 2x} = \dfrac{\cos x + \sin x}{\cos x - \sin x}$

27. $\sec 2u = \dfrac{\sec^2 u}{2 - \sec^2 u}$

28. $\dfrac{\sin 3x + \cos 3x}{\cos x - \sin x} = 1 + 2\sin 2x$

29. $\sin 4x = 4\sin x \cos x\,(1 - 2\sin^2 x)$

30. $\tan^2 x + \cos 2x = 1 - \cos 2x \tan^2 x$

31. $\sin 6x \tan 3x = 2\sin^2 3x$

32. $\cos 4x \sec^2 2x = 1 - \tan^2 2x$

In Exercises 33–42, find all solutions to the conditional equations in the interval $[0, 2\pi]$.

33. $\sin 2x = -\frac{1}{2}$

34. $\cos 2x = \frac{1}{2}\sqrt{3}$

35. $\sin 2x + \cos x = 0$

36. $\sin 2x - \sin x = 0$

37. $\cos 2x - \sin x = 0$

38. $\cos 2x + \sin x = 1$

39. $\cos 2x + \cos x = 0$

40. $\cos x + \sin 2x \sin x = 0$

41. $\tan 2x = \tan x$

42. $\cos 2x - 1 - \tan x = 0$

3.6 Oblique Triangles; The Law of Sines

An oblique triangle is a triangle that is not a right triangle. If at least three of the sides and/or angles are given, including at least one side, one can use trigonometry to find the remaining parts of the triangle. This is called **solving the triangle.** When doing this, the following law is often helpful.

(26) The Law of Sines In any triangle with sides a, b, c and opposite angles α, β, γ, respectively,

$$\frac{a}{\sin \alpha} = \frac{b}{\sin \beta} = \frac{c}{\sin \gamma}$$

Let us first see why this is true; then we shall use it to solve some triangles. First we draw the triangle, but in addition we drop an altitude from C and from B to points D and E, respectively. See Fig. 4. Suppose CD is of length r and BE is of length s. Then in triangle ACD, $\sin \alpha = \frac{r}{b}$, or $r = b \sin \alpha$, while in triangle BCD, $\sin \beta = \frac{r}{a}$, or $r = a \sin \beta$. Therefore, $b \sin \alpha = a \sin \beta$. Dividing both sides by $\sin \alpha \sin \beta$ yields

$$\frac{b \sin \alpha}{\sin \alpha \sin \beta} = \frac{a \sin \beta}{\sin \alpha \sin \beta} \quad \text{or} \quad \frac{b}{\sin \beta} = \frac{a}{\sin \alpha}$$

Similarly in triangle ABE, $\sin \alpha = \frac{s}{c}$, or $s = c \sin \alpha$, while in triangle CBE, $\frac{s}{a} = \sin(\angle BCE) = \sin(180° - \gamma) = \sin 180° \cos \gamma - \cos 180° \sin \gamma = (0) \cos \gamma - (-1)\sin \gamma = \sin \gamma$. Thus $s = a \sin \gamma$, and so $a \sin \gamma = c \sin \alpha$. Dividing both sides by $\sin \alpha \sin \gamma$ yields $\frac{a}{\sin \alpha} = \frac{c}{\sin \gamma}$. But we have already

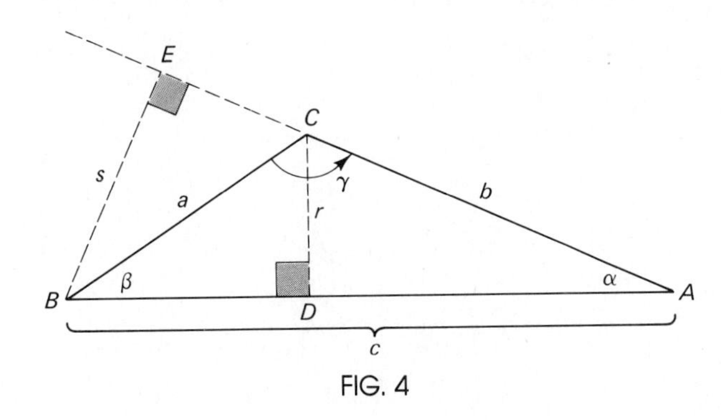

FIG. 4

shown that $\dfrac{a}{\sin\alpha} = \dfrac{b}{\sin\beta}$. Therefore

$$\frac{a}{\sin\alpha} = \frac{b}{\sin\beta} = \frac{c}{\sin\gamma} \qquad \blacksquare$$

Now that we have verified the Law of Sines, let us see how to use it to solve triangles.

ANGLE-SIDE-ANGLE (ASA) OR ANGLE-ANGLE-SIDE (AAS)

Suppose we are given two angles and one side. If α and β are the given angles, then since $\alpha + \beta + \gamma = 180°$, the third angle $\gamma = 180° - \alpha - \beta$. Since we know all three angles and we know one side, say side a, we can use the Law of Sines to find the other sides:

$$\frac{b}{\sin\beta} = \frac{a}{\sin\alpha} \quad \text{or} \quad b = \frac{a\sin\beta}{\sin\alpha}$$

while

$$\frac{c}{\sin\gamma} = \frac{a}{\sin\alpha} \quad \text{or} \quad c = \frac{a\sin\gamma}{\sin\alpha}$$

EXAMPLE 1 Solve the triangle ABC if $\alpha = 32°$, $\gamma = 103°$, and $b = 12$.

Solution Note that this is a case of ASA, as you can see in Fig. 5. The third angle is $\beta = 180° - 32° - 103° = 45°$. By the Law of Sines,

$$\frac{a}{\sin\alpha} = \frac{b}{\sin\beta} \quad \text{or} \quad a = \frac{b\sin\alpha}{\sin\beta} = \frac{12\sin 32°}{\sin 45°} \approx 8.99303$$

while

$$\frac{c}{\sin\gamma} = \frac{b}{\sin\beta} \quad \text{or} \quad c = \frac{b\sin\gamma}{\sin\beta} = \frac{12\sin 103°}{\sin 45°} \approx 16.5356 \qquad \blacksquare$$

FIG. 5

EXAMPLE 2 Solve the triangle ABC if $\alpha = 64°$, $\beta = 48°$ and $a = 3.1$.

Solution Note that this is a case of AAS, as you can see in Fig. 6. The third angle is

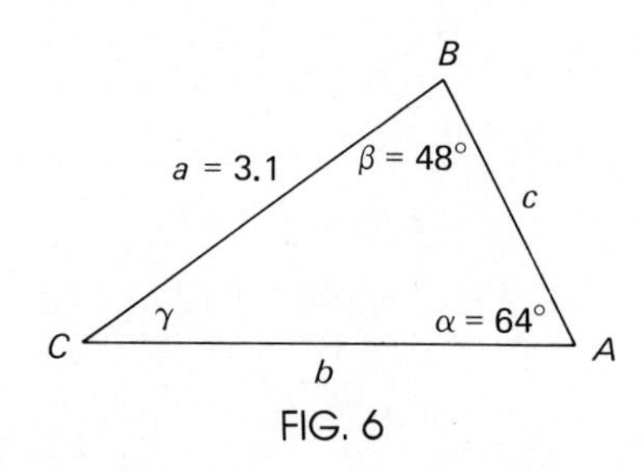

FIG. 6

$\gamma = 180° - 64° - 48° = 68°$. By the Law of Sines

$$\frac{a}{\sin \alpha} = \frac{b}{\sin \beta} \quad \text{or} \quad b = \frac{a \sin \beta}{\sin \alpha} = \frac{3.1 \sin 48°}{\sin 64°} \approx 2.56316$$

Similarly,

$$\frac{a}{\sin \alpha} = \frac{c}{\sin \gamma} \quad \text{or} \quad c = \frac{a \sin \gamma}{\sin \alpha} = \frac{3.1 \sin 68°}{\sin 64°} \approx 3.19792 \quad \blacksquare$$

APPLICATIONS

In some of the applications we worked in Sec. 1.5, the solutions were somewhat complicated. Sometimes the Law of Sines can be used to simplify these solutions.

EXAMPLE 3 (This is Example 4 of Sec. 1.5.) Suppose you are on level ground and you wish to measure the height of a mountain above ground level. From some point A along the level ground you determine that the angle of elevation of the peak is 0.95, while from a point B, which is 200 feet farther from the mountain, the angle of elevation of the peak is 0.92. Find the height. Assume the data are correct to two significant figures.

Solution First we draw a picture:

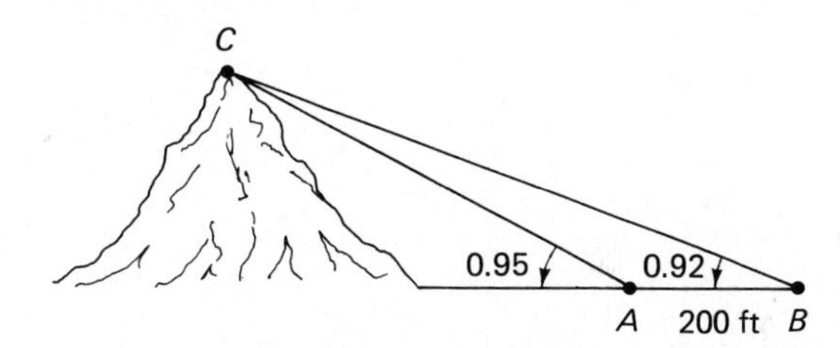

We see triangle ABC, but the height h is the length of an altitude from vertex C. We drop this altitude, forming right triangles ACD and BCD. In each right triangle we know an acute angle, but we do not know any of the sides.

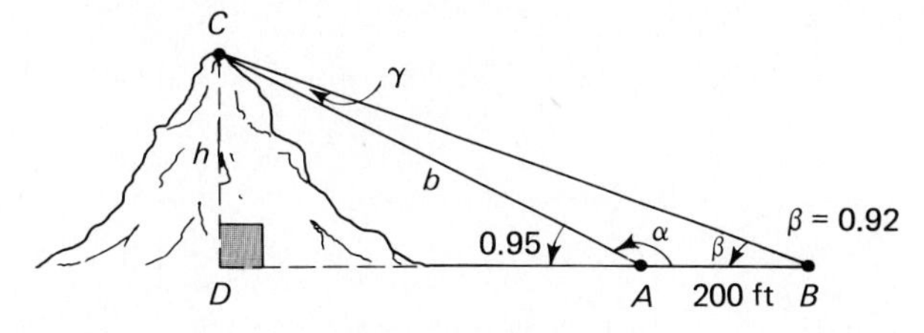

If we know the length of AC, which we shall call b, then we could easily find $h = b \sin 0.95$. We shall first find b with the Law of Sines. Now in triangle ABC,

$$\alpha = \pi - 0.95 \quad \text{and} \quad \gamma = \pi - \alpha - \beta = \pi - (\pi - 0.95) - 0.92 = 0.03$$

By the Law of Sines

$$\frac{b}{\sin \beta} = \frac{200}{\sin \gamma} \qquad \text{or} \qquad b = \frac{200 \sin 0.92}{\sin 0.03} \approx 5304.806$$

Therefore from triangle ACD,

$$h = b \sin 0.95 \approx 4315.01$$

Since the original data are correct to two significant figures, the answer is $h \approx 4300$ feet. ■

Angles are used in navigation to designate the course of a plane, boat, etc. Thus it is important to understand what is meant by the phrase "The object is on a course of 90°." In what direction is the object headed? North? East? South? West? Somewhere in between? The answer in this case is due east. This is because a navigator's compass is marked off in degrees in a clockwise direction, with 0° at due north (see diagram).

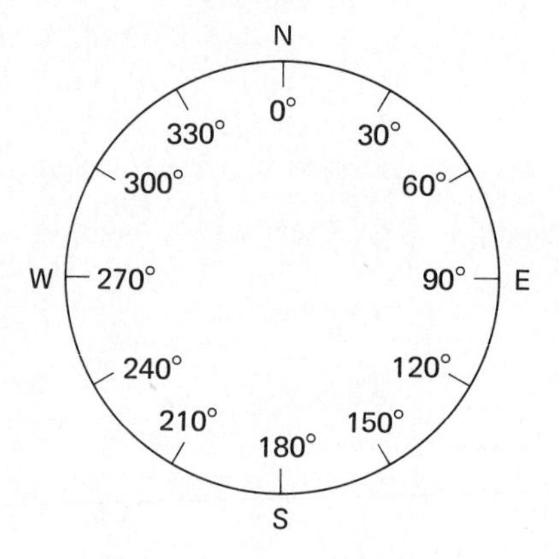

The following are examples of courses of 315° and 60°, respectively:

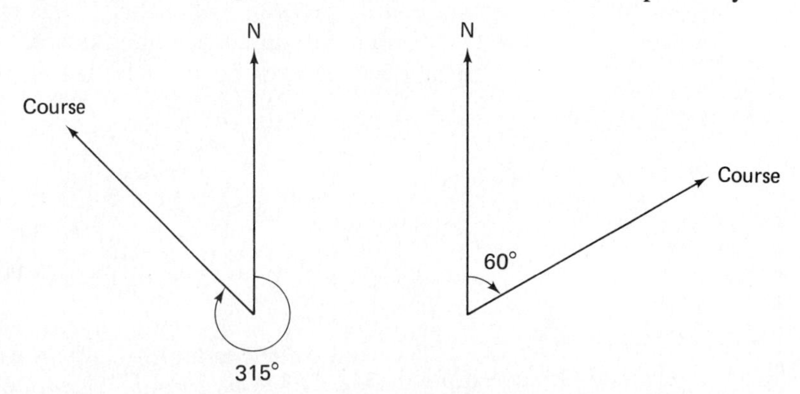

EXAMPLE 4 From a ship, port A is 43.6 nautical miles away on a course of 306°, while port B is on a course of 163°. If the ports are 87.5 nautical miles apart, how far is the ship from port B? Assume the data are correct to three significant figures.

Solution First we draw a picture. We let C denote the ship and label the angles as shown.

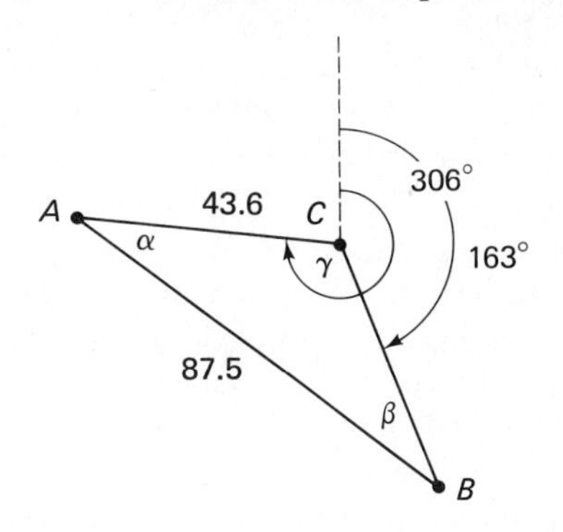

Then $\gamma = 306° - 163° = 143°$, and we have side-side-angle. By the Law of Sines,

$$\frac{87.5}{\sin 143°} = \frac{43.6}{\sin \beta} \quad \text{or} \quad \sin \beta = \frac{43.6 \sin 143°}{87.5} \approx 0.299987583$$

Thus $\beta \approx 17.450145°$, and $\alpha = 180° - \beta - \gamma \approx 19.549855°$. Therefore

$$\frac{87.5}{\sin 143°} = \frac{|BC|}{\sin \alpha} \quad \text{or} \quad |BC| = \frac{87.5 \sin \alpha}{\sin 143°} \approx 48.6526 \text{ nautical mi}$$

Since the initial data are accurate to three significant figures, the answer is 48.7 nautical mi. ■

EXERCISES

In Exercises 1–18 solve the triangle ABC having the given parts.

1. $a = 30$, $\beta = 60°$, $\gamma = 45°$

2. $b = 12$, $\alpha = 120°$, $\gamma = 30°$

3. $b = 25$, $\beta = 40°$, $\gamma = 20°$

4. $a = 40$, $\alpha = 30°$, $\beta = 80°$

5. $b = 50$, $\alpha = \frac{\pi}{4}$, $\gamma = \frac{\pi}{3}$

6. $a = 150$, $\alpha = \frac{2\pi}{3}$, $\beta = \frac{\pi}{6}$

7. $c = 22$, $\alpha = 112°$, $\gamma = 24°$

8. $b = 12$, $\alpha = 62°$, $\beta = 68°$

Ⓒ **9.** $\alpha = 36.12°$, $\beta = 61.27°$, $c = 407.1$

Ⓒ **10.** $\beta = 101°$, $\gamma = 26.34°$, $c = 31.04$

Ⓒ **11.** $\beta = 1.62$ rad, $\gamma = 0.53$ rad, $a = 96.5$

Ⓒ **12.** $\alpha = 1.21$ rad, $\beta = 0.97$ rad, $a = 625.5$

Ⓒ **13.** $\gamma = 0.34$ rad, $\alpha = 0.83$ rad, $c = 0.0145$

Ⓒ **14.** $\alpha = 46°$, $\gamma = 79.123°$, $b = 1.107$

Ⓒ **15.** $\gamma = 122.2°$, $\beta = 13.19°$, $b = 1492$

Ⓒ **16.** $\alpha = 1.4812$ rad, $\gamma = 0.8667$ rad, $a = 271.063$

Ⓒ **17.** $\beta = 88.42°$, $\alpha = 16.21°$, $b = 24.68$

Ⓒ **18.** $\alpha = 110°$, $\beta = 27.3°$, $c = 12.2$

19. A tree standing (vertically) on a slope inclined at an angle of 11° to the horizontal casts a shadow of length 51 feet up the slope. If the angle of elevation of the sun is 72.5°, how high is the tree?

20. A stretch of a river has straight banks. A man on one bank observes a tree on the other bank. His line of sight makes an angle of 51° with his bank. After walking 40 feet along his bank, he reaches a tree on his side of the river. The angle which his line of sight to the original tree makes with his bank has increased to 64°. He wants to stretch a rope between the two trees. How far apart are the trees?

21. A pole is slanting 13° away from vertical. To hold the pole up, a guy wire is attached to the top of the pole and anchored in the ground directly opposite the slant. If the pole is 96 feet long and the guy wire makes an angle of 47° with the (level) ground, how long is the guy wire?

Ⓒ**22.** A ramp 7.603 meters long is inclined from level ground to a loading dock. It makes an angle of 34.92° with the ground. If it is replaced with a ramp which makes an angle of 12.7° with the ground, how long is the new ramp?

Ⓒ**23.** Airport B is 200 miles due east of airport A. To reach airport C, a plane flies from A on a course of 346.35° or from B on a course of 302.2°. Find the distance from A to C and from B to C.

24. Airport B is 200 miles from airport A on a course of 52°. To reach airport C, a plane flies from A on a course of 346° or from B on a course of 302°. Find the distance from A to C and from B to C.

25. A ship leaves a small island on a course of 106°. The ship is traveling at 25 knots (nautical miles per hour). After 2 hours, the course is changed to 221°. Sometime later, the ship receives a message to return to the island. To do so, the ship turns to a heading of 33° 30′. What is the total length of the trip (in nautical miles)?

Ⓒ**26.** A plane has crashed and activated an emergency transmitter. The signal is being received by two rescue units, A and B. A is 8.63 kilometers due north of B. From the signal, the rescuers determine that they must take a course of 127.25° from A or 43.08° from B to reach the plane. How far is each rescue unit from the plane?

Ⓒ**27.** Two planes take off from the same airport. The first plane flies on a course of 220.1°. The second plane flies on a course of 154.4°. After the first plane flies 362.4 kilometers, the course from it to the second plane is 83.5°. How far is the second plane from the airport?

28. In the proof of the Law of Sines, redraw Fig. 4 so that γ is acute and see how the argument works then.

Exercise 29–31 refer to appropriate parts of triangle ABC.

29. Use the Law of Sines to prove

$$\frac{a \pm b}{b} = \frac{\sin \alpha \pm \sin \beta}{\sin \beta}$$

30. Prove $\dfrac{a - b}{a + b} = \dfrac{\sin \alpha - \sin \beta}{\sin \alpha + \sin \beta}$.

31. Verify the identity $(a - b) \cos \dfrac{\gamma}{2} = c \sin\left(\dfrac{\alpha - \beta}{2}\right)$, which is called *Mollweide's formula.*

3.7 The Ambiguous Case of the Law of Sines, SSA

If two sides and an angle opposite one of them are given, there can be complications in solving the triangle. In fact, the given data may determine more than one triangle or perhaps no triangle at all. The purpose of this section is to discuss this case thoroughly, giving both geometric and algebraic descriptions of the various possibilities.

We assume we are given sides AC of length b and CB of length a. In addition, assume the unknown side AB of length c lies along the x-axis. See Fig. 7 for two possibilities. Our problem now is where along the x-axis to place B. We break the problem into two cases, $0° < \alpha < 90°$ and $90° \leq \alpha < 180°$.

CASE $0° < \alpha < 90°$

For this case, there may be no triangle, one triangle, or two triangles that satisfy the data. In order to find B in Fig. 7(a), we attach to C a line segment of length a and let the other end swing freely in a circular arc. Then B lies at the

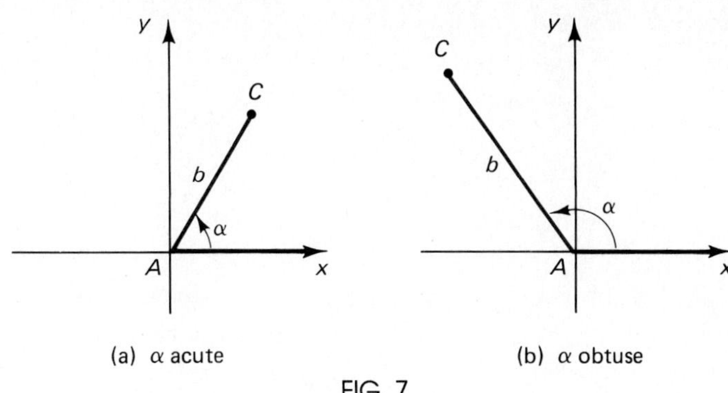

(a) α acute (b) α obtuse

FIG. 7

intersection of this arc and the positive x-axis. The various possibilities are described in the following table, where h is the length of the perpendicular dropped from C to the x-axis.

TABLE 1

Sketch	*Geometric description*	*Algebraic description*	*Number of possible triangles*
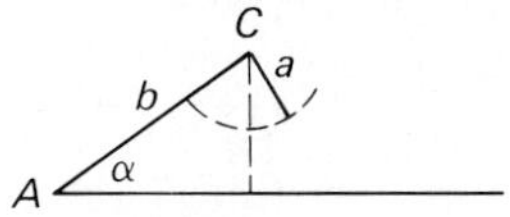	Arc does not intersect the x-axis	$a < h$	0
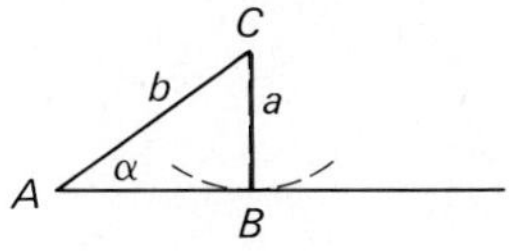	Arc is tangent to the x-axis	$a = h$	1
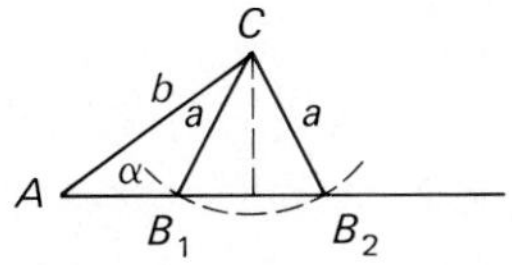	Arc intersects positive x-axis in two places	$h < a < b$	2
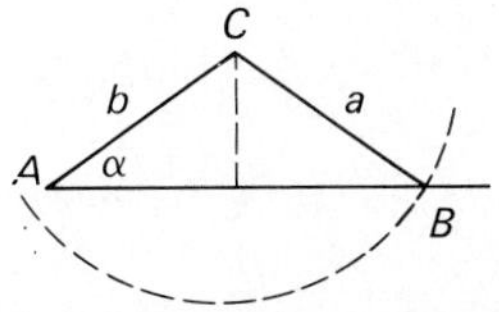	Arc intersects the positive x-axis in one place	$b \leq a$	1

EXAMPLE 1 Solve the triangle ABC if $a = 10$, $b = 12$, and $\alpha = 72°$.

Solution By the Law of Sines,

$$\frac{10}{\sin 72°} = \frac{12}{\sin \beta}$$

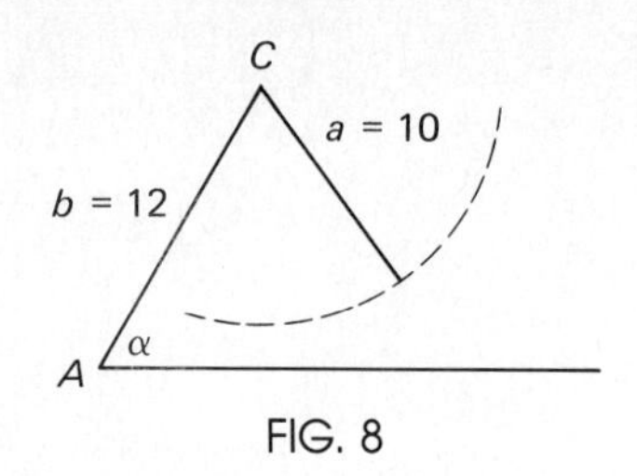

FIG. 8

Thus, $\sin \beta = \dfrac{12 \sin 72^\circ}{10} \approx 1.14127$. But there is no angle β with $\sin \beta$ greater than 1. Hence there is no triangle satisfying the data. ■

EXAMPLE 2 Solve the triangle ABC if $a = 10$, $b = 20$ and $\alpha = 30°$.

Solution By the Law of Sines,

$$\frac{10}{\sin 30^\circ} = \frac{20}{\sin \beta} \qquad \text{or} \qquad \sin \beta = \frac{20 \sin 30^\circ}{10} = 1$$

Therefore, β is exactly 90° and ABC is a right triangle. We thus have $\gamma = 60°$ and $c = 10\sqrt{3} \approx 17.3205$. ■

EXAMPLE 3 Solve the triangle ABC if $a = 10$, $b = 12$ and $\alpha = 51°$.

Solution By the Law of Sines,

$$\frac{10}{\sin 51^\circ} = \frac{12}{\sin \beta} \qquad \text{or} \qquad \sin \beta = \frac{12 \sin 51^\circ}{10} \approx 0.932575$$

We have two possibilities,

$$\beta = \beta_1 \approx \sin^{-1}(0.932575) \approx 68.8399^\circ$$

or

$$\beta = \beta_2 \approx 180^\circ - 68.8399^\circ \approx 111.160^\circ$$

If $\beta = \beta_1$, then $\gamma_1 \approx 180° - 51° - 68.8399° \approx 60.1601°$ and

$$c_1 = \frac{a \sin \gamma_1}{\sin \alpha} \approx 11.1616$$

If $\beta = \beta_2$, then $\gamma_2 \approx 180° - 51° - 111.160° \approx 17.840°$ and

$$c_2 = \frac{a \sin \gamma_2}{\sin \alpha} \approx 3.94212$$

The two solutions are in Fig. 9.

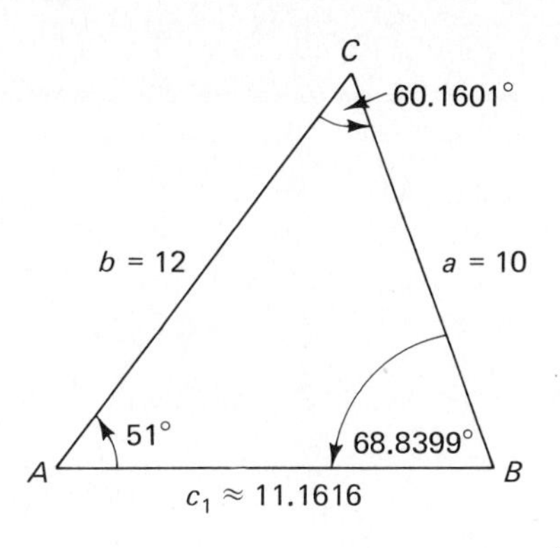

FIG. 9 ■

EXAMPLE 4 Solve the triangle ABC if $a = 12$, $b = 10$ and $\alpha = 51°$.

Solution We should see right away that the side opposite the given angle is greater than the side adjacent to the given angle (the case $a > b$ in Table 1). Hence there is a unique triangle satisfying the given data. But let us pretend we miss this fact and proceed to solve the triangle as in Example 3, so we can see how the algebra tells us which case we are in. By the Law of Sines,

$$\frac{12}{\sin 51°} = \frac{10}{\sin \beta} \qquad \text{or} \qquad \sin \beta = \frac{10 \sin 51°}{12} \approx 0.647622$$

We have two possibilities,

$$\beta = \beta_1 \approx \sin^{-1}(0.647622) \approx 40.3625°$$

or

$$\beta = \beta_2 \approx 180° - 40.3625° \approx 139.637°$$

If $\beta = \beta_1$, then $\gamma_1 \approx 180° - 51° - 40.3625° \approx 88.6375°$ and

$$c_1 = \frac{a \sin \gamma_1}{\sin \alpha} \approx 15.4367$$

If $\beta = \beta_2$, then $\gamma_2 \approx 180° - 51° - 139.637° = -10.637°$. Since triangles do not have negative angles, this case is impossible. Thus the algebra tells us there is only one triangle. When this happens, we should check to see that indeed $b < a$, so that only one triangle can be formed. ■

CASE $90° \leq \alpha < 180°$

For this case, there may be no triangle or one triangle, as illustrated in Table 2.

TABLE 2

Sketch	Geometric Description	Algebraic Description	Number of Possible Triangles
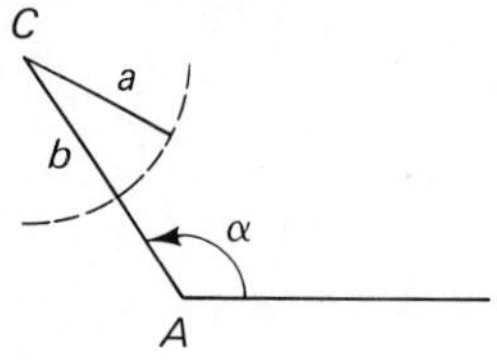	Arc does not intersect the positive x-axis	$a \leq b$	0
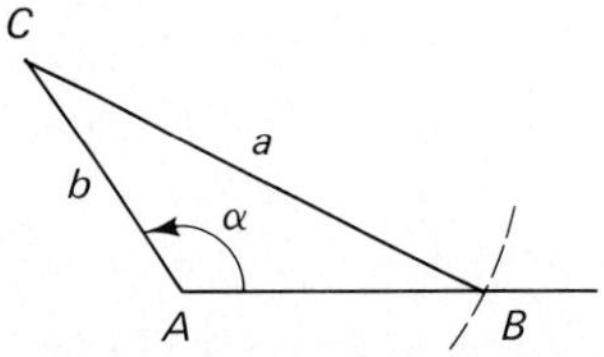	Arc intersects the positive x-axis in one place	$a > b$	1

EXAMPLE 5 Solve the triangle ABC if (a) $a = 12$, $b = 10$, and $\alpha = 110°$, (b) $a = 10$, $b = 12$, and $\alpha = 110°$.

Solution (a) First note that $\alpha \geq 90°$ and the side adjacent to α is smaller than the side opposite α. Hence there is a unique solution. By the Law of Sines,

$$\frac{12}{\sin 110°} = \frac{10}{\sin \beta} \quad \text{or} \quad \sin \beta = \frac{10 \sin 110°}{12} \approx 0.783077$$

Thus $\beta \approx \sin^{-1}(0.783077) \approx 51.5432°$, $\gamma = 180° - \alpha - \beta \approx 18.4568°$, and

$$c = \frac{12 \sin \gamma}{\sin 110°} \approx 4.04289$$

(b) We should note that $\alpha \geq 90°$ and the side adjacent to α is greater than the side opposite α. Hence there is no solution. If we missed this and started to solve it as in part (a), then after some (unnecessary) work, we would obtain a negative angle as we did in Example 4. ■

Warning: In the five examples so far, we always had α, a, and b given in the data. Of course, in general (and in the exercises) this will not be the case, and you must examine each problem carefully to determine what type of situation is described by the given data. Make sure to draw a diagram first and then take special care in labeling the sides and angles.

EXAMPLE 6 Solve the triangle ABC if $a = 10$, $b = 20$, and $\beta = 30°$.

Solution We recognize this as SSA and draw the diagram in Fig. 10, being very careful

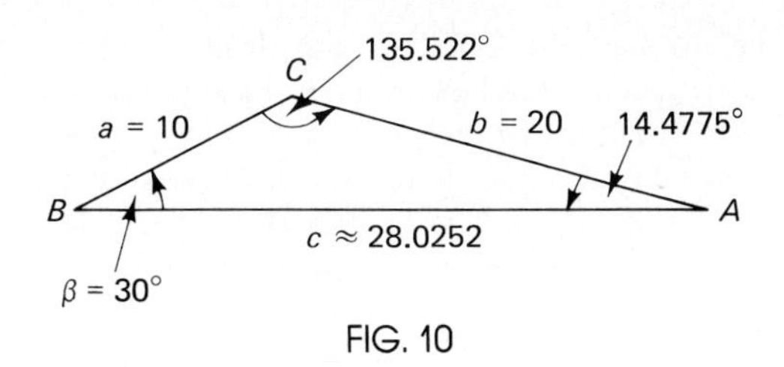

FIG. 10

to place b opposite the angle β. By the geometry (with the side opposite β being larger than the other given side), we solve the triangle as in Example 4.

$$\frac{a}{\sin \alpha} = \frac{b}{\sin \beta} \quad \text{or} \quad \sin \alpha = \frac{a \sin \beta}{b} = \frac{10\left(\frac{1}{2}\right)}{20} = \frac{1}{4}$$

Thus $\alpha = \sin^{-1}(0.25) \approx 14.4775°$, $\gamma = 180° - \alpha - \beta \approx 135.522°$, and

$$c = \frac{b \sin \gamma}{\sin \beta} \approx \frac{20 \sin(135.522°)}{\sin 30°} \approx 28.0252$$

Note that if we had been careless in labeling Fig. 10 and interchanged sides a and b, we would have obtained the problem we solved in Example 2. ■

EXERCISES

In Exercises 1–10, determine if the given data represent an ambiguous or unambiguous case. If ambiguous, without doing any computations, determine by inspection if possible whether there are 0, 1, or 2 solutions. Do not solve the given triangle.

1. $a = 5$, $\beta = 77°$, $\gamma = 50°$

2. $a = 12.1$, $\alpha = 28°$, $\beta = 112°$

3. $b = 9$, $c = 3$, $\beta = 20°$

4. $b = 8$, $c = 10$, $\gamma = 70°$

5. $a = 20$, $b = 10$, $\alpha = 100°$

6. $a = 15$, $b = 5$, $\beta = 120°$

7. $b = 18$, $\gamma = 19°$, $\alpha = 135°$

8. $a = 20$, $\beta = 50°$, $\gamma = 100°$

9. $a = 40$, $b = 15$, $\beta = 60°$

10. $a = 10$, $b = 20$, $\alpha = 45°$

In Exercises 11–28, solve the triangle ABC having the given parts, if possible. Find all solutions if there are more than one.

11. $a = 4$, $b = 6$, $\alpha = 30°$

12. $a = 10$, $c = 8$, $\alpha = 40°$

13. $b = 8$, $c = 10$, $\gamma = 120°$

14. $b = 5$, $c = 10$, $\beta = 120°$

15. $a = 10$, $b = 5$, $\alpha = 20°$

16. $a = 4$, $b = 3$, $\beta = 50°$

17. $a = 10$, $b = 5$, $\beta = 30°$

18. $\alpha = 110°$, $a = 27.12$, $c = 11.78$

ⓒ**19.** $\beta = 2.12$ rad, $a = 27.12$, $b = 11.78$

ⓒ**20.** $\alpha = 43°$, $a = 943.16$, $b = 476.5$

ⓒ**21.** $\gamma = 56°27'$, $b = 21.23$, $c = 32.21$

ⓒ**22.** $\beta = 0.634$ rad, $b = 637$, $c = 1239$

23. $\alpha = 0.4957$ rad, $c = 17$, $a = 9$

24. $\gamma = 2.46$ rad, $c = 1096$, $a = 548$

ⓒ**25.** $\alpha = 132.64°$, $a = 6.02$, $b = 23.51$

ⓒ**26.** $\beta = 68.142°$, $b = 113.715$, $c = 84.957$

ⓒ**27.** $\gamma = 0.3172$ rad, $b = 67.624$, $c = 61.1$

ⓒ**28.** $\gamma = 36°13'25''$, $a = 191.4$, $c = 134$

29. Sherlock Kemlock was investigating a waterfront robbery. A prime suspect, who was picked up in his motorboat in the harbor, claimed he had just been out fishing in his boat for the past several hours. The suspect claimed (see Fig. 11) that he forgot exactly when he left the harbor, but that he had trolled slowly north for a while, turned southeast and trolled for 1 hour until he was due east of the bay, and then headed directly west for half an hour, when he arrived at the bay only to be picked up by the harbor police. Since he was fishing, he was trolling at the same slow speed for the whole trip. Should Sherlock believe him?

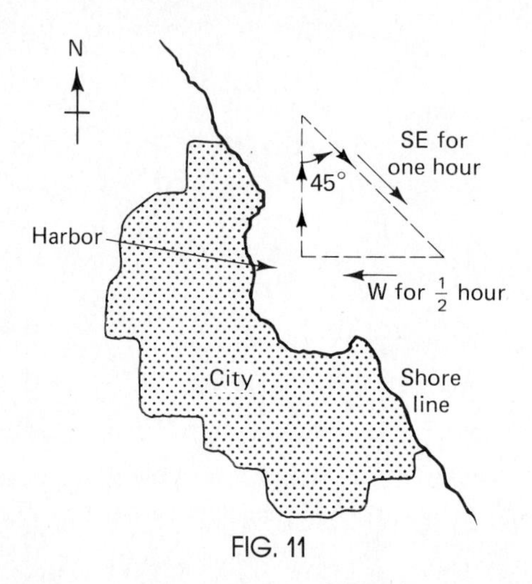

FIG. 11

30. A boat leaves a dock and travels due east. After a little while, the boat turns and heads southwest for one hour. The boat then turns and heads back to the dock, arriving in 50 minutes. If the boat traveled at a steady 12 mph for the whole trip, how long was the trip? Is there more than one possibility?

31. Points A and B are on opposite sides of a pond. A woman walks 67 feet from A straight to a point C (off to the side of the pond) and then turns and walks 91 feet straight to B. If she then observes that the angle between BA and BC is 38°, how far is it from A to B? (Can you determine the answer uniquely from the given information?)

32. Towns A, B, and C are connected by straight roads. It is 13.3 miles from A to B and 22.7 miles from B to C. At A, the roads to B and C meet at an angle of $63°24'$. How far is A from C?

3.8 The Law of Cosines

The Law of Sines is not sufficient to solve a triangle if we are given two sides and the included angle. For this we need another relationship between the sides and the angles. We derive such a relationship by placing the triangle ABC in the coordinate plane in such a way that A is at the origin and AB is along the positive x-axis. The two diagrams in Fig. 12 indicate this positioning of the triangle when α is obtuse and when α is acute. In either case $C = (x, y)$ is a point on the terminal side of α, an angle in standard position, and C is $r = b$ units from the origin. By the definition of the trigonometric functions, $\sin \alpha = \frac{y}{r} = \frac{y}{b}$, and $\cos \alpha = \frac{x}{r} = \frac{x}{b}$. Multiplying each equation by b yields $y = b \sin \alpha$ and $x = b \cos \alpha$. Since these are the coordinates of C, $C = (b \cos \alpha, b \sin \alpha)$. Since side AB has length c, B has coordinates $B = (c, 0)$.

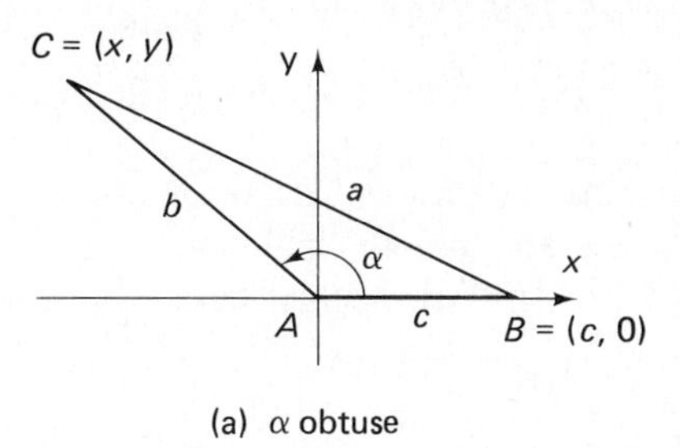

(a) α obtuse

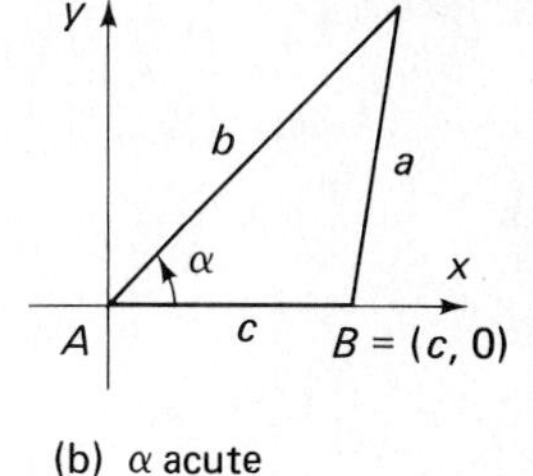

(b) α acute

FIG. 12

We obtain a relationship between a, b, c, and α by observing that the length of the side BC is not only a but is also the distance between B and C. Therefore

$$a = \sqrt{(b \cos \alpha - c)^2 + (b \sin \alpha)^2}$$

Squaring both sides yields

$$a^2 = (b \cos \alpha - c)^2 + (b \sin \alpha)^2$$

Multiplying out the right-hand side yields

$$\begin{aligned} a^2 &= (b \cos \alpha)^2 - 2bc \cos \alpha + c^2 + (b \sin \alpha)^2 \\ &= b^2 \cos^2 \alpha - 2bc \cos \alpha + c^2 + b^2 \sin^2 \alpha \\ &= b^2(\cos^2 \alpha + \sin^2 \alpha) + c^2 - 2bc \cos \alpha \end{aligned}$$

Since $\cos^2 \alpha + \sin^2 \alpha = 1$,

$$a^2 = b^2 + c^2 - 2bc \cos \alpha$$

The last equation is the relationship we were seeking. If we had placed B or C at the origin, we would have obtained

$$b^2 = a^2 + c^2 - 2ac \cos \beta \qquad \text{or} \qquad c^2 = a^2 + b^2 - 2ab \cos \gamma$$

Any one of these three equations is an instance of the Law of Cosines:

The Law of Cosines In a triangle, the square of the length of one side is equal to the sum of the squares of the lengths of the other two sides minus twice their product times the cosine of the angle between them. That is,

(27) $$a^2 = b^2 + c^2 - 2bc \cos \alpha$$

(28) $$b^2 = a^2 + c^2 - 2ac \cos \beta$$

(29) $$c^2 = a^2 + b^2 - 2ab \cos \gamma$$

The Law of Cosines is useful in solving triangles where you are given two sides and the included angle or three sides.

SIDE-ANGLE-SIDE (SAS)

If you are given two sides and the included angle (SAS), the method of solution is as follows:

First use the Law of Cosines in order to find the remaining side. Once you know three sides and one angle, you can use either the Law of Cosines or the Law of Sines to find a second angle. [The third is 180° (or π) minus the sum of the other two.] The Law of Sines is easier to use, but there is a possible problem.

If, say, a, b, c, and α are known and you try to find β, you get $\dfrac{a}{\sin \alpha} = \dfrac{b}{\sin \beta}$, or $\sin \beta = \dfrac{b}{a} \sin \alpha$. If β is acute, $\beta = \sin^{-1}\left[\dfrac{b}{a} \sin \alpha\right]$, but if β is obtuse, $\sin^{-1}\left[\dfrac{b}{a} \sin \alpha\right]$ is just the reference angle of β. Thus if you do not know ahead of time whether β is acute or obtuse, then you do not know whether $\sin^{-1}\left[\dfrac{b}{a} \sin \alpha\right]$ is β or the reference angle of β, and consequently the Law of Sines cannot be used. However, you can avoid this difficulty by making sure that you are looking for an acute angle. Since in a triangle the smaller angles are opposite the smaller sides and the smallest two angles are always acute, if you calculate the smaller of the two remaining angles, you will be sure it is acute. Summarizing,

> If three sides and one angle in a triangle are known and the Law of Sines is being used to find a second angle, make sure that angle is the smaller of the two remaining angles (i.e., opposite the smaller of the two remaining sides).

EXAMPLE 1 Solve the triangle ABC if $a = 6$, $c = 4$, and $\beta = 42°$.

Solution The remaining side is found by the Law of Cosines. We use equation (28) since a, c, and β are known, and we wish to find b:

$$b^2 = a^2 + c^2 - 2ac \cos \beta$$

$$= 6^2 + 4^2 - 2(6)(4) \cos 42° \approx 16.329048$$

Therefore, $b \approx 4.04092$. Now since $c < a$, we know that $\gamma < \alpha$ and that γ is acute. We compute γ by the Law of Sines:

$$\frac{c}{\sin \gamma} = \frac{b}{\sin \beta} \quad \text{or} \quad \sin \gamma = \frac{c \sin \beta}{b} \approx \frac{4 \sin 42°}{4.04092} \approx 0.662354$$

Since γ is acute, $\gamma \approx \sin^{-1}(0.662354) \approx 41.4797°$. Finally, the angle α is $\alpha = 180° - \beta - \gamma \approx 180° - 42° - 41.4797° \approx 96.5203°$. ■

EXAMPLE 2 Plane I leaves an airport at 1 P.M. and flies a straight course at 400 miles per hour. Plane II leaves the same airport at 1:30 P.M. and flies at 300 miles per hour on a straight course which makes an angle of 78° with that of plane I. How far apart are the planes at 3 P.M.? Assume all data are accurate to two significant figures.

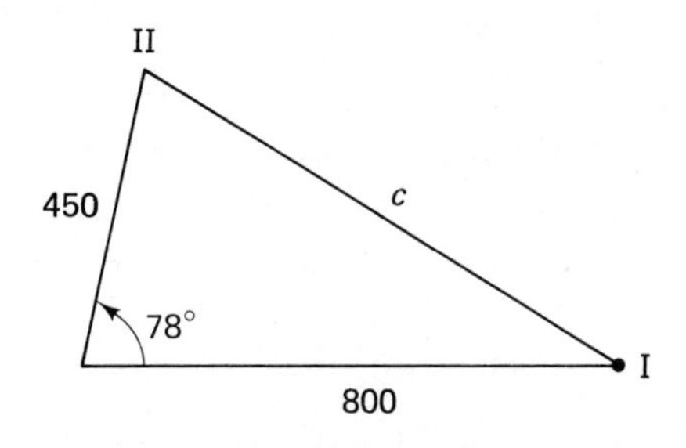

Solution At 3 P.M., plane I will have flown 800 miles, while plane II will have flown 450 miles. The distance between the planes at 3 P.M. is c in the triangle shown, where $a = 450$, $b = 800$, and $\gamma = 78°$. Thus by the Law of Cosines, equation (29),

$$\begin{aligned} c^2 &= a^2 + b^2 - 2ab \cos \gamma \\ &= (450)^2 + (800)^2 - 2(450)(800) \cos 78° \approx 692{,}803.58 \end{aligned}$$

Thus $c \approx \sqrt{692{,}803.58} \approx 832.348$ miles. Since the given data are accurate to two significant figures, the answer should be rounded to 830 miles. ■

SIDE-SIDE-SIDE (SSS)

If three sides of a triangle are given, we can use any of the three formulas for the Law of Cosines to find one angle. Then proceed as in the case SAS to find the remaining two angles.

EXAMPLE 3 Solve the triangle ABC if $a = 12$, $b = 7$, and $c = 11$.

Solution Let us find angle α first. We use the formula for the Law of Cosines in which α appears, namely (27), and solve the equation for $\cos \alpha$:

$$\begin{aligned} a^2 &= b^2 + c^2 - 2bc \cos \alpha \\ 2bc \cos \alpha &= b^2 + c^2 - a^2 \\ \cos \alpha &= \frac{b^2 + c^2 - a^2}{2bc} \\ &= \frac{7^2 + 11^2 - 12^2}{2(7)(11)} \\ &\approx 0.16883117 \end{aligned}$$

Therefore, $\alpha \approx \cos^{-1}(0.16883117) \approx 80.2801°$. Next we find the smaller angle β by the Law of Sines:

$$\sin \beta = \frac{b \sin \alpha}{a} \approx 0.574960$$

$$\beta \approx \sin^{-1}(0.574960) \approx 35.0968°$$

Therefore, $\gamma = 180° - \beta - \alpha \approx 64.6231°$. ■

EXERCISES

In Exercises 1–15, solve the triangle ABC having the given parts.

1. $a = 10$, $b = 4$, $\gamma = 60°$
2. $a = 400$, $c = 900$, $\beta = 102°$
3. $b = 50$, $c = 30$, $\alpha = 122°$
4. $b = 40$, $c = 30$, $\alpha = 80°$
5. $a = 1000$, $c = 600$, $\beta = 0.38$
6. $a = 60$, $b = 80$, $\gamma = 2.12$
7. $b = 14$, $c = 20$, $\alpha = 14°$
8. $a = 100$, $b = 400$, $\gamma = 140°$
9. $a = 6.39$, $c = 12.78$, $\beta = 40°$
10. [c] $a = 14.92$, $b = 7.49$, $c = 12.51$
11. [c] $a = 0.12$, $b = 0.67$, $c = 1.03$
12. $a = 5$, $b = 8$, $c = 11$
13. $a = 221$, $b = 284$, $c = 148$
14. [c] $a = 0.0035$, $b = 0.0024$, $c = 0.0017$
15. [c] $a = 2743$, $b = 967$, $c = 2102$
16. A triangular field has sides of length 147, 206, and 182 meters. Find the angles of the triangle.
17. A ship intends to sail from port A to port B, a distance of 350 miles. It sails at a constant speed of 35 miles per hour. After 2 hours, the captain discovers that they have been sailing off course by 18°. He then corrects the course. What is the total time of the trip?
18. A, B, and C are three airports. The distances from A to B and A to C are 348 and 296 miles, respectively. The angle between the flight paths from A to B and A to C is 71.4333°. How far is it from B to C?
19. Some pipe is to be laid directly through a small hill. From a point off to the side, a straight rope is run to the entry point, and another straight rope is run to the exit point. The first rope is 42 feet long, while the second is 33.6 feet long, and the two ropes meet at an angle of 56.245°. How long is the pipe, and at what angle with the first rope should it be driven so that it exits at the correct place?
20. A plane leaves an airport and flies at 360 miles per hour on a course of 103.25° for 1 hour and 33 minutes. Then it changes course to 236.75° and flies for 48 minutes. How far is it from the airport? If it wished to return, what would be its course?
21. [c] From one house you can go 1.68 kilometers along a straight road to a corner and then turn and go 2.35 kilometers along another straight road to a second house. The angle between the roads is 111°. There is a footpath that runs straight between the two houses. How much shorter is it to go on the footpath instead of the roads?
22. [c] A guy wire runs 15.3 meters from the top of a pole to the (level) ground. It is replaced by a 21.16-meter wire which runs from the top of the pole to a spot on the ground 8.47 meters farther from the pole than the old wire. What angle does the new wire make with the ground?
23. [c] A small electronic component is in the shape of a triangle with sides 6.23, 8.146, and 11.392 millimeters. Find the largest angle.
24. [c] Two boats leave a dock at the same time on straight courses. The angle between their courses is 51.2°. One boat is traveling at 14.7 kilometers per hour, while the other is traveling at 18.3 kilometers per hour. After 2 hours, how far apart are the boats?
25. [c] In Exercise 24, how far apart are the boats after 2 hours and 16 minutes?

3.9 The Area of a Triangle

It is well known that the area of a triangle is one-half the base times the altitude.

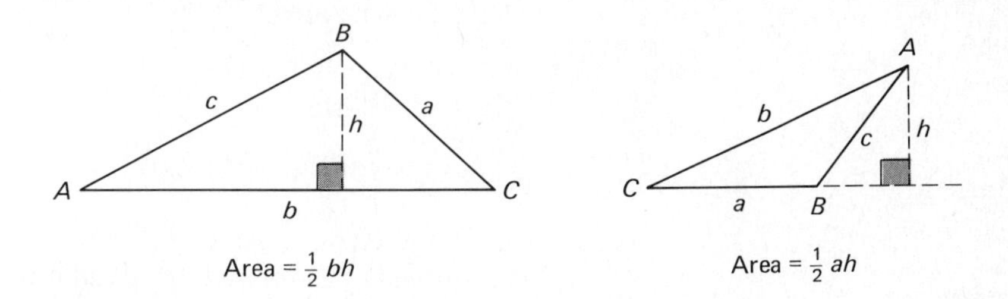

Suppose you are not given the height of a triangle, but you are given sufficient information to solve the triangle. In this section we shall see how to find the area of the triangle. We will use K for area (from the German, *Kontent*), since A usually denotes a vertex.

SAS

Suppose you are given two sides and the included angle of a triangle and you wish to find its area. Assume the given sides are a and c and the given angle is β. Choose one of the sides, say side c, as the base. Then the triangle looks like one of the three diagrams in Fig. 13:

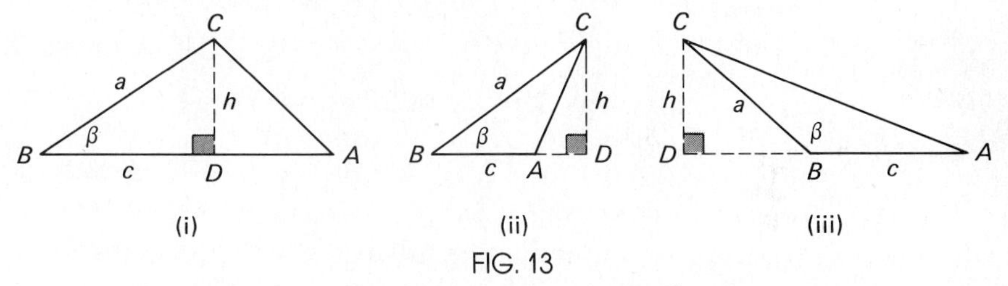

FIG. 13

In Figs. 13(i) and (ii), it is clear that $\sin \beta = \dfrac{h}{a}$, or $h = a \sin \beta$. In Fig. 13(iii), $\sin(180° - \beta) = \dfrac{h}{a}$; but $\sin(180° - \beta) = \sin 180° \cos \beta - \cos 180° \sin \beta = (0)\cos \beta - (-1)\sin \beta = \sin \beta$. Thus in any case $h = a \sin \beta$. Therefore the area is $K = \frac{1}{2}ch = \frac{1}{2}ca \sin \beta$.

> **(30)** If two sides a and c and the included angle β of a triangle are given, then the area is
>
> $$K = \tfrac{1}{2}\, ac \sin \beta$$

EXAMPLE 1 Triangle ABC has $a = 10$, $c = 14$, and $\beta = 30°$. Find its area.

Solution $K = \frac{1}{2}\, ac \sin \beta = \frac{1}{2}(10)(14)\sin 30° = 35.$ ■

EXAMPLE 2 ⓒ Triangle ABC has $b = 491.5$ centimeters, $c = 383.24$ centimeters, and $\alpha = 102.6°$. Find its area.

Solution The formula corresponding to (30) for these two sides and the included angle is $K = \frac{1}{2}bc \sin \alpha$. Thus

$$K = \tfrac{1}{2}(491.5)(383.24)\sin 102.6° \approx 91{,}913.0 \text{ sq cm} \quad ■$$

It is important to remember either formula (30) or the method used to derive the formula. Then if you are given *ASA, ASS, SSA,* or *SSS*, you can use the Law of Sines and/or the Law of Cosines until you have *SAS*, and then use (30). However, formulas have been derived so that if you have a lot of problems all of the same type (or instance *SSS*), you can compute the area directly. We now present two such formulas. The examples that follow demonstrate first how to use both these specialized formulas and then how to use (30) to solve the given problem.

> **(31)** If one side, say side c, and all three angles are known, then
>
> $$K = \tfrac{1}{2}c^2 \frac{\sin \alpha \sin \beta}{\sin \gamma}$$

Proof. By (30) we know

$$K = \tfrac{1}{2}ac \sin \beta$$

By the Law of Sines, we know

$$a = \frac{c \sin \alpha}{\sin \gamma}$$

Substituting the second equation in the first, we obtain

$$K = \tfrac{1}{2}\frac{c \sin \alpha}{\sin \gamma}\, c \sin \beta \qquad \text{or} \qquad K = \tfrac{1}{2}c^2 \frac{\sin \alpha \sin \beta}{\sin \gamma}$$

which is the result we wanted. ■

EXAMPLE 3 ⓒ (*AAS*) Find the area of triangle ABC if $a = 93.1$, $\alpha = 49°$, $\beta = 17°$.

Solution (Using (31)) The third angle is $\gamma = 180° - \alpha - \beta = 114°$. Formula (31) in this case becomes $K = \frac{1}{2}a^2 \frac{\sin \beta \sin \gamma}{\sin \alpha}$. Thus $K = \frac{1}{2}(93.1)^2 \frac{\sin 17° \sin 114°}{\sin 49°}$ ≈ 1533.75.

Solution (Using (30)) We have a side a and an adjacent angle β. We need the next side, side c. First we find the opposite angle as before, $\gamma = 180° - \alpha - \beta = 114°$, and then we use the Law of Sines to find c: $\frac{c}{\sin \gamma} = \frac{a}{\sin \alpha}$, or $c = \frac{a \sin \gamma}{\sin \alpha} =$

$93.1 \dfrac{\sin 114°}{\sin 49°} \approx 112.69366$. Then use (30):

$$K = \tfrac{1}{2}ac \sin \beta \approx \tfrac{1}{2}(93.1)(112.69366)\sin 17° \approx 1533.75 \quad \blacksquare$$

The next formula is usually attributed to Heron (or Hero) of Alexandria (*c.* 250 B.C.)

(32) Heron's Formula If the three sides are known, and $s = \frac{1}{2}(a + b + c)$, ($s$ is called the **semiperimeter**), then

$$K = \sqrt{s(s - a)(s - b)(s - c)}$$

EXAMPLE 4 (SSS) Find the area of triangle ABC if $a = 8$, $b = 6$, and $c = 4$.

Solution (Using (32)) First we compute $s = \frac{1}{2}(a + b + c) = \frac{1}{2}(8 + 6 + 4) = 9$. Then

$$\begin{aligned} K &= \sqrt{s(s - a)(s - b)(s - c)} \\ &= \sqrt{9(1)(3)(5)} \\ &= 3\sqrt{15} \approx 11.6190 \end{aligned}$$

Solution (Using (30)) We can use the Law of Cosines to find any angle. We arbitrarily pick α. Then with its two adjacent sides we have SAS.

$$\cos \alpha = \frac{b^2 + c^2 - a^2}{2bc} = \frac{36 + 16 - 64}{48} = \frac{-12}{48} = -\frac{1}{4}$$

Thus $\alpha = \cos^{-1}(-0.25) \approx 104.478°$. Then

$$K = \tfrac{1}{2}bc \sin \alpha \approx \tfrac{1}{2}(6)(4) \sin 104.478° \approx 11.6190 \quad \blacksquare$$

The proof of Heron's formula is not difficult and provides a nice concatenation of several trigonometric and algebraic principles. But it is a little long and messy, so we just present an outline here and leave the details to the exercises.

By (30) we know $K = \frac{1}{2}ac \sin \beta$ or

(33) $$K^2 = \frac{1}{4}a^2c^2 \sin^2 \beta$$

By the Fundamental Identities,

(34) $$\sin^2 \beta = 1 - \cos^2 \beta = (1 - \cos \beta)(1 + \cos \beta)$$

By the Law of Cosines,

(35) $$\cos \beta = \frac{a^2 + c^2 - b^2}{2ac}$$

Substituting (35) into (34), then (34) into (33), and a page of careful algebra yields Heron's formula. See Exercises 35–38.

EXERCISES

In Exercises 1–20, find the area of a triangle with the given measurements:

1. $a = 40$, $b = 20$, $\gamma = 30°$

2. $a = 5$, $c = 2$, $\beta = 45°$

3. $a = 6$, $c = 12$, $\beta = \frac{2\pi}{3}$

4. $b = 13$, $c = 10$, $\alpha = \frac{\pi}{3}$

5. $b = 20$, $\alpha = 30°$, $\gamma = 45°$

6. $c = 10$, $\alpha = 60°$, $\beta = 45°$

7. $a = 16$, $\beta = 39.2°$, $\gamma = 81.3°$

8. $a = 40$, $\alpha = 92°10'$, $\beta = 66°40'$

9. $b = 747$, $\alpha = 0.70$, $\beta = 1.2$

10. $c = 2$, $\beta = 84°$, $\gamma = 17°$

[C] **11.** $a = 35$, $b = 40$, $\beta = 63°$

[C] **12.** $b = 181.63$, $c = 209.2$, $\gamma = 136°$

[C] **13.** $b = 59.2$, $c = 73.5$, $\beta = 48°$

[C] **14.** $a = 91$, $c = 36$, $\alpha = 82°$

[C] **15.** $a = 8.962$, $c = 8.418$, $\gamma = 1°$

[C] **16.** $a = 0.0847$, $b = 0.2306$, $\alpha = 9.5°$

[C] **17.** $a = 0.0847$, $b = 0.2306$, $c = 0.1913$

[C] **18.** $a = 6$, $b = 9$, $c = 10$

[C] **19.** $a = 24.638$, $b = 12.259$, $c = 30.025$

[C] **20.** $a = 97.125$, $b = 118.625$, $c = 158.25$

21. $K = 50$, $a = 40$, $\beta = 30°$. Find b and c.

22. $K = 8$, $b = 1$, $\alpha = 45°$. Find a and c.

23. $K = 10$, $\alpha = 40°$, $\beta = 60°$, $\gamma = 80°$. Find a, b, and c.

24. $K = 12$, $\alpha = 30°$, $\beta = 80°$, $\gamma = 70°$. Find a, b, and c.

25. A field is in the shape of a triangle with sides of lengths 160 yards, 230 yards, and 150 yards. If one acre is 4840 square yards, find the area of the field in acres.

26. A triangular surface is to be painted. Two of the sides meet in an angle of 73°30′ and these sides have lengths 190 feet and 90 feet. What is the area of the surface? If one gallon of paint covers 120 square feet, how much paint is needed?

27. A triangular field is to be fertilized. One bag of fertilizer covers 1000 square yards. If one side has length 124 yards and the other two sides meet the given side in angles of 54° and 49°, how much fertilizer is needed?

28. A flat piece of metal is in the shape of a parallelogram. If two of the sides have lengths 12 centimeters and 8 centimeters and the shorter diagonal has length 10 centimeters, find the area.

29. A plane is down somewhere between islands A, B, and C. If B is on a course of 26° from A, C is on a course of 191° from B, A is on a course of 348° from C, and the distance from A to B is 9 miles, how many square miles must be searched for survivors?

30. Find the length of the shortest side of a triangle whose area is 400 square inches and whose angles are 18°, 64°, 98°.

31. A triangular floor with sides of lengths 6 yards, 4 yards, and 5 yards is going to be carpeted by carpet which costs \$12.95 per square yard. How much will the carpet cost?

32. A triangular field has area 2800 square yards. A side of length 110 yards meets a second side in an angle of 45°. How long is the second side? How long is the third side?

33. A triangular piece of wood has area 3750 square centimeters. The angles are 16°, 52°, and 112°, respectively. Find the length of the longest side.

34. A triangular surface has been painted using 4.3 cans of paint. Each can of paint covers 13.2 square meters. Two of the sides of the triangle have lengths 9.6 and 15.4 meters. What is the length of the third side? (*Hint:* First use (30) to find the angle opposite that side, then use the Law of Cosines.)

In Exercises 35–38, we derive Heron's formula.

35. Use the Law of Cosines (see Eq. 35) to show

$$1 + \cos\beta = \frac{(a + c)^2 - b^2}{2ac} = \frac{(a + c - b)(a + c + b)}{2ac}$$

36. Similarly to Exercise 35, show

$$1 - \cos \beta = \frac{b^2 - (a - c)^2}{2ac} = \frac{(b - a + c)(b + a - c)}{2ac}$$

37. Use Exercises 35 and 36 to show

$$\sin^2 \beta = \frac{(a + b - c)(a + c - b)(b + c - a)(a + b + c)}{4a^2c^2}$$

$$= \frac{(a + b + c - 2c)(a + b + c - 2b)(a + b + c - 2a)(a + b + c)}{4a^2c^2}$$

$$= \frac{4}{a^2c^2}(s - c)(s - b)(s - a)s$$

38. Use Exercise 37 to show

$$K^2 = (s - c)(s - b)(s - a)s \quad \text{or} \quad K = \sqrt{s(s - a)(s - b)(s - c)}$$

3.10 Vectors

Many familiar concepts have numbers associated with them, such as distance, area, volume, etc. However, other concepts not only have numbers associated with them but direction as well. For instance, the velocity of an object is its speed together with the direction it is heading (e.g., the velocity of a car may be 50 miles per hour [mph] due north). Other concepts that involve both a magnitude and a direction include acceleration and force.

> **(36) Definition** A quantity that has both a magnitude and a direction is called a **vector.**

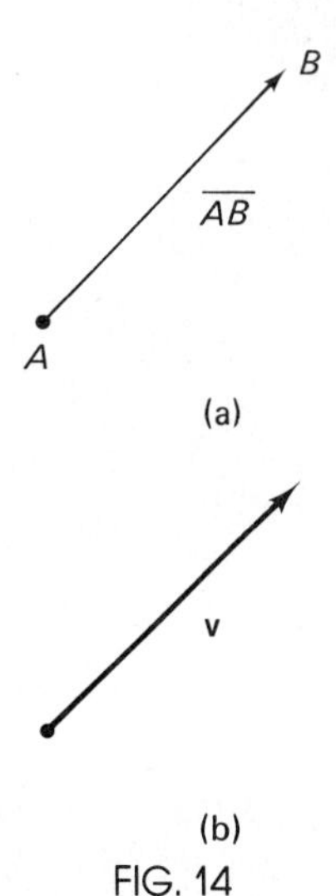

FIG. 14

Note: It is sometimes convenient to use ordered pairs in order to denote vectors, where the first entry represents magnitude and the second entry represents direction. For example, the velocity of the airplane is (250 mph, due west).

When working with vectors, it often helps to represent them geometrically. Now when we want to represent a number geometrically, we usually draw a line segment. Similarly, we usually represent a vector by drawing an *arrow* or *directed line segment* whose length has the given magnitude *and* which is pointed in the given direction. If the arrow goes from a point A (called the **initial point**) to the point B (called the **terminal point**), we denote the vector by $\overline{AB}$. See Fig. 14(a). If the initial and terminal points are not specified, we may denote vectors by small Roman boldface letters, such as **v**. See Fig. 14(b). The length of the arrow is the **length** or **magnitude** of the vector and is denoted by putting vertical lines on either side of the vector (like absolute values). Thus the lengths of the vectors in Fig. 14 are $|\overline{AB}|$ and $|\mathbf{v}|$.

There is one vector which is special, namely, the vector of zero length.

This is called the **zero vector.** It is denoted by **0**, and you may assume it has any direction that you wish.

We say that two vectors are **equal,** or that two arrows represent the same vector, if they have the same magnitude *and* the same direction. For example, in Fig. 14, $\overline{AB} = \mathbf{v}$. A consequence of this definition of equality is that when you want to represent a vector geometrically, you may put the initial point anywhere that is convenient (and then draw an arrow of the appropriate length and direction).

Now consider what happens when two forces act on the same thing. For example, suppose you are rowing a boat toward the north at 4 mph across a river which is flowing toward the east at 3 mph. See Fig. 15. Then the boat actually moves as though it were being propelled at a speed and in a direction which is some combination of the speed and direction of the water and the speed and direction of the rowing. This combination is called the **vector sum** of the two vectors.

If two vectors **v** and **w** are not in the same or opposite directions, the rule for their sum is as follows: We first draw the arrows corresponding to **v** and **w**

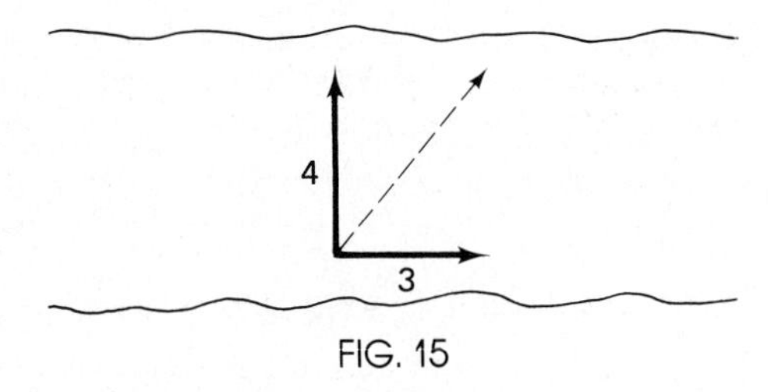

FIG. 15

with the same initial point O. See Fig. 16(a). Thus $\mathbf{v} = \overline{OA}$ and $\mathbf{w} = \overline{OB}$. Then we complete this to a parallelogram $OACB$ and draw in the diagonal $\overline{OC}$. See Fig. 16(b).

> **(37) Definition** The vector sum of $\overline{OA}$ and $\overline{OB}$ is the vector $\overline{OC}$. See Fig. 16.

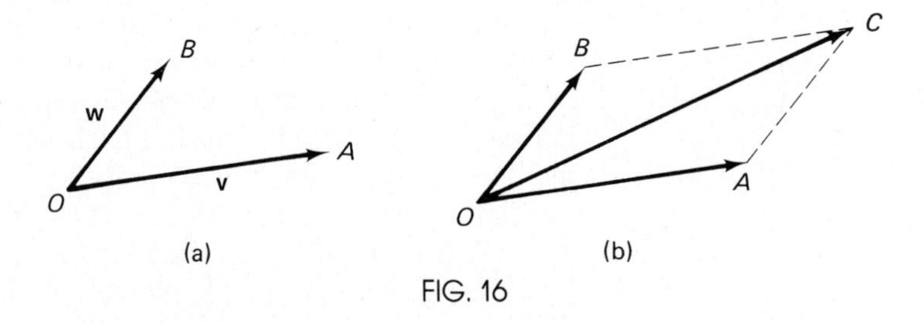

FIG. 16

Note that since opposite sides of a parallelogram are equal and parallel, the vectors $\overline{OA}$ and $\overline{BC}$ are equal, as are $\overline{OB}$ and $\overline{AC}$. Thus we could have defined the vector sum of **v** and **w** by saying the following: First draw any vector $\overline{OA}$ equal to **v**. Then using A as the initial point, draw the vector $\overline{AC} = \mathbf{w}$. Then $\overline{OC}$ is the sum of **v** and **w**. This is what we use in the special cases when **v** and **w** are in the same or opposite directions.

> If two vectors have the same direction, then their sum has that same direction, and the magnitude of the sum is the sum of the two magnitudes.
>
> If two vectors have opposite directions, then either
>
> **1.** Their lengths are equal, and their sum is the zero vector, or
> **2.** Their lengths are unequal, and their sum has the direction of the vector with larger magnitude and a magnitude which is the larger minus the smaller of the magnitudes of the given vectors.

From this it follows that the zero vector **0** acts as an additive identity:

$$\mathbf{0} + \mathbf{v} = \mathbf{v} + \mathbf{0} = \mathbf{v}$$

For, we can think of **0** as having zero length and the direction of **v** (recall that we may think of **0** as having any direction that is convenient).

The general case of adding two vectors was defined using a parallelogram, and we shall need the following properties:

> **(38)** In a parallelogram,
>
> **i.** Opposite sides are equal.
> **ii.** Opposite angles are equal.
> **iii.** The sum of two adjacent angles is 180°

Suppose we want to compute the vector sum of two arbitrary vectors. First, we shall assume that the vectors are described by giving their lengths and their directions on a navigator's compass. In this context, Example 1 illustrates how to find the sum using the Laws of Sines and Cosines.

EXAMPLE 1 Suppose a small plane is flying on a heading of 253° (the heading is the direction the plane is pointed), and suppose there is a wind of 25 mph in the direction of 150°. If the airspeed of the plane is 115 mph (the airspeed is the speed relative to the air), find the ground speed and actual course of the plane. Assume the data are accurate to three significant figures.

Solution First we draw a diagram indicating the given vectors, and to keep the diagram uncluttered, we also draw a separate parallelogram. See Fig. 17. There are two things to find: the length $|\overline{OC}|$ and the angle $\angle NOC$. We find the length $|\overline{OC}|$ first. Referring to Fig. 17(b), since $OBCA$ is a parallelogram, $|\overline{BC}| = |\overline{OA}| = 25$. If we knew angle β, we could easily use the Law of Cosines to compute $|\overline{OC}|$. We do know that $150° + \alpha = 253°$ [see Fig. 17(a)] so that $\alpha =$

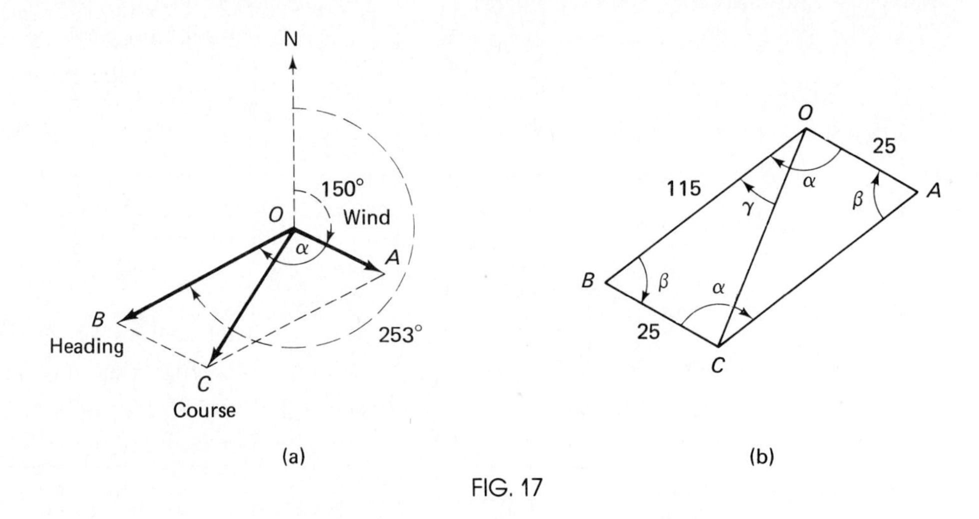

FIG. 17

$253° - 150° = 103°$. Now by (38iii), $\alpha + \beta = 180°$, so $\beta = 180° - 103° = 77°$. By the Law of Cosines,

$$|\overline{OC}|^2 = 115^2 + 25^2 - 2(115)(25)\cos 77° \approx 12{,}556.531$$

Thus, $|\overline{OC}| \approx 112.056$ is the ground speed (in mph) of the plane.

If we knew angle γ, then we could easily find the heading since $\gamma + \angle NOC = 253°$. By the Law of Sines,

$$\frac{\sin\gamma}{|\overline{BC}|} = \frac{\sin\beta}{|\overline{OC}|}$$

so
$$\sin\gamma = \frac{|\overline{BC}|\sin\beta}{|\overline{OC}|} \approx \frac{25\sin 77°}{112.0559} \approx 0.2173848$$

Hence
$$\gamma \approx \sin^{-1}(0.2173848) \approx 12.5555°$$

and the actual course is

$$253° - 12.5555° \approx 240.445° \approx 240° \quad \blacksquare$$

Some problems require a little *vector algebra,* so we should point out that vector addition can be seen to satisfy the same properties of addition as the real numbers. These properties will be useful in Example 2.

EXAMPLE 2 Port B is 200 miles from port A on a course of 60°. The current is due east (i.e., in a direction of 90°) at 6 mph. Determine the heading and water speed of a ship in order for that ship to sail directly from A to B in 8 hours. Assume the data are accurate to two significant figures.

Solution In this problem, we are given one vector $\mathbf{v}$ (6 mph, 90°), and we want to find a second vector $\mathbf{x}$, so that the vector sum is $\mathbf{w}$ ($\frac{200}{8} = 25$ mph, 60°). Thus we want to solve

$$\mathbf{v} + \mathbf{x} = \mathbf{w}$$

for **x**. See Fig. 18(a). Since vector addition behaves like addition of numbers, we can solve this for **x**, obtaining

$$\mathbf{x} = \mathbf{w} + (-\mathbf{v})$$

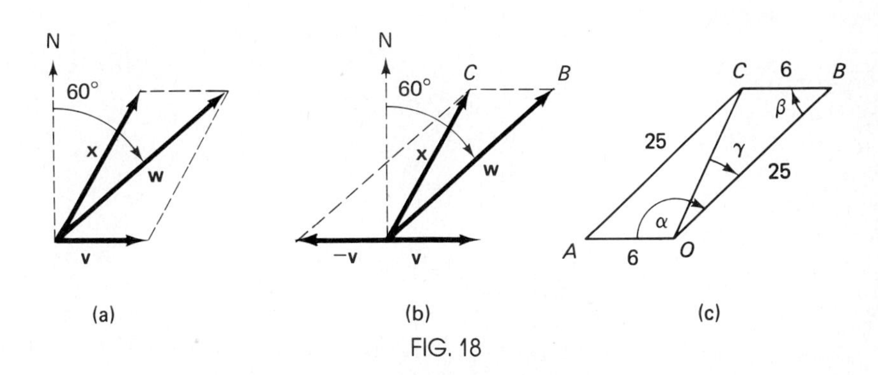

FIG. 18

where $-\mathbf{v}$ is (6 mph, 270°). See Fig. 18(b). Writing this as an addition problem allows us to follow the pattern of Example 1. We first see that $\alpha = 150°$ $(= 90° + 60°)$. Since $\alpha + \beta = 180°$, $\beta = 180° - 150° = 30°$. By the Law of Cosines,

$$|\overline{OC}|^2 = 25^2 + 6^2 - 2(25)(6)\cos 30° \approx 401.1924$$

so that

$$|\overline{OC}| \approx 20.02979 \approx 20$$

is the water speed in mph.

To find γ, by the Law of Sines,

$$\sin\gamma = \frac{|\overline{CB}|\sin\beta}{|\overline{OC}|} \approx \frac{6\sin 30°}{20.02979} \approx 0.1497769$$

so that

$$\gamma \approx \sin^{-1}(0.1497768) \approx 8.614°$$

By Fig. 18(b), $\angle NOC + \gamma = 60°$, so that the heading is $\angle NOC \approx 60° - 8.614° \approx 51.386° \approx 51°$. ■

If **u, v,** and **w** are vectors and **w = u + v,** then **u** and **v** are called **components** of **w** (in the sense that they "make up" **w**). Thus each vector **w** has many pairs of components. There are many problems in which you are given one vector **w**, and you are asked to find one or both of its components with certain properties. This process is called **resolving** the vector.

EXAMPLE 3 A 50-pound block of ice is on a ramp inclined at 20°. What force pushing up the ramp is necessary to keep the block from sliding? What is the actual force caused by the weight of the ice against the ramp? (We assume friction is negligible.) Data are correct to two significant figures.

Solution The weight of the block is a vector $\overline{OC}$ which points vertically downward. The vector $\overline{OC}$ can be resolved into two vectors: $\overline{OB}$, which is perpendicular to the ramp and represents the actual force of the block against the ramp, and $\overline{OA}$, which is parallel to the ramp pointing downwards and represents the force of the block down the ramp. Thus we want to compute $|\overline{OA}|$ and $|\overline{OB}|$. See Fig. 19. Note that $\overline{OA}$ and $\overline{OB}$ are perpendicular. Since ODC is a right triangle, $\beta = 90° - 20° = 70°$. Since OAC is a right triangle,

$$|\overline{OA}| = |\overline{OC}| \cos \beta = 50 \cos 70° \approx 17.101 \text{ lb} \approx 17 \text{ lb}$$

Since $OACB$ is a rectangle,

$$|\overline{OB}| = |\overline{AC}| = |\overline{OC}| \sin \beta = 50 \sin 70° \approx 46.9846 \text{ lb} \approx 47 \text{ lb}$$

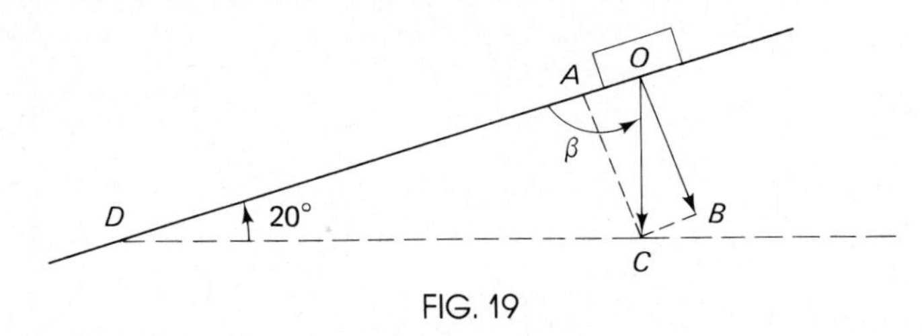

FIG. 19

■

EXERCISES

In Exercises 1–8, find the sum.

	FIRST VECTOR		SECOND VECTOR	
	Magnitude	**Direction**	**Magnitude**	**Direction**
1.	50	90°	25	60°
2.	1046	45°	837	135°
[c] **3.**	6.28	215°	4.65	334°
[c] **4.**	96.256	196.3°	12.31	83.05°
[c] **5.**	11.14	33.9°	25.91	351°
[c] **6.**	31.47	171°	42.81	259.3°
7.	819	106.51°	819	286.51°
8.	64.8	303°	100.2	303°

9. A man can row his boat at 3 mph. He is in a stream flowing from north to south at 4 mph. How fast and in what direction will he travel if he rows
a. Directly downstream?
b. Directly upstream?
c. Due east (across the stream)?
d. Northeast (45° on a navigator's compass)?
e. Southeast (135°)?

10. A duck is flying directly north at 25 km/hr (kilometers per hour). How fast and in what direction will the bird actually travel if the wind is blowing at 30 km/hr
a. Directly north?
b. Directly south?

c. From the west?
d. Toward the southeast (135°)?
e. Toward the northwest (315°)?

11. Assume that a portion of the Mississippi River is 1 mile wide and that the current is 6 mph. Points A and B are directly opposite each other on the banks. If a man starts from A rowing at 4 mph and keeps his boat parallel to $\overline{AB}$,
a. What is his actual velocity?
b. How far downriver from B will he land?
c. How long will it take him to land?
d. How long would it take him if there were no current?

12. An airplane flies over the northwest corner of Colorado heading due south at 200 mph with a crosswind of 25 mph blowing directly east. If it is 275 miles between the northern and southern borders of Colorado,
a. What is the actual velocity of the plane?
b. How far east of the southwestern corner of Colorado will the plane cross the border?
c. How long will it take the plane to cross the southern border of Colorado?
d. How long would it take the plane to reach the southern border if there were no wind?

13. Referring to Exercise 11, suppose the man puts a motor on the boat which propels it at 9 mph.
a. In what direction should he head in order to actually land at B?
b. How long would it take him to arrive at B?
c. Would it be possible for him to arrive at B by rowing at 4 mph?

14. Referring to Exercise 12, suppose the pilot wanted to fly directly to the southwestern corner of Colorado.
a. In what direction should he head?
b. How long will it take him to reach the corner?

15. A plane is flying at an airspeed of 400 mph with a heading of 328°. There is a wind of 40 mph in the direction of 243°. Find the course and ground speed of the plane.

16. A ship is sailing at 21 km/hr on a heading of 107° through a crosscurrent of 3 mph in the direction of 21°. Find the actual speed and direction of the ship.

17. There is a 50-km/hr wind in the direction of 36°. A pilot wishes to fly on a course of 121° with a ground speed of 400 km/hr. Determine her heading and airspeed.

18. Two people want to take their motorboat to an island 20 miles away on a heading of 349°. If there is a 2 mph crosscurrent in the direction of 40°, in what direction and at what speed should they head if they want to take 2 hours for the trip?

19. A patrol boat leaves its base on a heading of 17° traveling at 17 km/hr. After 2 hours it turns to a new heading of 310° and slows to 12 km/hr. After 45 minutes on this new heading, how far and in what direction is the boat from its base?

20. An airplane flies due north at 200 mph. After 3 hours, it turns and flies southeast (135°) at 210 mph. After an hour on this new heading, how far is it from its starting point?

21. A 38-kilogram weight is on a ramp inclined at 27°. If there is no friction, what force, parallel to the ramp, is necessary to keep the weight at rest?

22. A 2000-pound car is on a hill inclined at 15°. How much force, pushing up the hill, is necessary to keep the car from rolling?

23. A 50-pound weight on an incline is being held at rest by a force of 21 pounds pushing up the incline. What is the angle of inclination?

24. A force of 48 pounds is required to keep a weight of 70 pounds from sliding down an inclined plane. What is the angle of inclination?

25. A manual lawnmower requires a force of 10 pounds parallel to the ground in order to move it through the grass. If the handle is inclined at 41° to the ground, how much force is necessary along the handle? See Fig. 20(a).

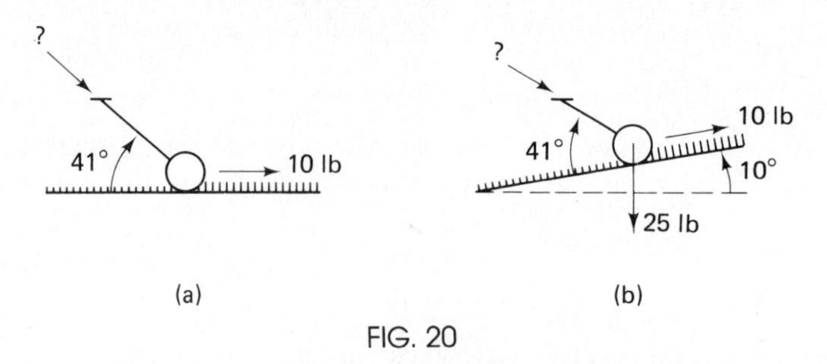

FIG. 20

26. If the lawnmower of Exercise 25 weighs 25 pounds and is pushed up a hill of 10°, how much force exerted along the handle is necessary to keep it moving through the grass? The handle is still inclined 41° to the slope, and you must take into account both the 10 pounds from Exercise 25 and the weight of the lawnmower. See Fig. 20(b).

3.11 Polar Coordinates

You are all familiar with the way we coordinatize the plane with a Cartesian (rectangular) coordinate system. We draw a horizontal number line (the x-axis) and a vertical line (the y-axis) as shown in Fig. 21. Associated with each point P, there is a unique ordered pair of numbers (x, y), called the coordinates of P. Conversely, with each pair (x, y) we associate the unique point P which has (x, y) as coordinates.

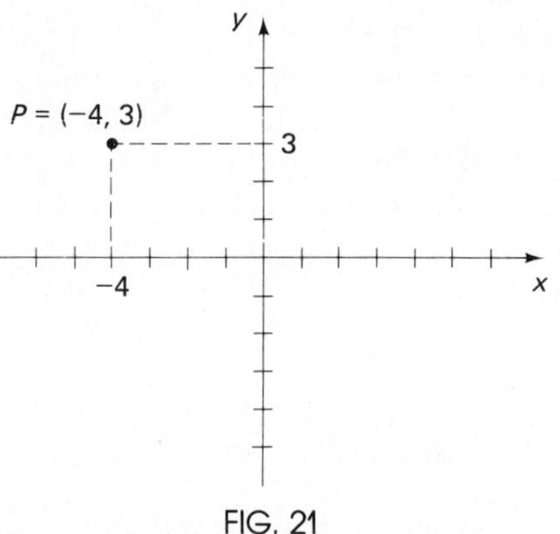

FIG. 21

In this section we introduce another system of coordinates, the system of **polar coordinates.** We start with a point O, called the **origin,** or **pole,** and a horizontal half line with endpoint O, called the **polar axis.** We think of the polar axis as being the nonnegative half of a number line. This gives us a scale so that we can measure distance in the plane.

(39) Definition We say that a point P has polar coordinates (r, θ), $r \geq 0$, if θ has the polar axis as initial side and

i. P is on the terminal side of θ.

ii. $|OP| = r$ (i.e., the distance from the origin to P is r).

See Fig. 22(a).

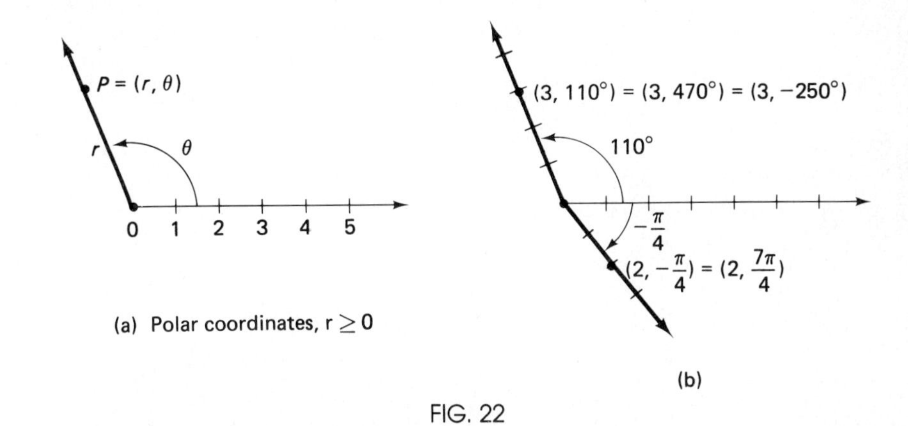

(a) Polar coordinates, r ≥ 0

(b)

FIG. 22

The angle θ is considered to be an angle in standard position; it is positive for counterclockwise rotations and negative for clockwise rotations. Since we know that any angles which differ by complete rotations have the same terminal side, (r, θ) and $(r, \theta + 360°n)$ are different polar coordinates for the same point. [Of course θ could be measured in radians, in which case we write (r, θ) and $(r, \theta + 2n\pi)$.] Some examples of points in polar coordinates and a few of the many different ways these coordinates may be written are illustrated in Fig. 22(b). The origin O has coordinates $(0, \theta)$ for any angle θ.

It is convenient for applications to allow r to be negative.

> **(40) Definition** We say that a point P has polar coordinates (r, θ), $r < 0$, if we draw the terminal side of θ and then extend that line $|r|$ units on the other side of the origin to P. See Fig. 23(a, b).

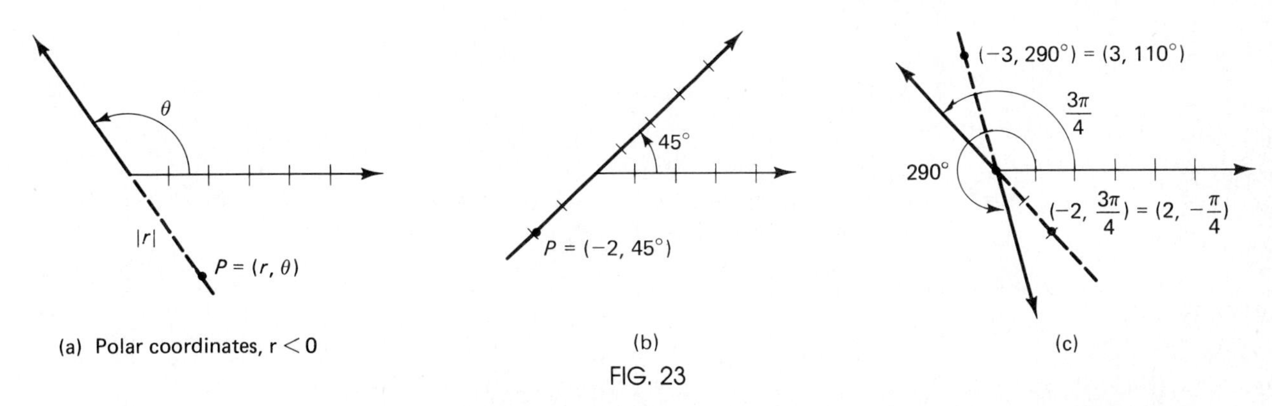

(a) Polar coordinates, r < 0

(b)

(c)

FIG. 23

In Fig. 23(c), we illustrate how polar coordinates of the points in Fig. 22(b) can be written with r negative. Notice that if P has polar coordinates (r, θ), $r < 0$, then $(|r|, \theta + 180°)$ are also polar coordinates for P.

There are many times when it is very useful to be able to relate polar coordinates and Cartesian coordinates. To do this, we let the origins coincide and let the polar axis be the positive x-axis. See Fig. 24.

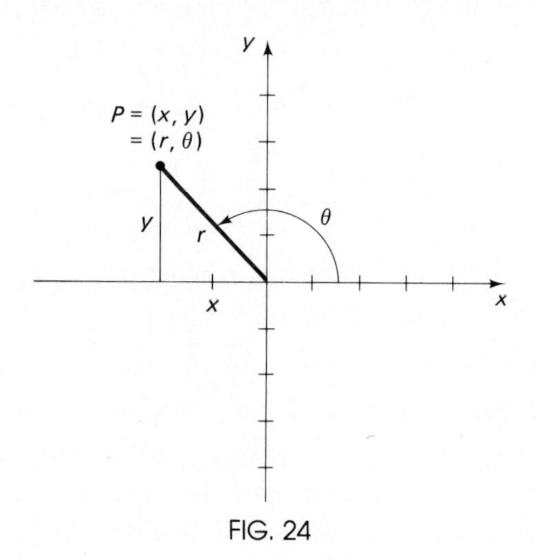

FIG. 24

Given the Cartesian coordinates (x, y) of a point P, we can obtain polar coordinates (r, θ) by setting $r = \sqrt{x^2 + y^2}$ and finding θ from any of the following equations (which you will recognize as the defining equations of the three trigonometric functions):

(41)

$$\left.\begin{aligned} \sin\theta &= \frac{y}{r} \\ \cos\theta &= \frac{x}{r} \end{aligned}\right\} \quad \text{where } r = \sqrt{x^2 + y^2}$$

$$\tan\theta = \frac{y}{x}, \qquad x \neq 0$$

You just have to keep in mind what quadrant θ is in.

EXAMPLE 1 Find polar coordinates (r, θ) of the points with Cartesian coordinates (a) $(-2, 3)$ and (b) $(-2, -3)$.

Solution (a) Set $r = \sqrt{(-2)^2 + 3^2} = \sqrt{13}$. Since arccosine gives angles in Q_1 or Q_2 and $(-2, 3)$ is in Q_2, we can use $\cos\theta = \frac{x}{r} = \frac{-2}{\sqrt{13}}$ and solve $\theta = \cos^{-1}\left(\frac{-2}{\sqrt{13}}\right) \approx 123.69°$. Thus, $(r, \theta) \approx (\sqrt{13}, 123.69°)$.

(b) Set $r = \sqrt{(-2)^2 + (-3)^2} = \sqrt{13}$. Since no inverse trigonometric function gives an angle in Q_3, we use our knowledge of reference angles to find θ. If ϕ is the reference angle, $\tan \phi = \frac{y}{x} = \frac{-3}{-2} = 1.5$. Thus, $\phi = \tan^{-1}(1.5) \approx 56.3099°$. Since θ is in Q_3, $\theta = \phi + 180° \approx 236.310°$. ■

Given the polar coordinates (r, θ) of a point P, we can easily obtain the Cartesian coordinates (x, y). If $r \geq 0$, we just solve the first two equations in (41) for x and y, obtaining

$$\textbf{(42)} \qquad x = r\cos\theta \quad \text{and} \quad y = r\sin\theta \quad \text{if } r \geq 0$$

If $r < 0$, then $(|r|, \theta + \pi)$ are also polar coordinates for P, so we can use these in (42) to obtain

$$x = |r|\cos(\theta + \pi) \quad \text{and} \quad y = |r|\sin(\theta + \pi) \quad \text{if } r < 0$$

However, these can be simplified, since if $r < 0$,

$$\begin{aligned} x &= |r|\cos(\theta + \pi) = |r|(\cos\theta\cos\pi - \sin\theta\sin\pi) \\ &= -|r|\cos\theta = r\cos\theta \end{aligned}$$

and

$$\begin{aligned} y &= |r|\sin(\theta + \pi) = |r|(\sin\theta\cos\pi + \cos\theta\sin\pi) \\ &= -|r|\sin\theta = r\sin\theta \end{aligned}$$

In other words, (42) holds for all r.

$$\textbf{(43)} \qquad x = r\cos\theta \quad \text{and} \quad y = r\sin\theta$$

EXAMPLE 2 Find the Cartesian coordinates (x, y) of the point with polar coordinates $(-2, 4.8 \text{ rad})$.

Solution Observe that the angle is in radians:

$$x = -2\cos(4.8) \approx -0.174998$$

$$y = -2\sin(4.8) \approx 1.99233 \quad ■$$

We now consider the graphs of some equations in polar coordinates.

EXAMPLE 3 Sketch the graph of $\quad r = 5$.

Solution The set of all points of distance 5 from the origin (and no restriction on θ) is a circle of radius 5 centered at the origin. See Fig. 25. This circle also has

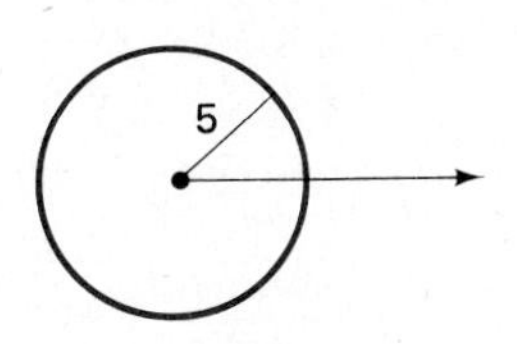

The circle $r = 5$

FIG. 25

equation $r = -5$, since $(-5, \theta)$ and $(5, \theta + \pi)$ are different polar coordinates for the same point. ■

EXAMPLE 4 Sketch the graph of $\theta = 130°$.

Solution The graph of $\theta = 130°$ is all points on the terminal side of θ $(r \geq 0)$ together with all points opposite the origin from these $(r < 0)$. Thus it is the straight line in Fig. 26.

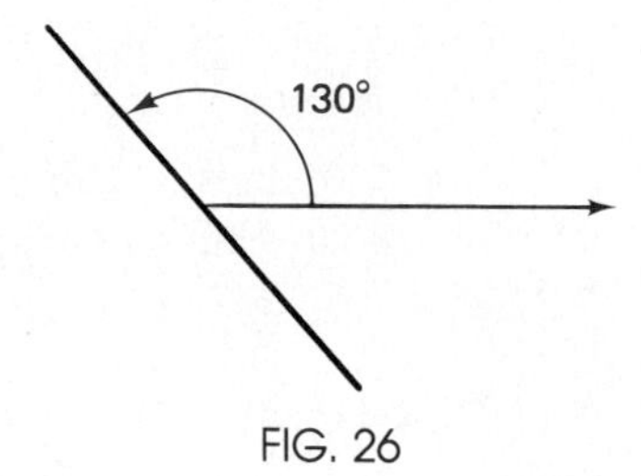

FIG. 26

■

EXAMPLE 5 Convert $r = \cos\theta + 2\sin\theta$ to Cartesian coordinates.

Solution The easiest way is to multiply both sides by r, obtaining $r^2 = r\cos\theta + 2r\sin\theta$. Since $r^2 = x^2 + y^2$ and $x = r\cos\theta$, $y = r\sin\theta$, this is

$$x^2 + y^2 = x + 2y \quad ■$$

EXAMPLE 6 Sketch the graph of $r = 2(1 - \cos\theta)$.

Solution We first observe that since $\cos(-\theta) = \cos\theta$, the point (r, θ) is on the graph precisely when the point $(r, -\theta)$ is on the graph. In other words, the graph is symmetric about the x-axis. Thus if we sketch the graph when $0 \leq \theta \leq \pi$ and reflect the result in the x-axis, we obtain the whole graph. We plot the points

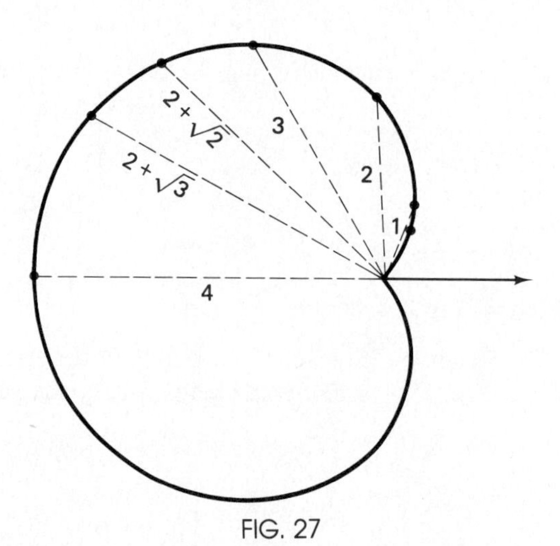

FIG. 27

from the following table, although we could use a calculator:

θ	0	$\frac{\pi}{6}$	$\frac{\pi}{4}$	$\frac{\pi}{3}$	$\frac{\pi}{2}$	$\frac{2\pi}{3}$	$\frac{3\pi}{4}$	$\frac{5\pi}{6}$	π
$r = 2(1 - \cos\theta)$	0	$2 - \sqrt{3}$	$2 - \sqrt{2}$	1	2	3	$2 + \sqrt{2}$	$2 + \sqrt{3}$	4

This curve is called a cardioid (which means heart shaped). ■

It is often useful to construct a different type of table than the table given in Example 6, one that enables us to see the general shape of the curve by indicating how r changes as θ changes over various intervals. The following is an illustration of such a table for the preceding problem.

θ	$\cos\theta$	$1 - \cos\theta$	$r = 2(1 - \cos\theta)$
$0 \to \frac{1}{2}\pi$	$1 \to 0$	$0 \to 1$	$0 \to 2$
$\frac{\pi}{2} \to \pi$	$0 \to -1$	$1 \to 2$	$2 \to 4$
$\pi \to \frac{3}{2}\pi$	$-1 \to 0$	$2 \to 1$	$4 \to 2$
$\frac{3}{2}\pi \to 2\pi$	$0 \to 1$	$1 \to 0$	$2 \to 0$

We interpret the first line of this table as follows:

As θ goes from 0 to $\frac{\pi}{2}$

$\cos\theta$ goes from 1 to 0

$1 - \cos\theta$ goes from 0 to 1

Hence

$$r = 2(1 - \cos\theta) \text{ goes from 0 to 2}$$

The remaining lines are interpreted similarly.

EXERCISES

In Exercises 1–10, plot the points with the given polar coordinates.

1. $(2, 40°)$

2. $(-3, 120°)$

3. $(-1, 18°)$

4. $(1, 198°)$

5. $(2.3, -400°)$

6. $\left(3, \frac{\pi}{2} \text{ rad}\right)$

7. $\left(-1, \frac{7\pi}{6} \text{ rad}\right)$

8. $(-3.5, 12\pi \text{ rad})$

9. $(5, 5 \text{ rad})$

10. $(2, 15 \text{ rad})$

In Exercises 11–22, find a set of polar coordinates for the points with the given Cartesian coordinates.

11. $(-3, 0)$
12. $(0, -4)$
13. $(2, 2\sqrt{3})$
14. $(-5\sqrt{3}, 5)$
15. $(-4, 4)$
16. $(-7, -24)$
17. $(5, -12)$
18. $(3\sqrt{3}, 3)$
19. $(-6, 8)$
20. $(-3, -4)$
21. $(1, -3)$
22. $(-1, 3)$

In Exercises 23–30, find Cartesian coordinates for the points with the given polar coordinates.

23. $(2, 120°)$
24. $(-3, 45°)$
25. $(-6, -12°)$
26. $(4.6, 300°)$
27. $(0, 1.63 \text{ rad})$
28. $\left(12.3, \frac{5\pi}{6} \text{ rad}\right)$
29. $(-2, 2 \text{ rad})$
30. $(-3, 27.415 \text{ rad})$

In Exercises 31–36, convert the equations to polar coordinates.

31. $x = 7$
32. $x^2 + y^2 = 9$
33. $y = 2x$
34. $(x + 2)^2 + y^2 = 4$
35. $\frac{x^2}{4} + y^2 = 1$
36. $x^2 - y^2 = 1$

In Exercises 37–44, convert the equations to Cartesian coordinates.

37. $r = 3$
38. $r = -4$
39. $\theta = \frac{3\pi}{4}$
40. $r \cos\theta = -4$
41. $r \sin\theta = 7$
42. $r = \cos\theta$
43. $r^2 \cos\theta = 3 \sin\theta - \cos\theta$
44. $r = 3 \sin\theta + 1$

In Exercises 45–56, sketch the graph of the given equation.

45. $r = 2$
46. $r = -3$
47. $\theta = 100°$
48. $\theta = \frac{\pi}{4}$
49. $r \cos\theta = 3$
50. $r \sin\theta = 2$
51. $r = 1 - \cos\theta$
52. $r = 1 + \sin\theta$
53. $r = 2 \sin\theta - 1$
54. $r = 1 - 2\cos\theta$
55. $r = \sin 2\theta$
56. $r = \sin 3\theta$

REVIEW EXERCISES

In Exercises 1–10, verify the identities.

1. $\cos\theta + \tan\theta\sin\theta = \sec\theta$

2. $\dfrac{\sin x - \cos x}{\csc x - \sec x} = -\dfrac{1}{\csc x \sec x}$

3. $\sec^4\theta - \tan^4\theta = \dfrac{1 + \sin^2\theta}{\cos^2\theta}$

4. $\dfrac{\tan u + \cot u}{\tan u - \cot u} = \dfrac{-1}{\cos 2u}$

5. $\dfrac{1}{1 - \sin x} - \dfrac{1}{1 + \sin x} = 2\tan x \sec x$

6. $\sin\left(\theta - \dfrac{\pi}{2}\right) = -\cos\theta$

7. $\tan\left(x + \dfrac{5\pi}{4}\right) = \dfrac{1 + \tan x}{1 - \tan x}$

8. $\cos\left(\dfrac{2\pi}{3} - x\right) = \dfrac{1}{2}(\sqrt{3}\sin x - \cos x)$

9. $\sin 4x = 4\sin x\cos^3 x - 4\cos x\sin^3 x$

10. $\dfrac{\sin^2 2x}{1 + \cos 2x} = 2\sin^2 x$

In Exercises 11–16, find all solutions.

11. $2\sin x + 1 = 0$

12. $2\cos x\csc x = \cos x$

13. $3\sin x\cos x = \sin x$

14. $\tan^2 x + 5\tan x + 6 = 0$

15. $\cos 2x - \cos x = 0$

16. $\sin 2x - \tan x = 0$

17. If $\sin u = -\frac{5}{13}$ and $\cos v = \frac{4}{5}$, with u in Q_3 and v in Q_4, find $\sin(u + v)$, $\cos(u + v)$, and $\tan(u + v)$.

18. Use an addition formula to find the exact value of $\sin\dfrac{7\pi}{12}$.

In Exercises 19–23, solve the triangle ABC having the given parts.

19. $\alpha = 24°$, $\beta = 85°$, $c = 108$

20. $\gamma = \dfrac{\pi}{5}$, $a = 27$, $c = 15$

21. $\gamma = \dfrac{\pi}{5}$, $a = 27$, $c = 18$

22. $a = 4065$, $b = 4849$, $c = 6507$

23. $\alpha = 0.98$, $b = 0.926$, $c = 0.305$

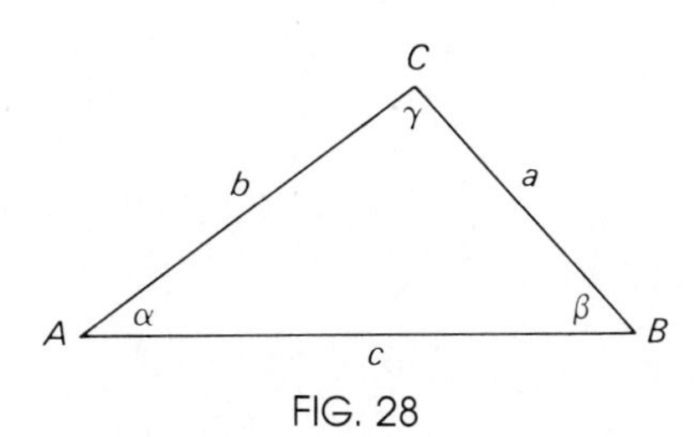

FIG. 28

24. From one observer, the angle of elevation of a balloon is 38°. From an observer 1431 meters away on the opposite side, the angle of elevation of the balloon is 55°. How far is the balloon from each observer?

25. A plane leaves an airport and flies 83.65 miles on one course and then turns and flies 98.15 miles on a course of 26.2°. At that point, the plane then turns to a course of 167.9° in order to head back to the airport. How far is it from the airport?

26. A triangular wooden object is being built. Two sides of lengths 157.3 and 126.9 centimeters are to meet in an angle of 84°. How long should the third side be?
27. A golf ball is 186 yards from the hole. The ball is then hit (straight) 172.4 yards and ends up 31.3 yards from the hole. What is the angle between the line of flight of the ball and a direct line to the hole?

In Exercises 28 and 29, find the sum $\mathbf{u} + \mathbf{v}$.

28. **u** has magnitude 60, direction 150°; **v** has magnitude 40, direction 225°.
29. **u** has magnitude 117.51, direction 321.2°; **v** has magnitude 139.08, direction 84.25°.
30. A river is 0.5 kilometers wide and the current is 8 km/hr. Points A and B are directly opposite each other on the banks. If a boat propelled at 6 km/hr leaves A and is always parallel to $\overline{AB}$,
    a. What is its actual velocity?
    b. How far downriver from B will it land?
    c. How far will it travel?
    d. How long will it take to land?
31. A small plane is flying at an airspeed of 150 mph with a heading of 247°. There is a wind of 35 mph in the direction of 132°. Find the course and ground speed of the plane.
32. There is a 50-mph wind in the direction of 346°. A pilot wishes to fly on a course of 321° with a ground speed of 250 mph. Determine the heading and airspeed.

CHAPTER 4

Complex Numbers

Complex numbers first arose in trying to understand the Cardan-Tartaglia formula for solving cubics. For example, Cardan (1501–1576) knew (for other reasons) that there was only one positive solution to the equation $x^3 = 15x + 4$ and that 4 worked by direct substitution. However, the formulas, which worked well for many equations, gave $x = \sqrt[3]{2 + \sqrt{-121}} + \sqrt[3]{2 - \sqrt{-121}}$ as that only positive solution. Historically, the ancient Greeks had denied the existence of negative numbers. By Cardan's time, negative numbers were on shaky ground, though made plausible by the number line. Cardan used negative numbers but called them "numeri ficti." Roots of negative numbers were "unimaginable." Several years later, an Italian algebraist Rafael Bombelli (1526–1573) had a "wild idea" and wrote the answer as $\sqrt[3]{2 + 11\sqrt{-1}} + \sqrt[3]{2 - 11\sqrt{-1}}$, and then he was able to show that this is $2 + \sqrt{-1} + 2 - \sqrt{-1} = 4$. However, $\sqrt{-1}$ was still regarded as imaginary for the next two centuries. In 1797, Caspar Wessel (1745–1818) invented the complex plane. This went virtually unnoticed until 30 years later when it was used extensively by Carl Friedrich Gauss (1777–1855). This geometric interpretation brought about the acceptance of roots of negative numbers, but the use of the word "imaginary" persists.

4.1 The Complex Numbers

In most practical problems we encounter, we are interested in those solutions that are real numbers. However, there are situations where it is useful to have solutions even though they are not real numbers. For these situations, we define a number system larger than the reals which contains these solutions.

For example, consider the equation $x^2 = -1$. There is no real number whose square is -1, so this equation has no real solutions. Let us invent a solution and call it i, so that $i^2 = -1$. Thus, i is a solution to the equation $x^2 = -1$, and we would like i to be in some larger number system. This new number system must contain the reals $\mathbb{R}$ (in the same sense that the reals contain the rationals, and the rationals contain the integers) and the number i. Being a number system (containing the reals), it should possess the operations of addition and multiplication which must obey all the fundamental laws, such as the associative, commutative, and distributive laws. For example, since $i^2 = -1$,

$$(-i)^2 = (-i)(-i) = (-1)i(-1)i = (-1)(-1)ii = 1 \cdot i^2 = -1$$

so $-i$ is also a solution to $x^2 = -1$ in the new number system.

This new number system is called the system of **complex numbers,** and the set of all complex numbers is denoted by $\mathbb{C}$. Let us see what complex numbers look like. Suppose b is in $\mathbb{R}$; then it is also in $\mathbb{C}$ along with i so that $b \cdot i = bi$ is in $\mathbb{C}$. If a is another real number, a is also in $\mathbb{C}$ so $a + bi$ is in $\mathbb{C}$. Thus, $\mathbb{C}$ must contain all numbers of the form $a + bi$, a and b real. In fact, $\mathbb{C}$ is the set of all numbers of this form.

Definition The set of **complex numbers** is $\mathbb{C} = \{a + bi \mid a \text{ and } b \text{ are real numbers}\}$. The form $a + bi$ is called the **standard form** of a complex number.

Note that the set of all complex numbers $a + bi$ with $b = 0$ is the set of real numbers. Because of this, in a complex number $z = a + bi$, a is called the **real part** of z. In $z = a + bi$, b is called the **imaginary part** of z. (Note that b, *not* bi, is the imaginary part of $a + bi$.) If $a = 0$, z is called **pure imaginary.**

Before we can define operations like addition and multiplication of complex numbers, we need to understand what it means for two complex numbers to be equal.

> **Definition of Equality of Complex Numbers** If a, b, c, d are real,
> $$a + bi = c + di \quad \text{if and only if} \quad a = c \quad \text{and} \quad b = d.$$

This definition can be used to solve certain types of equations.

EXAMPLE 1 Solve for x and y real: $\quad 3x + 4i = -6 + 3yi.$

Solution We recognize that this is an equation of the form

$$a + bi = c + di$$

where $a = 3x$, $b = 4$, $c = -6$, $d = 3y$. Since x and y are *real* numbers, a, b, c, and d are all real numbers. Using the definition of equality of complex numbers, we have

$$a = c \quad \text{and} \quad b = d$$

or

$$3x = -6 \quad \text{and} \quad 4 = 3y$$

Solving, we obtain $x = -2$ and $y = \frac{4}{3}$. ■

We now turn to the problem of defining the sum and product of any two complex numbers. If two complex numbers are in standard form, we would like their sum and product also to be in standard form. We start with $z = a + bi$ and $w = c + di$, a, b, c, and d real, and try to add and multiply them as though they were real numbers but keeping in mind that $i^2 = -1$:

$$z + w = (a + bi) + (c + di) = a + c + bi + di = (a + c) + (b + d)i$$

$$\begin{aligned} z \cdot w = (a + bi) \cdot (c + di) &= ac + adi + bci + bdi^2 \\ &= ac + (ad + bc)i + bd(-1) \\ &= (ac - bd) + (ad + bc)i \end{aligned}$$

Thus we see that addition and multiplication can be defined in $\mathbb{C}$ as follows:

> **Definition** If $z = a + bi$ and $w = c + di$, a, b, c, and d real, then
> $$z + w = (a + c) + (b + d)i$$
> $$z \cdot w = (ac - bd) + (ad + bc)i$$

However, this definition need not be memorized to do computations. In practice, to compute $z + w$ or $z \cdot w$, just proceed in the "natural" way, replacing i^2 by -1 when it occurs.

EXAMPLE 2 Find (a) $(5 + 3i) + (-7 + 6i)$; (b) $(5 + 3i)(-7 + 6i)$

Solution (a) $(5 + 3i) + (-7 + 6i) = 5 - 7 + 3i + 6i = -2 + 9i$

(b) $(5 + 3i)(-7 + 6i) = -35 + 30i - 21i + 18i^2 = -35 + 9i - 18$

$= -53 + 9i$ ■

We can also subtract in the "natural" way:

$$(5 + 3i) - (-7 + 6i) = 5 + 3i + 7 - 6i = 12 - 3i$$

One type of multiplication is special:

$$(a + bi)(a - bi) = a^2 - (bi)^2 = a^2 - b^2i^2 = a^2 + b^2$$

The result $a^2 + b^2$ is zero if $a = b = 0$, but otherwise it is a *positive real* number. Thus the complex numbers $a + bi$ and $a - bi$ have a special relationship.

> **Definition** If $z = a + bi$, we define the **complex conjugate** $\bar{z}$ of z to be the complex number $\bar{z} = a - bi$.

EXAMPLE 3 (a) $\overline{5 + 3i} = 5 - 3i$; (b) $\overline{5 - 3i} = 5 + 3i$;

(c) $\overline{-7 + 6i} = -7 - 6i$; (d) $\bar{2} = \overline{2 + 0i} = 2 - 0i = 2$ ■

Example 3 leads us to the observation that for $z = a + bi$, we have $z = \bar{z}$ (that is, $a + bi = a - bi$) if and only if $a = a$ and $b = -b$; i.e., $b = 0$. Thus

> $z = \bar{z}$ if and only if z is a real number.

From the preceding, we see that if $z = a + bi$, then

> $$z\bar{z} = (a + bi)(a - bi) = a^2 + b^2$$

Let us determine some further properties of conjugation. Suppose $z = a + bi$ and $w = c + di$. Then

$$\begin{aligned}\overline{z + w} &= \overline{(a + bi) + (c + di)} = \overline{(a + c) + (b + d)i} \\ &= (a + c) - (b + d)i = a + c - bi - di \\ &= (a - bi) + (c - di) = \bar{z} + \bar{w}\end{aligned}$$

Similarly,

$$\begin{aligned}\overline{zw} &= \overline{(a + bi)(c + di)} = \overline{(ac - bd) + (ad + bc)i} \\ &= (ac - bd) - (ad + bc)i = [ac - (-b)(-d)] + [a(-d) + (-b)c]i \\ &= (a - bi)(c - di) = \bar{z}\bar{w}\end{aligned}$$

Thus we have seen that two properties of complex conjugation are

$$\overline{z + w} = \overline{z} + \overline{w} \qquad \text{and} \qquad \overline{z \cdot w} = \overline{z} \cdot \overline{w}$$

EXAMPLE 4 $\overline{(5 + 3i) + (-7 + 6i)} = \overline{-2 + 9i} = -2 - 9i$, while $\overline{(5 + 3i)} + \overline{(-7 + 6i)} = (5 - 3i) + (-7 - 6i) = -2 - 9i$. ■

EXAMPLE 5 $\overline{(5 + 3i)(-7 + 6i)} = \overline{(-35 + 30i - 21i - 18)} = \overline{-53 + 9i} = -53 - 9i$, while $\overline{(5 + 3i)}\,\overline{(-7 + 6i)} = (5 - 3i)(-7 - 6i) = -35 - 30i + 21i - 18 = -53 - 9i$. ■

The complex conjugate is particularly useful when doing division of complex numbers. To divide $z = a + bi$ by a nonzero complex number $w = c + di$ and then put the answer in standard form, we merely multiply numerator and denominator of $\frac{z}{w}$ by the complex conjugate $\overline{w}$ of the denominator:

$$\frac{z}{w} = \frac{z}{w} \cdot \frac{\overline{w}}{\overline{w}} = \frac{a + bi}{c + di} \cdot \frac{c - di}{c - di} = \frac{(a + bi)(c - di)}{c^2 + d^2}$$

$$= \frac{ac + bd}{c^2 + d^2} + \frac{-ad + bc}{c^2 + d^2} i$$

EXAMPLE 6

$$\frac{5 + 3i}{-7 + 6i} = \frac{(5 + 3i)(-7 - 6i)}{(-7 + 6i)(-7 - 6i)} = \frac{-35 - 30i - 21i + 18}{(-7)^2 + 6^2}$$

$$= -\frac{17}{85} - \frac{51}{85} i = -\frac{1}{5} - \frac{3}{5} i \quad ■$$

Computing powers of i is an interesting problem. Starting with the first four powers of i, we see that

$$i^1 = i \qquad i^2 = -1$$

$$i^3 = i^2 \cdot i = (-1)i = -i \qquad i^4 = i^2 \cdot i^2 = (-1)(-1) = 1$$

Then the cycle starts again for the next four powers:

$$i^5 = i^4 \cdot i = 1 \cdot i = i, \qquad i^6 = i^4 \cdot i^2 = 1 \cdot (-1) = -1,$$

$$i^7 = i^4 \cdot i^3 = 1 \cdot (-i) = -i, \qquad i^8 = i^4 \cdot i^4 = 1 \cdot 1 = 1$$

This cycle repeats every four powers, so to compute i to any positive integral power, we simply split off all the multiples of 4 in the exponent and compute what remains.

EXAMPLE 7 Find (a) i^{49}; (b) i^{91}; (c) $i^{487,123}$.

Solution (a) $i^{49} = i^{48+1} = (i^4)^{12} \cdot i = 1^{12} \cdot i = i.$

(b) $i^{91} = i^{88+3} = (i^4)^{22} \cdot i^3 = 1^{22} \cdot (-i) = -i.$

(c) Dividing 487,123 by 4, we get 121,780.75. This means that 487,123 divided by 4 is 121,780 with a remainder of 3. That is,

$$487{,}123 = 4 \cdot 121{,}780 + 3$$

Thus

$$i^{487{,}123} = i^{4 \cdot 121{,}780+3} = (i^4)^{121{,}780} i^3 = 1^{121{,}780} \cdot (-i) = -i \quad \blacksquare$$

EXERCISES

In Exercises 1–48, simplify. Write the answer in the form a + bi.

1. $\overline{3 + i}$

2. $\overline{4 - 2i}$

3. $\overline{5}$

4. $\overline{2i}$

5. $(3 + 2i) + (4 + 5i)$

6. $(7 + 9i) + (3 - 8i)$

7. $(11 - 4i) + (4 - 3i)$

8. $(-2 - 7i) + (-5 + 15i)$

9. $(4 + 3i) + \overline{(-4 + 3i)}$

10. $(-9 - 2i) + \overline{(9 - 3i)}$

11. $(2 - 3i) - (-4 - 2i)$

12. $(18 - 2i) - (13 - 8i)$

13. $-(-3 + 8i) - \overline{(-7 + 6i)}$

14. $-\overline{(4 - 5i)} - (-7 - 9i)$

15. $5 - (2 - 3i)$

16. $-9 - (-7 + 2i)$

17. $7i - (3 + 9i)$

18. $(15 + 2i) - 8i$

19. $(-5 - 2i) - (3 + 3i)$

20. $(14 + 5i) + (3 - 8i)$

21. $i - (4 + 12i)$

22. $(-11 - 11i) - (4 - 2i)$

23. $(2 - 3i)(4 + i)$

24. $(7 - i)(6 + 4i)$

25. $\overline{(1 + 2i)(4 - i)}$

26. $\overline{(7 + 4i)(5 - i)}$

27. $i(3 - i)$

28. $\overline{(3 - 2i)}4i$

29. $2(4 + 7i) + i(1 - i)$

30. $(3i + 1)(2i) + (2i)(3 - i)$

31. $(2 - 3i)^2$

32. $(7 + 2i)^2$

33. i^{17}

34. i^{30}

35. i^{3600}

36. $i^{51{,}271}$

37. $i^{29} + i^{21} + i$

38. $i^{4219} - i^{8121} + i^{3802}$

39. $\dfrac{2 + i}{3 + 2i}$

40. $\dfrac{4 - i}{2 - 5i}$

41. $\dfrac{1 + 3i}{4 - i}$

42. $\dfrac{1 - i}{1 + i}$

43. $\dfrac{1}{1 + 2i}$

44. $\dfrac{1}{3 - i}$

45. $\dfrac{4 + 3i}{i}$

46. $\dfrac{2 - i}{3i}$

47. $\dfrac{1}{i}$

48. $-\dfrac{3}{2i}$

In Exercises 49–58, solve for x and y real.

49. $4x + 7i = -8 + 3yi$

50. $7 - 2yi = 18x + 3i$

51. $(x + yi)i + 2 = 3x + y - 4i$

52. $(x - y) + (x + y)i = 4 - y - i$

53. $2^x + \frac{1}{9}i = 8 + (\log_3 y)i$

54. $\log 2 + 3e^{2x}i = 10^y + 6i$

55. $8^y + 27i = 2 + 9^x i$

56. $5 + (\log_4 y)i = \log_5 x + 2i$

57. $\sin x + \frac{1}{2}i = 1 + (\cos y)i$

58. $-1 + (\tan^{-1} x)i = \tan y + \sqrt{3}i$

4.2 Complex Roots of Equations

In this section we shall study equations in which some of the solutions are complex numbers. We begin with linear equations. Linear equations cannot have nonreal solutions unless they have nonreal coefficients. We shall see that this is not true for higher-degree equations.

LINEAR EQUATIONS

Recall that the general strategy for solving a linear equation in x is to do the following:

1. Get all the x's on one side and everything else on the other.
2. Combine similar terms.
3. Divide by the coefficient of x, and simplify.

This same general strategy works for solving a linear equation involving complex numbers.

EXAMPLE 1 Solve for x complex, putting the answer in the form $a + bi$:

$$(2 - 3i)x + 4 - i = (6 + 3i)x + 7 + 4i$$

Solution First, get all the x's on one side and everything else on the other:

$$(2 - 3i)x - (6 + 3i)x = -4 + i + 7 + 4i$$

Next, combine similar terms:

$$(-4 - 6i)x = 3 + 5i$$

Now, divide by the coefficient of x:

$$x = \frac{3 + 5i}{-4 - 6i}$$

This is the answer, but we want it expressed in the form $a + bi$. So we carry out the division as in the previous section, by multiplying top and bottom by the conjugate of the denominator:

$$x = \frac{3 + 5i}{-4 - 6i} \frac{-4 + 6i}{-4 + 6i} = \frac{-12 - 2i + 30i^2}{16 + 36}$$

$$= \frac{-42 - 2i}{52} = -\frac{21}{26} - \frac{1}{26}i \quad \blacksquare$$

QUADRATIC EQUATIONS

We now turn to quadratic equations and begin with the equation $x^2 = d$. Recall that if d is a positive real number, the equation $x^2 = d$ has two solutions, $x = +\sqrt{d}$ and $x = -\sqrt{d}$. The symbol $\sqrt{d}$ represents $+\sqrt{d}$ and is called the **principal square root** of the positive real number d.

Now consider the equation $x^2 = -d$, where d is again a positive real number. This equation has no real solutions, but it does have two complex solutions, $x = \sqrt{d}\,i$ and $x = -\sqrt{d}\,i$. To see this, first observe that

$$(x - \sqrt{d}\,i)(x + \sqrt{d}\,i) = x^2 - (\sqrt{d})^2(i)^2 = x^2 - d(-1) = x^2 + d$$

Thus we can solve $x^2 = -d$ by factoring:

$$x^2 + d = 0, \qquad (x + \sqrt{d}\,i)(x - \sqrt{d}\,i) = 0$$

$$x = \sqrt{d}\,i \qquad \text{or} \qquad x = -\sqrt{d}\,i$$

Therefore, in the same way that we solve $x^2 = d$, $d > 0$, by writing $x = \pm\sqrt{d}$, we can now solve $x^2 = -d$, $d > 0$, by writing $x = \pm\sqrt{d}\,i$.

EXAMPLE 2 Solve (a) $x^2 = -16$ and (b) $x^2 = -17$.

Solution (a) $x^2 = -16$, $\quad x = \pm\sqrt{16}\,i$, $\quad x = \pm 4i$.
(b) $x^2 = -17$, $\quad x = \pm\sqrt{17}\,i$, $\quad x \approx \pm 4.12311i$. ■

We also extend the definition of principal root, following the pattern in the real case:

> **Definition** If d is a positive real number, then the **principal square root** of $-d$ is denoted by $\sqrt{-d}$ and is defined to be the complex number $\sqrt{d}\,i$.

Warning: If a and b are positive real numbers, $\sqrt{-a}\,\sqrt{-b}$ is *not* the same as $\sqrt{(-a)(-b)}$.

EXAMPLE 3 Simplify (a) $\sqrt{-4}\,\sqrt{-9}$ and (b) $\sqrt{(-4)(-9)}$.

Solution (a) $\sqrt{-4}\,\sqrt{-9} = (2i)(3i) = 6i^2 = -6$.
(b) $\sqrt{(-4)(-9)} = \sqrt{36} = +6$. Hence $\sqrt{-4}\,\sqrt{-9} \neq \sqrt{(-4)(-9)}$.
To avoid errors, always convert expressions involving radicals to the form $a + bi$ before any algebraic manipulations. ■

EXAMPLE 4 Simplify $(2 - \sqrt{-75})(-3 + \sqrt{-3})$.

Solution

$$\begin{aligned}(2 - \sqrt{-75})(-3 + \sqrt{-3}) &= (2 - 5\sqrt{3}\,i)(-3 + \sqrt{3}\,i)\\ &= -6 + 17\sqrt{3}\,i - 5(\sqrt{3})^2 i^2\\ &= -6 + 17\sqrt{3}\,i - 15(-1) = 9 + 17\sqrt{3}\,i\end{aligned}$$ ■

We are now ready to solve the general quadratic equation $ax^2 + bx + c = 0$, $a \neq 0$, a, b, c real numbers. The solutions are given by the quadratic formula:

$$x = \frac{-b + \sqrt{b^2 - 4ac}}{2a} \quad \text{and} \quad x = \frac{-b - \sqrt{b^2 - 4ac}}{2a}$$

When $b^2 - 4ac$ is less than zero, the equation $ax^2 + bx + c = 0$ has no real roots. But since we now know square roots of negative real numbers are complex numbers, we see that the quadratic formula gives us two complex solutions. The formula can be evaluated by the methods already given in this section.

EXAMPLE 5 Solve $3x^2 + 2x + 1 = 0$.

Solution By the quadratic formula we have

$$\begin{aligned}x &= \frac{-2 \pm \sqrt{2^2 - 4(3)(1)}}{2(3)} = \frac{-2 \pm \sqrt{-8}}{6}\\ &= \frac{-2 \pm 2\sqrt{2}\,i}{6} = -\frac{1}{3} \pm \frac{\sqrt{2}}{3}i\end{aligned}$$ ■

Remark. These answers can be verified by direct substitution. See Exercise 22.

Note that the two roots in Example 5 are conjugates of each other. This illustrates a general principle concerning complex roots. Suppose you are solving an equation of the form

$$a_n x^n + a_{n-1}x^{n-1} + \cdots + a_1 x + a_0 = 0, \qquad a_n \neq 0$$

It follows from a theorem called the Fundamental Theorem of Algebra that this equation has exactly n roots. If all the coefficients are *real* numbers, then whenever one complex number is a root, its conjugate is also a root. Thus there are always an even number of roots that are nonreal, and they can be grouped in pairs, each with its conjugate. This general principle will be illustrated in some of the following examples.

Sometimes higher-degree equations can be solved by factoring the expressions into linear and quadratic factors and handling these individually.

EXAMPLE 6 Solve $x^3 + 1 = 0$.

Solution $x^3 + 1 = 0$. Factoring, we obtain $(x + 1)(x^2 - x + 1) = 0$. Thus, $x + 1 = 0$ or $x^2 - x + 1 = 0$. Solving, we obtain

$$x = -1 \qquad \text{or} \qquad x = \frac{-(-1) \pm \sqrt{(-1)^2 - 4(1)(1)}}{2} = \frac{1}{2} \pm \frac{\sqrt{3}}{2}i$$

Therefore the solutions to $x^3 + 1 = 0$ are

$$x = -1, \qquad \frac{1}{2} + \frac{\sqrt{3}}{2}i, \qquad \frac{1}{2} - \frac{\sqrt{3}}{2}i$$

Note that the two nonreal roots are conjugates. ■

The quadratic formula can also be used to solve quadratic equations with complex coefficients.

EXAMPLE 7 Solve $\quad 2x^2 + 3ix + 2 = 0.$

Solution By the quadratic formula we have

$$x = \frac{-3i \pm \sqrt{(3i)^2 - 4(2)(2)}}{2(2)} = \frac{-3i \pm \sqrt{-25}}{4}$$

$$= \frac{-3i \pm 5i}{4} = -2i, \frac{1}{2}i$$

Note that this example shows that when some of the coefficients are not real, it is not necessary for conjugates of roots to be roots. ■

Equations in which $b^2 - 4ac$ is a complex number can also be solved. However, we must put off this discussion until the next two sections, where we develop techniques for finding square roots of complex numbers.

If you start with several complex numbers, it is straightforward to construct an equation with these numbers as roots. You simply multiply the appropriate linear factors together.

EXAMPLE 8 Find an equation with roots $1, -i, 2i$.

Solution First form $(x - 1)(x + i)(x - 2i) = 0$ and then multiply the factors together:

$$(x - 1)(x^2 - ix + 2) = 0$$

$$x^3 + (-1 - i)x^2 + (2 + i)x - 2 = 0$$ ■

The interesting observation to make is that if all the complex roots appear in conjugate pairs, then the resulting equation has all real coefficients. In such situations, when multiplying the factors together, first multiply together the factors with conjugate roots, and do this by regrouping as illustrated in Example 9.

EXAMPLE 9 Find an equation with roots $-2, 1 + i, 1 - i$.

Solution

$$(x + 2)[x - (1 + i)][x - (1 - i)] = 0$$

$$(x + 2)[(x - 1) - i][(x - 1) + i] = 0$$

$$\begin{aligned}(x + 2)[(x - 1)^2 - i^2] &= 0\\ (x + 2)(x^2 - 2x + 1 + 1) &= 0\\ (x + 2)(x^2 - 2x + 2) &= 0\\ x^3 - 2x + 4 &= 0 \quad \blacksquare\end{aligned}$$

EXERCISES

In Exercises 1–20, simplify. Write the answer in the form $a + bi$.

1. $\sqrt{-4}$

2. $-\sqrt{-16}$

3. $-\sqrt{-20}$

4. $\sqrt{-1200}$

5. $\sqrt{-25} + \sqrt{-36}$

6. $\sqrt{-16} + \sqrt{-4}$

7. $\sqrt{-100}\sqrt{-81}$

8. $\sqrt{-144}\sqrt{-9}$

9. $-\sqrt{-12}\sqrt{-75}$

10. $-\sqrt{-20}\sqrt{-45}$

11. $\dfrac{\sqrt{-169}}{\sqrt{-4}}$

12. $\dfrac{-\sqrt{-36}}{\sqrt{-4}}$

13. $(3 + \sqrt{-4})(2 - \sqrt{-9})$

14. $(-7 - \sqrt{-9})(8 + \sqrt{-25})$

15. $(2 - \sqrt{-12})(5 - \sqrt{-75})$

16. $(-4 + \sqrt{-45})(-3 + \sqrt{-80})$

17. $\sqrt{-4}(5 - \sqrt{-9})$

18. $\sqrt{-8}(\sqrt{-48} + \sqrt{40})$

19. $(\sqrt{-8})^3$

20. $(\sqrt{-12})^3 + (\sqrt{-12})^5$

In Exercises 21–24, show by direct substitution that the answers satisfy the given equation.

21. $x^2 + x + 1 = 0,\ x = -\frac{1}{2} \pm \frac{1}{2}\sqrt{-3}\,i$

22. $3x^2 + 2x + 1 = 0,\ x = -\frac{1}{3} \pm \frac{1}{3}\sqrt{2}\,i$

23. $2x^2 + 3ix + 2 = 0,\ x = -2i,\ \frac{1}{2}i$

24. $x^3 + 1 = 0,\ x = -1,\ \frac{1}{2} \pm \frac{1}{2}\sqrt{3}\,i$

In Exercises 25–64, solve the given equation, putting all answers in the form $a + bi$.

25. $3ix + 4 = 5$

26. $48x - 1 = 6ix$

27. $(2 + i)x + 2i = 7i.$

28. $(2 - 2i)x + 4 = -6$

29. $4ix - 5 = 0$

30. $(7 + i)x - 8 = (7 + 3i)x$

31. $(-1 - 2i)x - 4i = 5 + i$

32. $(4 - i)x - 6 = 8 + 3i$

33. $6ix + 2 - i = (3 + 2i)x - 4i$

34. $(2 - i)x + 3 = 4 - i - (3 + 2i)x$

35. $2x - 3i = ix - 4$

36. $ix - 4 = 3x + 2i$

37. $x^2 - 2x + 2 = 0$

38. $x^2 + 2x + 5 = 0$

39. $2x^2 - 3x + 2 = 0$

40. $3x^2 - x + 2 = 0$

41. $3x^2 + 4x + 2 = 0$

42. $-2x^2 + x - 4 = 0$

43. $x^3 - 1 = 0$

44. $x^3 + 8 = 0$

45. $x^3 - 125 = 0$

46. $x^3 + 64 = 0$

47. $x^4 - 16 = 0$

48. $x^4 - \frac{1}{81} = 0$

49. $x^6 - 64 = 0$

50. $x^3 + \frac{1}{8} = 0$

51. $2x^4 + 3x^2 + 1 = 0$

52. $4x^4 + 21x^2 + 27 = 0$

53. $x^4 + x^3 + x^2 = 0$

54. $9x^4 + 13x^2 + 4 = 0$

55. $x^2 + ix + 2 = 0$

56. $2x^2 - ix + 3 = 0$

57. $x^2 + (1 - 2i)x - i = 0$

58. $3x^2 + (2 + 3i)x + i = 0$

59. $2x^2 + (2 - 2i)x - i = 0$

60. $9x^2 + (3 + 3i)x + \frac{1}{2}i = 0$

[C] 61. $0.21x^2 - 0.35x + 0.49 = 0$

[C] 62. $813.2x^2 + 141.7x + 731.3 = 0$

[C] 63. $4.191x^2 - 3.198x + 1.092 = 0$

[C] 64. $x^3 + 5.128 = 0$

In Exercises 65–76, find an equation having the given numbers as roots.

65. $2i, -2i$

66. $3 + i, 3 - i$

67. $-2 + 3i, -2 - 3i$

68. $\frac{1}{3} - \frac{1}{2}i, \frac{1}{3} + \frac{1}{2}i$

69. $\sqrt{3}i, -\sqrt{3}i$

70. $1 + \sqrt{2}i, 1 - \sqrt{2}i$

71. $1, i, -3i$

72. $4, -2i, -i$

73. $-1, 1 - i, 1 + i$

74. $3, -3 + i, -3 - i$

75. $i, 2 + i, 2 - i$

76. $-i, 1 - 2i, 1 + 2i$

4.3 Trigonometric Form of Complex Numbers

We have previously said that two complex numbers $a + bi$ and $c + di$ were equal if and only if $a = c$ and $b = d$. Similarly, two ordered pairs (a, b) and (c, d) are equal if and only if $a = c$ and $b = d$. Thus it is natural to associate the complex number $a + bi$ with the ordered pair (a, b). This correspondence is *one-to-one*. That is, each complex number corresponds to exactly one ordered pair of real numbers and vice versa.

Now given an ordered pair (a, b) of real numbers, we normally think of that ordered pair as representing the point in the plane that has coordinates $x = a$ and $y = b$. Thus using the preceding correspondence, we may view each complex number $a + bi$ geometrically as the point (a, b) in the Cartesian plane. See Fig. 1.

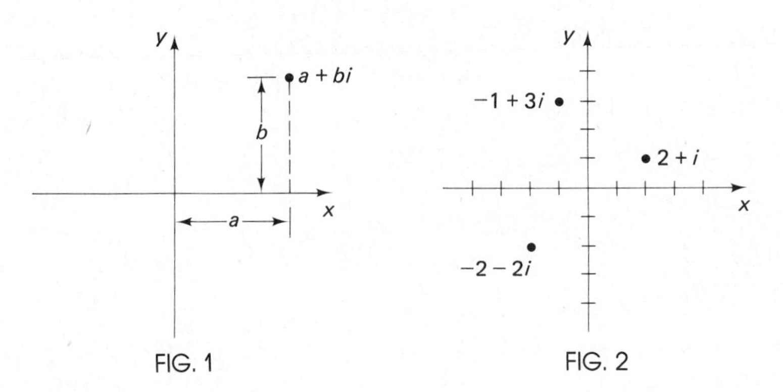

FIG. 1 FIG. 2

EXAMPLE 1 Represent $2 + i$, $-1 + 3i$, and $-2 - 2i$ geometrically.

Solution See Fig. 2. ■

When the plane is viewed this way, it is usually referred to as the **complex plane.** Since the numbers $a + 0i$, i.e., the real numbers, are the points on the x-axis, the x-axis is called the **real axis.** Similarly, the y-axis is called the **imaginary axis** since points there correspond to pure imaginary numbers.

In the complex plane, it is sometimes useful to view a complex number $z = a + bi$ as a vector from the origin to (a, b). For instance, if $z = -4 + i$ and $w = 3 + 2i$, then $z + w = -1 + 3i$. Now if we consider z and w as vectors $\mathbf{z}$ and $\mathbf{w}$ and add them by vector addition (using parallelograms), we see that adding z and w as vectors coincides with adding them as complex numbers. See Fig. 3. This illustrates the following general principle:

Addition of complex numbers may be described geometrically as addition of vectors.

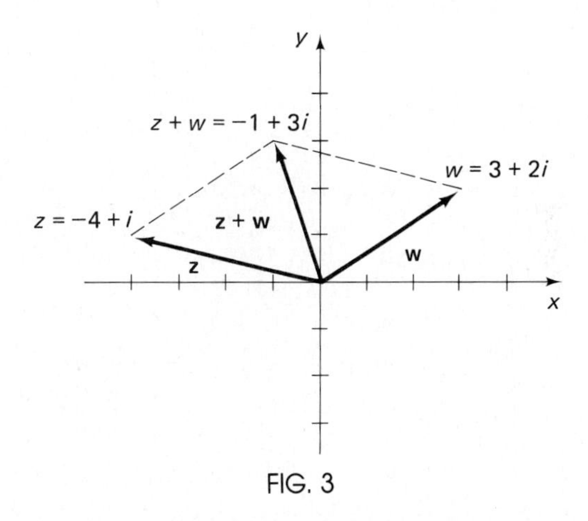

FIG. 3

To get a nice geometric description of multiplication of complex numbers, we look at a different form of complex numbers, obtained as follows. Suppose $z = a + bi$ is plotted in the complex plane. Draw a line from the point (a, b) to the origin. See Fig. 4. The length of this line is $r = \sqrt{a^2 + b^2}$, the distance

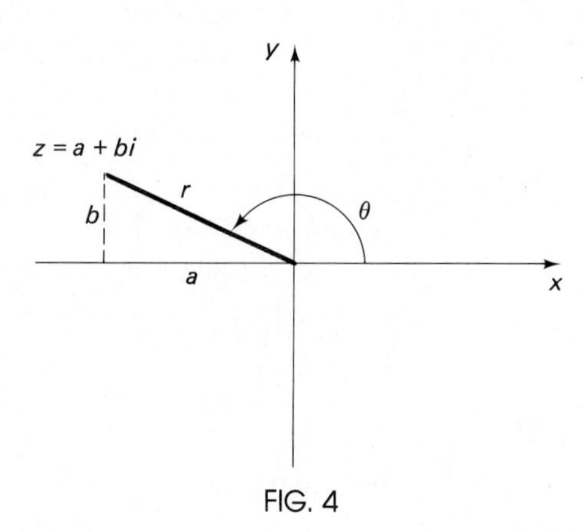

FIG. 4

from the point to the origin. Let θ be any angle in standard position for which (a, b) is on the terminal side. Then by the definition of the trigonometric functions, $\cos\theta = \frac{a}{r}$ and $\sin\theta = \frac{b}{r}$. It follows that $a = r\cos\theta$ and $b = r\sin\theta$. Therefore, $z = a + bi = r\cos\theta + (r\sin\theta)i$, or

$$z = r(\cos\theta + i\sin\theta)$$

We write $i\sin\theta$ instead of $(\sin\theta)i$ to eliminate parentheses and because writing $\sin\theta i$ makes it appear that the angle is θi.

The expression $r(\cos\theta + i\sin\theta)$ is called the **trigonometric form** or **polar form** of z. [Those students who have had polar coordinates will recognize r and θ as the polar coordinates of (a, b).] The trigonometric form is *not* unique, since θ is not unique. For example,

$$1 + i = \sqrt{2}(\cos\tfrac{1}{4}\pi + i\sin\tfrac{1}{4}\pi) = \sqrt{2}(\cos\tfrac{9}{4}\pi + i\sin\tfrac{9}{4}\pi) = \cdots$$

In addition, if $z = 0$, then $r = 0$, and θ can be *any* angle. If $z = r(\cos\theta + i\sin\theta)$, the angle θ is called the **argument** of z, and r is called the **absolute value** (or **modulus**) of z. Just as in the reals, the absolute value of z is the distance between z and the origin and is denoted by $|z|$. Since $r = \sqrt{a^2 + b^2}$ and $z\bar{z} = a^2 + b^2$,

$$|z| = \sqrt{z\bar{z}}$$

Since the expression $\cos\theta + i\sin\theta$ is somewhat cumbersome to write, we usually shorten this to $\operatorname{cis}\theta$, the abbreviation *cis* being obtained from $\underline{c}$os θ + $\underline{i}$ $\underline{s}$in θ.

Notation: $\operatorname{cis}\theta = \cos\theta + i\sin\theta$

Thus the trigonometric form is abbreviated by

$$z = r(\cos\theta + i\sin\theta) = r\operatorname{cis}\theta$$

In the first few examples, we shall write complex numbers in the trigonometric form both as $r(\cos\theta + i\sin\theta)$ and as the shortened version $r\operatorname{cis}\theta$. After that, we shall use the longer form only where necessary for clarity.

EXAMPLE 2 Put the numbers $z_1 = 1 + i$, $z_2 = \sqrt{3} - i$, and $z_3 = -3 + 4i$ in trigonometric form.

Solution See Fig. 5. For z_1: $r = \sqrt{1 + 1} = \sqrt{2}$; $\cos\theta = \frac{1}{\sqrt{2}}$, and $\sin\theta = \frac{1}{\sqrt{2}}$. One such angle θ is $\theta = 45°$. Thus,

$$z_1 = \sqrt{2}(\cos 45° + i\sin 45°) = \sqrt{2}\operatorname{cis} 45°$$

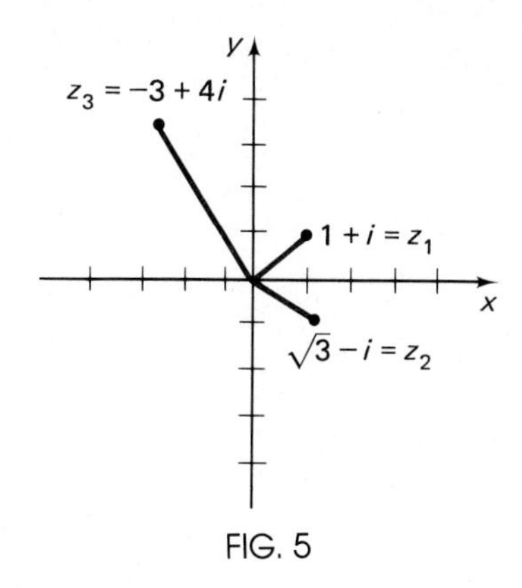

FIG. 5

For z_2: $r = \sqrt{(\sqrt{3})^2 + (-1)^2} = \sqrt{4} = 2$; $\cos\theta = \frac{1}{2}\sqrt{3}$, and $\sin\theta = -\frac{1}{2}$. Choose $\theta = -\dfrac{\pi}{6}$. Then

$$z_2 = 2\left[\cos\left(-\frac{\pi}{6}\right) + i\sin\left(-\frac{\pi}{6}\right)\right] = 2\operatorname{cis}\left(-\frac{\pi}{6}\right)$$

For z_3: $r = \sqrt{(-3)^2 + 4^2} = \sqrt{25} = 5$; $\cos\theta = -\frac{3}{5}$, and $\sin\theta = \frac{4}{5}$. One such angle is $\theta = \cos^{-1}(-\frac{3}{5}) \approx 126.870°$. Thus,

$$z_3 \approx 5(\cos 126.870° + i\sin 126.870°) = 5\operatorname{cis} 126.870° \quad \blacksquare$$

The virtue of the trigonometric form of complex numbers is that it allows us to simplify multiplication and division. Suppose z and w are two complex numbers in trigonometric form, $z = r\operatorname{cis}\theta = r(\cos\theta + i\sin\theta)$ and $w = s\operatorname{cis}\phi = s(\cos\phi + i\sin\phi)$. Then, using the rules of multiplication of complex numbers, we obtain

$$\begin{aligned} zw &= rs(\cos\theta + i\sin\theta)(\cos\phi + i\sin\phi) \\ &= rs[(\cos\theta\cos\phi - \sin\theta\sin\phi) + i(\sin\theta\cos\phi + \cos\theta\sin\phi)] \end{aligned}$$

The quantities in parentheses can be simplified using the addition formulas:

$$\cos\theta\cos\phi - \sin\theta\sin\phi = \cos(\theta + \phi)$$
$$\sin\theta\cos\phi + \cos\theta\sin\phi = \sin(\theta + \phi)$$

Therefore, $zw = rs[\cos(\theta + \phi) + i\sin(\theta + \phi)] = rs\operatorname{cis}(\theta + \phi)$, a number again in trigonometric form.

Multiplication of complex numbers in trigonometric form:

$$(r\operatorname{cis}\theta)(s\operatorname{cis}\phi) = rs\operatorname{cis}(\theta + \phi)$$

What this says is *when multiplying complex numbers, multiply the absolute values and add the arguments.*

EXAMPLE 3 $(3 \text{ cis } 41°)(2 \text{ cis } 197°) = 6 \text{ cis } 238°$. See Fig. 6(a). ■

EXAMPLE 4 $(3 \text{ cis } 108°)^2 = 9 \text{ cis } 216°$. See Fig. 6(b). ■

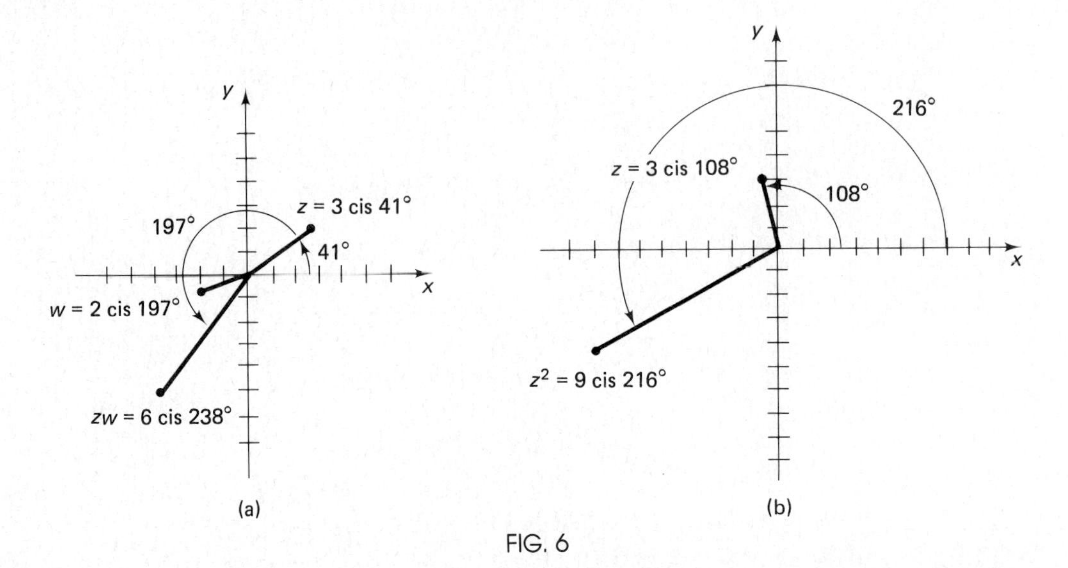

FIG. 6

The rule for (multiplicative) inverses is quite interesting. If $z = r \text{ cis } \theta = r(\cos \theta + i \sin \theta)$,

$$\begin{aligned}\frac{1}{z} &= \frac{1}{r(\cos \theta + i \sin \theta)} \\ &= \frac{1}{r} \frac{1}{\cos \theta + i \sin \theta} \frac{\cos \theta - i \sin \theta}{\cos \theta - i \sin \theta} \\ &= \frac{1}{r} \frac{\cos \theta - i \sin \theta}{\cos^2 \theta + \sin^2 \theta} \\ &= r^{-1}[\cos(-\theta) + i \sin(-\theta)] = r^{-1} \text{ cis}(-\theta)\end{aligned}$$

The last line follows from the identities $\sin^2 \theta + \cos^2 \theta = 1$, $\cos(-\theta) = \cos \theta$, and $\sin(-\theta) = -\sin \theta$. Hence we have shown the following:

> The inverse of complex numbers in trigonometric form:
>
> $$(r \text{ cis } \theta)^{-1} = r^{-1} \text{ cis}(-\theta)$$

What this says is *to find the inverse of a complex number, you find the inverse of its absolute value and the negative of its argument.*

EXAMPLE 5 $(0.5 \text{ cis } 40°)^{-1} = 2 \text{ cis}(-40°)$. See Fig. 7(a). ■

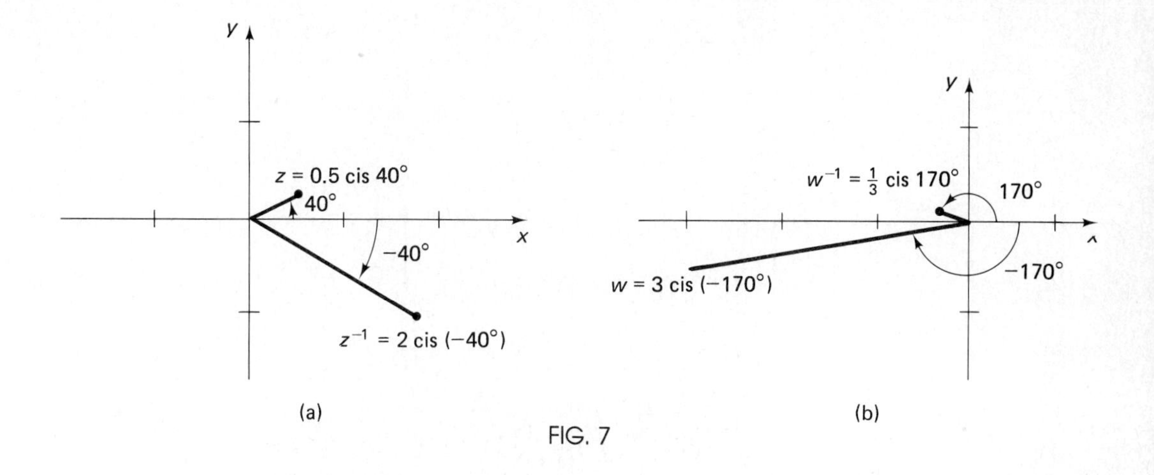

FIG. 7

EXAMPLE 6 $[3\ \text{cis}(-170°)]^{-1} = \frac{1}{3}\text{cis}\ 170°$. See Fig. 7(b). ■

Now using the multiplication and inverse rules, we see that if $z = r\ \text{cis}\ \theta$, $w = s\ \text{cis}\ \phi$,

$$\frac{z}{w} = zw^{-1} = (r\ \text{cis}\ \theta)\left[\frac{1}{s}\text{cis}(-\phi)\right]$$

$$= \frac{r}{s}\text{cis}(\theta - \phi)$$

Therefore,

> Division of complex numbers in trigonometric form:
>
> $$\frac{r\ \text{cis}\ \theta}{s\ \text{cis}\ \phi} = \frac{r}{s}\ \text{cis}(\theta - \phi)$$

What this says is *to divide complex numbers, you divide the absolute values and subtract the arguments.*

EXAMPLE 7

$$\frac{8\ \text{cis}\ 41°}{2\ \text{cis}\ 147°} = \frac{8}{2}\ \text{cis}(41° - 147°) = 4\ \text{cis}(-106°)$$

See Fig. 8 on the next page. ■

EXAMPLE 8

$$\frac{4\ \text{cis}\ 3.17}{8\ \text{cis}\ 1.49} = 0.5\ \text{cis}\ 1.68 \quad ■$$

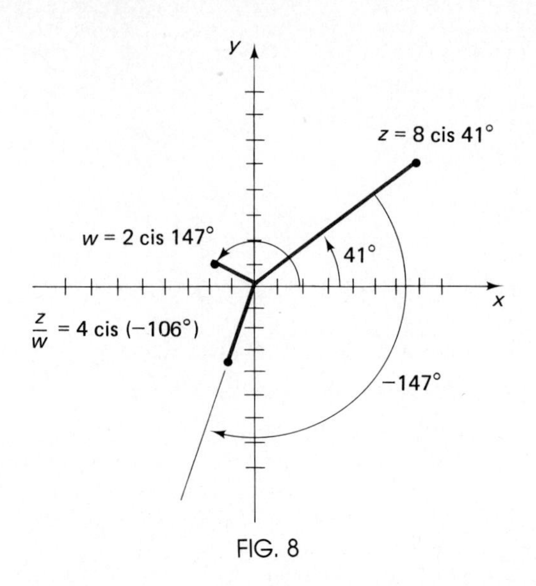

FIG. 8

EXERCISES

In Exercises 1–5, plot each complex number, and then find its absolute value and put it into trigonometric form (long and short version).

1. $z_1 = 1 + i$, $z_2 = -1 + i$, $z_3 = -1 - i$, $z_4 = 1 - i$
2. $z_1 = 1 + \sqrt{3}i$, $z_2 = -1 + \sqrt{3}i$, $z_3 = -1 - \sqrt{3}i$, $z_4 = 1 - \sqrt{3}i$
3. $z_1 = \sqrt{3} + i$, $z_2 = -\sqrt{3} + i$, $z_3 = -\sqrt{3} - i$, $z_4 = \sqrt{3} - i$
4. $z_1 = 2 - 2i$, $z_2 = 4 - 4i$
5. $z_1 = 6 + 2i$, $z_2 = -3 + i$, $z_3 = -9 + 3i$, $z_4 = 6 - 2i$

In Exercises 6–10, change each complex number from trigonometric form to the form $a + bi$.

6. 3 cis 120°
7. 6 cis 210°
8. 2 cis 27°
9. 2 cis 1.73
10. $8 \text{ cis } \frac{7\pi}{4}$

In Exercises 11–14, multiply. Sketch the problem graphically.

11. (3 cis 98°)(6 cis 321°)
12. $\left(80 \text{ cis } \frac{\pi}{5}\right)\left(0.03 \text{ cis } \frac{\pi}{3}\right)$
ⓒ**13.** (11.516 cis 4.32)(64.3 cis 5.8)
ⓒ**14.** (3.914 cis 21°42′)(13.47 cis 261.3°)

In Exercises 15–18, find the inverse of the given number. Sketch.

15. 2 cis 70°
16. $1.5 \text{ cis}\left(\frac{5\pi}{6}\right)$
17. $0.6 \text{ cis}\left(\frac{-\pi}{5}\right)$
18. 1.2 cis(−160°)

In Exercises 19–22, divide. Sketch.

19. $\dfrac{6 \text{ cis } 284^\circ}{3 \text{ cis } 107^\circ}$

20. $\dfrac{3 \text{ cis } 107^\circ}{6 \text{ cis } 284^\circ}$

[C] **21.** $\dfrac{61.28 \text{ cis}\left(\frac{2\pi}{3}\right)}{46.5 \text{ cis}\left(\frac{4\pi}{5}\right)}$

[C] **22.** $\dfrac{4916 \text{ cis } 0.71}{739 \text{ cis } 2.03}$

23. Compute **(a)** $(1 + i)(\sqrt{3} - i)(-4 + 4i)(-1 - \sqrt{3}\,i)$ and **(b)** $\dfrac{(1 + i)(-1 - \sqrt{3}\,i)}{(\sqrt{3} - i)(-4 + 4i)}$ directly. Then change each of the four complex numbers to trigonometric form, and recompute the answers.

24. Show that $\text{cis } \theta = \text{cis}(\theta + 2n\pi)$, n an integer.

4.4 Powers and Roots of Complex Numbers

Having a complex number in trigonometric form is particularly useful if you want to raise the number to an integer power. Let us look at some small powers of $r \text{ cis } \theta$:

$$
\begin{aligned}
(r \text{ cis } \theta)^2 &= (r \text{ cis } \theta)(r \text{ cis } \theta) \\
&= r \cdot r \text{ cis}(\theta + \theta) \\
&= r^2 \text{ cis } 2\theta \\
(r \text{ cis } \theta)^3 &= (r \text{ cis } \theta)^2(r \text{ cis } \theta) \\
&= (r^2 \text{ cis } 2\theta)(r \text{ cis } \theta) \\
&= r^2 \cdot r \text{ cis}(2\theta + \theta) \\
&= r^3 \text{ cis } 3\theta \\
(r \text{ cis } \theta)^4 &= (r \text{ cis } \theta)^3(r \text{ cis } \theta) \\
&= (r^3 \text{ cis } 3\theta)(r \text{ cis } \theta) \\
&= r^4 \text{ cis } 4\theta
\end{aligned}
$$

These examples are following a very simple pattern:

> **(1)** If n is a positive integer,
> $$(r \text{ cis } \theta)^n = r^n \text{ cis } n\theta$$

This formula can be shown to be true for all positive integers n by applying the principle of mathematical induction. We now extend this to all integers.

To be consistent with the rules for powers of real numbers, we define any nonzero complex number raised to the zero power to be 1:

> **Definition** $\quad (a + bi)^0 = 1 \quad$ if $a + bi \neq 0$

Since the trigonometric form of 1 is $1(\cos 0 + i \sin 0)$, formula (1) holds with $n = 0$ whenever $r(\cos \theta + i \sin \theta) \neq 0$, i.e., whenever $r \neq 0$:

$$\begin{aligned}(r \text{ cis } \theta)^0 &= r^0 \text{ cis } 0\theta \\ &= r^0 (\cos 0\theta + i \sin 0\theta) \\ &= 1(\cos 0 + i \sin 0) = 1\end{aligned}$$

Let us look at negative integer powers. From the previous section, we know that

$$(r \text{ cis } \theta)^{-1} = r^{-1} \text{ cis}(-\theta) = r^{-1} \text{ cis}[(-1)\theta]$$

More generally, if n is any *positive* integer and (1) holds,

$$\begin{aligned}(r \text{ cis } \theta)^{-n} &= [(r \text{ cis } \theta)^{-1}]^n \\ &= [r^{-1} \text{ cis}(-\theta)]^n \\ &= (r^{-1})^n \text{ cis}[n(-\theta)] \qquad \text{[by (1)]} \\ &= r^{-n} \text{ cis}[(-n)\theta]\end{aligned}$$

Thus we have the following formula, called **De Moivre's Theorem:**

De Moivre's Theorem For any integer n, $\quad (r \text{ cis } \theta)^n = r^n \text{ cis } n\theta$

EXAMPLE 1

$$\begin{aligned}(2 \text{ cis } 10°)^5 &= 2^5 \text{ cis } 50° \\ (2 \text{ cis } 10°)^{-7} &= 2^{-7} \text{ cis}(-70°) \\ \left(3.2 \text{ cis } \frac{3\pi}{13}\right)^{10} &= (3.2)^{10} \text{ cis } \frac{30\pi}{13} \qquad \blacksquare\end{aligned}$$

The advantage of De Moivre's Theorem becomes apparent if one tries to calculate powers of complex numbers directly. It would take a lot of work to compute $(-\sqrt{3} + i)^{53}$ directly. However, $-\sqrt{3} + i$ can be put in trigonometric form, $-\sqrt{3} + i = 2 \text{ cis } 150°$. Thus

$$\begin{aligned}(-\sqrt{3} + i)^{53} &= (2 \text{ cis } 150°)^{53} \\ &= 2^{53} \text{ cis}[53(150)°] \\ &= 2^{53} \text{ cis } 7950° \\ &= 2^{53} \text{ cis } 30°, \qquad \text{since } 7950° = 30° + 22(360°) \\ &= 2^{53}(\cos 30° + i \sin 30°) \\ &= 2^{53}\left(\frac{\sqrt{3}}{2} + \frac{1}{2}i\right) = 2^{52}\sqrt{3} + 2^{52}i\end{aligned}$$

In the preceding example, even if you have a calculator you should go from

$$\cos 30° + i \sin 30° \qquad \text{to} \qquad \frac{\sqrt{3}}{2} + \frac{1}{2}i$$

without using a calculator. It is usually better to avoid using a calculator in such situations, because the answer is in a more compact form that can be handled more easily in later computations. However, there are situations that are quite tedious to do without a calculator or computer.

EXAMPLE 2 Ⓒ Find $(1.102 - \sqrt[3]{2}\,i)^{12}$.

Solution First put $z = 1.102 - \sqrt[3]{2}\,i$ in trigonometric form:

$$|z| = r = \sqrt{(1.102)^2 + (\sqrt[3]{2})^2} \approx 1.6738593$$

Since $\cos\theta = \dfrac{1.102}{r}$ is positive and $\sin\theta = -\dfrac{\sqrt[3]{2}}{r}$ is negative, θ is in Q_4 with a reference angle of $\cos^{-1}\left(\dfrac{1.102}{r}\right) \approx 0.852160$ rad; i.e., $\theta \approx -0.852160$. Therefore

$$\begin{aligned} z^{12} &\approx [1.6738593 \operatorname{cis}(-0.852160)]^{12} \\ &\approx (1.6738593)^{12} \operatorname{cis}[12(-0.852160)] \\ &\approx 483.757[\cos(-10.2259) + i \sin(-10.2259)] \\ &\approx -336.647 + 347.404i \qquad \blacksquare \end{aligned}$$

We started our discussion of the complex numbers by stating that we wanted to be able to solve the equation $x^2 + 1 = 0$. Not only can this equation be solved in the complex numbers, but so can any polynomial equation with complex coefficients. This is not obvious, and the proof requires some rather deep mathematics. Moreover, it follows from a theorem called the Fundamental Theorem of Algebra that any polynomial equation of degree n has precisely n roots. One such equation is $x^n - c = 0$, $c \neq 0$, or, equivalently,

$$\textbf{(2)} \qquad x^n = s \operatorname{cis} \phi, \qquad s \neq 0$$

where $c = s \operatorname{cis} \phi$ is expressed in polar form. Now we shall see how De Moivre's Theorem can be employed to find the n solutions to equation (2), i.e., to find the n distinct nth roots of the complex number $c = s \operatorname{cis} \phi$.

Suppose $r \operatorname{cis} \theta$ is one solution to (2). Then

$$(r \operatorname{cis} \theta)^n = s \operatorname{cis} \phi \qquad \text{or} \qquad r^n \operatorname{cis} n\theta = s \operatorname{cis} \phi$$

Written out in the nonabbreviated trigonometric form, this is

$$r^n(\cos n\theta + i \sin n\theta) = s(\cos \phi + i \sin \phi)$$

For this equality to hold, we must have $r^n = s$, $\cos n\theta = \cos \phi$, and $\sin n\theta = \sin \phi$. Now if r and s are positive real numbers and $r^n = s$, then $r = s^{1/n}$. Moreover, if $\cos n\theta = \cos \phi$ and $\sin n\theta = \sin \phi$, then $n\theta$ and ϕ must have the same terminal side. In other words, $n\theta = \phi + 360° k$ for some integer k. Thus, $\theta = \dfrac{\phi}{n} + \dfrac{360° k}{n}$. Therefore we have seen that for any integer k, $z_k = s^{1/n} \operatorname{cis}\left(\dfrac{\phi}{n} + \dfrac{360° k}{n}\right)$ is a solution to equation (2): $x^n = s \operatorname{cis} \phi$. Of course, if the angles are in radians, then $z_k = s^{1/n} \operatorname{cis}\left(\dfrac{\phi}{n} + \dfrac{2k\pi}{n}\right)$.

EXAMPLE 3 Solve $x^4 = 1$.

Solution We write the number as $1 = 1(\cos 0° + i \sin 0°)$. By the preceding discussion, the solutions are

$$z_k = 1^{1/4} \operatorname{cis}\left(0° + \frac{k}{4} 360°\right) = \operatorname{cis}(90° k)$$

where k is an integer. Let us look at some of them:

$z_0 = \operatorname{cis} 0° = 1$, $z_1 = \operatorname{cis} 90° = i$, $z_2 = \operatorname{cis} 180° = -1$, $z_3 = \operatorname{cis} 270° = -i$, $z_4 = \operatorname{cis} 360° = z_0$, $z_5 = \operatorname{cis} 450° = z_1$, etc.

You can see the first four solutions are all distinct and thereafter they repeat. This illustrates our statement that a polynomial equation of degree n has n roots. ■

Of course, we could have solved Example 3 more simply by rewriting the equation as $x^4 - 1 = 0$ and factoring

$$x^4 - 1 = (x^2 - 1)(x^2 + 1) = (x - 1)(x + 1)(x - i)(x + i)$$

Thus the roots are ± 1, $\pm i$, which checks the computations in Example 3. The next example cannot be done so simply.

EXAMPLE 4 Solve $x^5 = -16\sqrt{3} + 16i$.

Solution We rewrite the right side in trigonometric form:

$$-16\sqrt{3} + 16i = 32(-\tfrac{1}{2}\sqrt{3} + \tfrac{1}{2}i) = 32(\cos 150° + i \sin 150°)$$

By the discussion before Example 3, the solutions are

$$z_k = 32^{1/5} \operatorname{cis}\left(\frac{150°}{5} + \frac{k}{5} 360°\right) = 2 \operatorname{cis}(30° + 72°k)$$

where k is an integer. The first several are

$$z_0 = 2 \operatorname{cis} 30° = 2\left(\frac{\sqrt{3}}{2} + \frac{1}{2}i\right) = \sqrt{3} + i$$

$$z_1 = 2\ \text{cis}(30° + 72°) = 2\ \text{cis}(102°) \approx -0.415823 + 1.95630i$$

$$z_2 = 2\ \text{cis}(30° + 144°) = 2\ \text{cis}(174°) \approx -1.98904 + 0.209057i$$

$$z_3 = 2\ \text{cis}(30° + 216°) = 2\ \text{cis}(246°) \approx -0.813473 - 1.82709i$$

$$z_4 = 2\ \text{cis}(30° + 288°) = 2\ \text{cis}(318°) \approx 1.48629 - 1.33826i$$

$$z_5 = 2\ \text{cis}(30° + 360°) = 2\ \text{cis}(30°) = z_0$$

$$z_6 = 2\ \text{cis}(30° + 360° + 72°) = 2\ \text{cis}\ 102° = z_1, \quad \text{etc.}$$

Here, the first five solutions z_0, z_1, z_2, z_3, z_4 are all distinct and the others repeat one of these five. ■

Examples 3 and 4 illustrate the general way we can find the n distinct solutions to the equation $x^n = c$.

> **(3)** The n distinct solutions to $x^n = s\ \text{cis}\ \phi$ are
>
> $$z_k = s^{1/n}\ \text{cis}\left(\frac{\phi}{n} + \frac{360°\, k}{n}\right) \qquad \text{if } \phi \text{ is in degrees}$$
>
> or
>
> $$z_k = s^{1/n}\ \text{cis}\left(\frac{\phi}{n} + \frac{2k\pi}{n}\right) \qquad \text{if } \phi \text{ is in radians}$$
>
> for $k = 0, 1, 2, \ldots, n - 1$.

EXAMPLE 5 Find all solutions to $x^6 = \sqrt{3} - i$.

Solution Since $\sqrt{3} - i = 2\ \text{cis}\left(-\frac{\pi}{6}\right)$, the solutions are

$$z_k = 2^{1/6}\text{cis}\left(\frac{-\pi/6}{6} + \frac{2k\pi}{6}\right) \qquad \text{for } k = 0, 1, 2, 3, 4, 5$$

They are

$$z_0 = 2^{1/6}\ \text{cis}\left(-\frac{\pi}{36}\right) \approx 1.11819 - 0.0978290i$$

$$z_1 = 2^{1/6}\ \text{cis}\left(-\frac{\pi}{36} + \frac{\pi}{3}\right) = 2^{1/6}\ \text{cis}\left(\frac{11\pi}{36}\right) \approx 0.643818 + 0.919467i$$

$$z_2 = 2^{1/6}\ \text{cis}\left(-\frac{\pi}{36} + \frac{2\pi}{3}\right) = 2^{1/6}\ \text{cis}\left(\frac{23\pi}{36}\right) \approx -0.474373 + 1.01730i$$

$$z_3 = 2^{1/6}\ \text{cis}\left(-\frac{\pi}{36} + \pi\right) = 2^{1/6}\ \text{cis}\left(\frac{35\pi}{36}\right) \approx -1.11819 + 0.0978290i$$

$$z_4 = 2^{1/6}\,\text{cis}\left(-\frac{\pi}{36} + \frac{4\pi}{3}\right) = 2^{1/6}\,\text{cis}\left(\frac{47\pi}{36}\right) \approx -0.643818 - 0.919467i$$

$$z_5 = 2^{1/6}\,\text{cis}\left(-\frac{\pi}{36} + \frac{5\pi}{3}\right) = 2^{1/6}\,\text{cis}\left(\frac{59\pi}{36}\right) \approx 0.474373 - 1.01730i$$

For many purposes, it is preferable to leave the answer in the form $z_0 = 2^{1/6}\,\text{cis}(-\frac{\pi}{36})$, etc. However, computing the answer out, using either tables or a calculator, shows us the relative sizes of the numbers and makes it obvious that the last three answers are negatives of the first three. ■

Let us plot the points $z_0, z_1, z_2, z_3, z_4, z_5$ from Example 5 in the complex plane. They all have the same absolute value, $2^{1/6}$. Thus they all lie on a circle of radius $2^{1/6}$ with center at the origin. First plot z_0; it has argument $\frac{\phi}{n} = -\frac{\pi}{36} = -5°$. Then the points $z_0, z_1, \ldots, z_5$ divide the circumference of the circle into $n = 6$ equal arcs, each subtending an angle of $\frac{2\pi}{n} = \frac{\pi}{3} = 60°$.

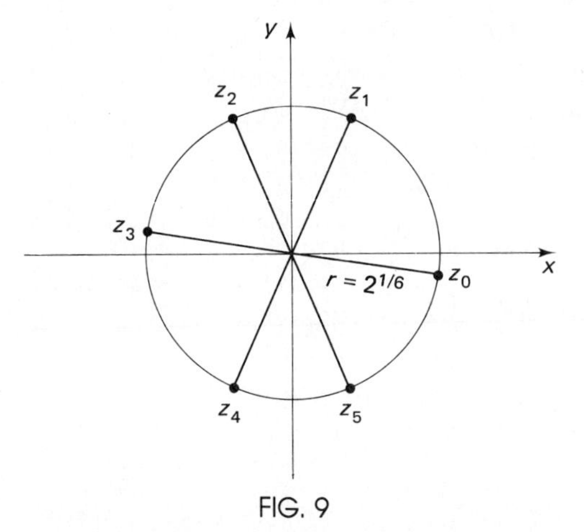

FIG. 9

Such a geometric picture is always possible. The n roots of $x^n = s(\cos\phi + i\sin\phi)$ all lie on a circle about the origin of radius $s^{1/n}$. They divide the circumference into n equal arcs, each subtending an angle of $\frac{360°}{n}$ (or $\frac{2\pi}{n}$ rad) starting from z_0, which has an argument of $\frac{\phi}{n}$.

Now that we know how to find roots of complex numbers, we can solve any quadratic equation $ax^2 + bx + c = 0$, $a \neq 0$, with complex coefficients. Previously when we used the quadratic formula

$$x = \frac{-b \pm \sqrt{b^2 - 4ac}}{2a}$$

we considered examples where $b^2 - 4ac$ was a real number. But we could not handle the case where $b^2 - 4ac$ was complex but not real, because we did not know how to take square roots of such numbers. Now we use the techniques we have just learned. We should first observe that for square roots, the formulas in (3) give

$$z_0 = s^{1/2} \operatorname{cis}\left(\frac{\phi}{2}\right), \qquad z_1 = s^{1/2} \operatorname{cis}\left(\frac{\phi}{2} + 180°\right) = -s^{1/2} \operatorname{cis}\left(\frac{\phi}{2}\right)$$

since $\cos(\theta + 180°) = -\cos\theta$ and $\sin(\theta + 180°) = -\sin\theta$.

Thus the two square roots are $\pm z_0$, which should not be surprising.

EXAMPLE 6 Solve $\quad ix^2 - (5 + 3i)x + 7 - 4i = 0.$

Solution By the quadratic formula

$$x = \frac{5 + 3i \pm \sqrt{[-(5 + 3i)]^2 - 4(i)(7 - 4i)}}{2i}$$

$$= \frac{5 + 3i \pm \sqrt{25 + 30i - 9 - 28i - 16}}{2i}$$

$$= \frac{5 + 3i \pm \sqrt{2i}}{2i}$$

Now $\quad 2i = 2 \operatorname{cis} 90° \quad$ so

$$\pm\sqrt{2i} = \pm(2 \operatorname{cis} 90°)^{1/2} = \pm\sqrt{2} \operatorname{cis} 45° = \pm\sqrt{2}\left(\frac{1}{\sqrt{2}} + \frac{1}{\sqrt{2}}i\right)$$

$$= \pm(1 + i)$$

Therefore $\quad x = \dfrac{5 + 3i \pm (1 + i)}{2i}$

$$= \frac{6 + 4i}{2i}, \frac{4 + 2i}{2i} = 2 - 3i, 1 - 2i \quad ■$$

EXERCISES

In Exercises 1–6, simplify using De Moivre's theorem.

1. $(3 \operatorname{cis} 102°)^6$

2. $(8 \operatorname{cis} 37°)^{100}$

3. $(2 \operatorname{cis} 213°)^{-8}$

4. $\left(6.4 \operatorname{cis}\frac{\pi}{5}\right)^{11}$

5. $\left[3.7\left(\cos\frac{\pi}{3} - i \sin\frac{\pi}{3}\right)\right]^5$

6. $(1.31 \operatorname{cis} 8.92)^{20}$

In Exercises 7–14, simplify to the form $a + bi$.

7. $\left(\frac{1}{2} - \frac{\sqrt{3}}{2}i\right)^{35}$

8. $\left(-\frac{\sqrt{2}}{2} - \frac{\sqrt{2}}{2}i\right)^{49}$

9. $\left(-\frac{\sqrt{2}}{2}+\frac{\sqrt{2}}{2}i\right)^{93}$

10. $\left(-\frac{\sqrt{3}}{2}+\frac{1}{2}i\right)^{207}$

[C] 11. $(1.3 - 1.82i)^8$

[C] 12. $(3 + 4i)^7$

[C] 13. $(2.3 + 1.037i)^9$

[C] 14. $(-0.604 - 1.425i)^5$

In Exercises 15–22, find all n distinct roots. Put the answer in both polar form and in rectangular form, a + bi.

15. $x^5 = 1$

16. $x^6 = 1$

17. $x^3 = 4 - 4\sqrt{3}i$

18. $x^5 = -1 - i$

19. $x^6 = 64i$

20. $x^7 = 5 + i$

21. $x^5 = -3 + 2i$

[C] 22. $x^2 = -7.301 - 13.86i$

23. Construct the fifth roots of $1 + i$ and of 32 geometrically.

[C] 24. Find all solutions to $(3 - i)z^7 - 2 + 4i = 0$.

In Exercises 25–32, solve for x.

25. $8x^2 - 4\sqrt{2}x + (1 + 2i) = 0$

26. $ix^2 - 4x - (9 + 4i) = 0$

27. $8x^2 - 4\sqrt{3}x + (\frac{1}{2} - \sqrt{3}i) = 0$

28. $8x^2 - 4\sqrt{3}ix - (\frac{1}{2} - \sqrt{3}i) = 0$

[C] 29. $x^2 + (5 - 3i)x + 4 + 2i = 0$

[C] 30. $3x^2 - (3 - i)x + \frac{5}{6} = 0$

[C] 31. $ix^2 + (1 + i)x - 2 = 0$

[C] 32. $4ix^2 - 5x - 1 + i = 0$

4.5 An Application

Up to this point, we have not discussed any applications of complex numbers. This is because the applications of complex numbers involve fairly sophisticated mathematical and physical concepts. Since many students appreciate having some idea of the way in which the mathematics they are learning is useful, we will briefly indicate how complex numbers help in describing and understanding elementary electrical circuits.

We discussed simple circuits in the applications section of the chapter on general trigonometry. However, using only trigonometry we could not adequately generalize Ohm's Law to circuits containing a condenser, because of the phase shift. We now indicate how this is done with complex numbers. Before continuing, you might briefly review that section.

We first redescribe the *EMF* function for an *AC* circuit. Previously this was given as

$$E = a\sin(\omega t)$$

By changing the zero point, $t = 0$, we can equivalently use

$$E = a\cos(\omega t)$$

We now pretend that this is the real part of a complex function and that really

$$E = a(\cos(\omega t) + i\sin(\omega t)) = a\,\text{cis}(\omega t)$$

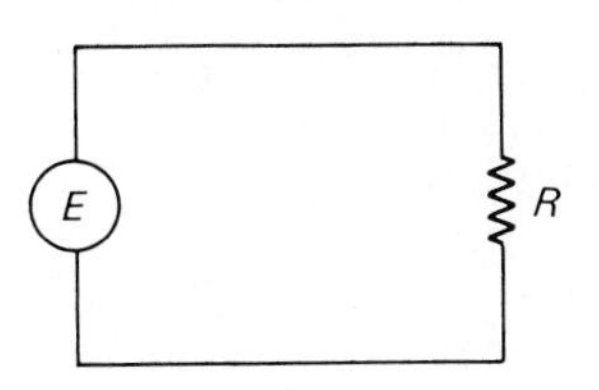

The imaginary part is, well, imaginary, but in fact it affects the results. If the circuit has just a resistor, R, then

$$I = \frac{a}{R}(\cos \omega t + i \sin \omega t) = \frac{a}{R}\operatorname{cis}(\omega t)$$

and Ohm's Law $E = IR$ holds. The imaginary part has no observable effect. If we now have a circuit with just a condenser, then the condenser throws the current 90° ahead of the EMF. This rotation of 90° can be accomplished mathematically by multiplying by i. Thus we now say the current is

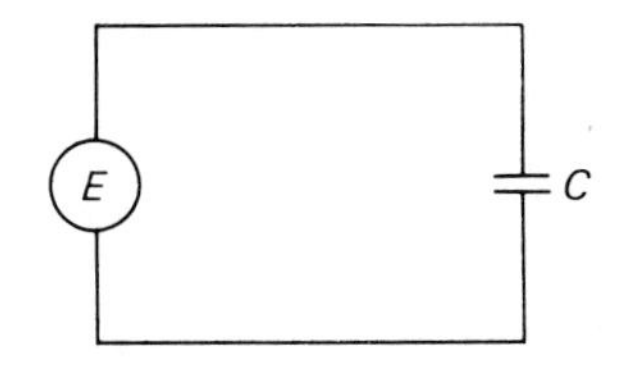

$$I = a\omega C(\cos \omega t + i \sin \omega t)i = a\omega Ci \operatorname{cis}(\omega t)$$

and the capacitance is pure imaginary,

$$\frac{1}{\omega Ci} = -\frac{1}{\omega C}i$$

Now Ohm's Law reads

$$E = I\left(-\frac{1}{\omega C}i\right)$$

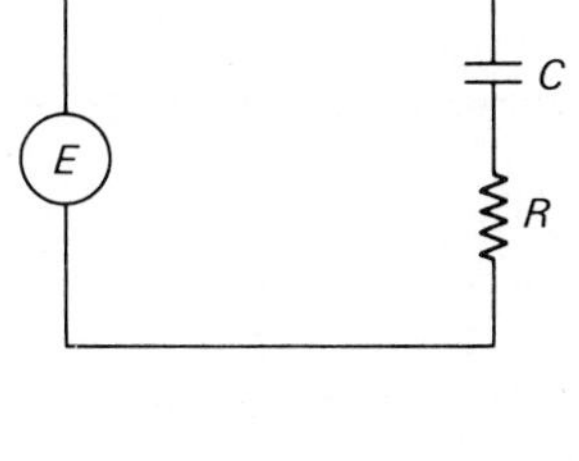

What now begins to be spectacular is what happens when we have both a resistor and a condenser "in series," i.e., one right after the other. Before, we said Ohm's Law is almost

$$E = IZ, \qquad Z = \sqrt{R^2 + \left(\frac{1}{\omega C}\right)^2}$$

except for a phase shift of $\phi = \tan^{-1}\left(\frac{1}{\omega CR}\right)$. We now consider Z to be a complex number $Z = a + bi$, which is the sum of R (real) and $-\frac{1}{\omega C}i$ (pure imaginary),

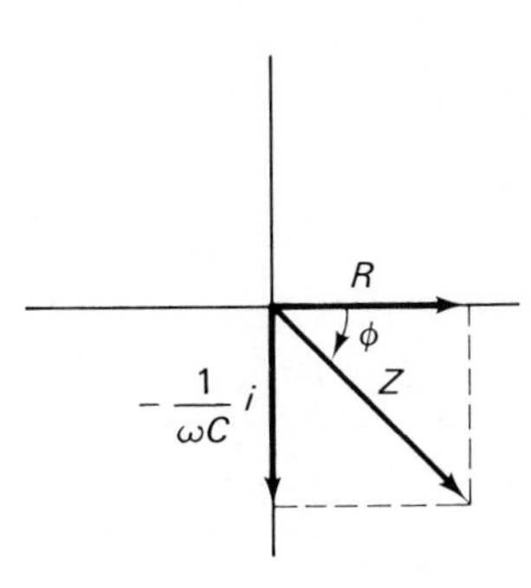

$$Z = R - \frac{1}{\omega C}i$$

The magnitude of Z is $\sqrt{a^2 + b^2} = \sqrt{R^2 + \left(-\frac{1}{\omega C}\right)^2}$ which is what we obtained before. The argument of Z is

$$\phi = \tan^{-1}\left(\frac{b}{a}\right) = \tan^{-1}\left(\frac{-1/\omega C}{R}\right) = -\tan^{-1}\left(\frac{1}{\omega CR}\right)$$

Thus, in polar form,

$$Z = \sqrt{R^2 + \left(\frac{1}{\omega C}\right)^2}\operatorname{cis}(-\phi), \qquad \phi = \tan^{-1}\left(\frac{1}{\omega CR}\right)$$

Finally, we take Ohm's Law to be

$$E = IZ$$

and solve for I, obtaining

$$I = EZ^{-1} = a \text{ cis } \omega t \left(\sqrt{R^2 + \left(\frac{1}{\omega C}\right)^2} \text{ cis}(-\phi) \right)^{-1}$$

$$= \frac{a}{\sqrt{R^2 + \left(\frac{1}{\omega C}\right)^2}} \text{ cis } \omega t \text{ cis } \phi$$

$$= \frac{a}{\sqrt{R^2 + \left(\frac{1}{\omega C}\right)^2}} \text{ cis}(\omega t + \phi)$$

Thus you see mathematically that the current I *must* have a phase shift forward of ϕ, which is exactly what happens physically.

REVIEW EXERCISES

In Exercises 1–16, write in the form $a + bi$.

1. $(3 - 2i) + \overline{(4 - 7i)}$
2. $(-2 + 3i)\overline{(5 + 6i)}$
3. $(-5 + \sqrt{-9})(2 - \sqrt{-4})$
4. $(3 - 4i)^2$
5. $i^{3219} - i^{427} + i^{18}$
6. $\dfrac{3 - i}{4 + 2i}$
7. $\dfrac{1 - i}{2i}$
8. $\sqrt{-81}\sqrt{-27}$
9. $(\sqrt{-27})^3$
10. $\dfrac{2 + \sqrt{-4}}{3 - \sqrt{-8}}$
11. $4 \text{ cis } 45°$
12. $3 \text{ cis}\left(-\dfrac{\pi}{2}\right)$
13. $(2 \text{ cis } 3)(3 \text{ cis } 2)$
14. $(4 \text{ cis } 30°)^{-1}$
15. $(3 \text{ cis } 150°)^{10}$
16. $\left[16 \text{ cis}\left(-\dfrac{\pi}{2}\right)\right]^{-4}$

In Exercises 17 and 18, solve for x and y real.

17. $3 - 2yi = 9^x - 7i$
18. $\cos x - \pi i = 0.5 + i \tan y$

In Exercises 19–25, solve.

19. $4ix - 5 = 6x - i$
20. $3ix - 2 = 0$
21. $x^3 - 8 = 0$
22. $2x^4 - 3x^2 + 1 = 0$
23. $3x^2 + (2 - 3i)x - i = 0$
24. $ix^2 + (2 - 3i)x - 2 - i = 0$
25. $x^5 = 4 - i$

CHAPTER 5

Exponential and Logarithmic Functions

Knowledge of exponents goes back to the very beginnings of mathematics. Tables of exponents have been found among old Babylonian tablets, and in ancient Mesopotamia, they even knew the compound interest formula. Logarithms, on the other hand, were invented fairly recently, as a tool to transform large, tedious multiplication problems (mainly from astronomy) into relatively easy addition problems. In 1614, John Napier (1550–1617) published his *Mirifici logarithmorum canonis descriptio* ("A Description of the Marvelous Rule of Logarithms") in which he virtually developed what we would call today the system of logarithms to the base $\frac{1}{e}$, though he did it with a geometric description of certain ratios. Napier coined the word *logarithm* from two Greek words, *logos* ("ratio") and *arithmos* ("number"). Henry Briggs (1561–1639), first in collaboration with Napier and then on his own, modified Napier's definition and developed the common logarithm as a more easily applied computing tool. John Bürgi (1552–1632) independently developed what was almost the natural logarithm system. He probably did this earlier than Napier but only published his version in 1620. Logarithms were to remain an invaluable computing tool until the advent of calculators and computers.

The relationship between logarithms and exponentials that we find so useful for problem solving today went unsuspected for over 60 years and was not fully exploited until Leonhard Euler (1707–1783), who was also the first person to use the letter e as the base for the natural logarithm.

5.1 Exponential Functions

From Chapter 1, we have various definitions of a^x, depending on the nature of x:

$a^n = a \cdots a$ (n factors)	if n is a positive integer
$a^0 = 1$	if $a \neq 0$. 0^0 is undefined
$a^{-n} = \dfrac{1}{a \cdots a}$ (n factors of a)	if n is a positive integer
$a^{1/m} = \sqrt[m]{a}$	m a positive integer
$a^{n/m} = (\sqrt[m]{a})^n$	m, n integers, $m > 0$, $\dfrac{n}{m}$ in lowest terms

Of course, a^{-n} is undefined if $a = 0$, and $a^{n/m}$ is not a real number if a is negative and m is even and n is not. It is natural to ask what is the meaning of a^x if x is irrational. For instance, what does $2^{\sqrt{2}}$ mean? We know that $\sqrt{2}$ has an infinite decimal expansion

$$\sqrt{2} = 1.4142135\ldots$$

Then each number of the sequence

$$1, \quad 1.4, \quad 1.41, \quad 1.414, \quad 1.4142\ldots$$

is a rational number (since $1.4 = \frac{14}{10}$, $1.41 = \frac{141}{100}$, etc.) which is an approximation to $\sqrt{2}$; the more decimal places, the better the approximation.

These rational numbers can be used to form a new sequence

$$2^1, \quad 2^{1.4}, \quad 2^{1.41}, \quad 2^{1.414}, \quad 2^{1.4142}, \ldots$$

Using some techniques in calculus, it can be shown that this second sequence approaches a unique real number; we define that real number to be $2^{\sqrt{2}}$. Whenever $a \geq 0$, we can use the same techniques to define a^x for any irrational number x; we can use it only for $a \geq 0$ because of the difficulty with even roots of negative numbers.

Under this definition, if x is an irrational number, then a^x "fits in" exactly where it should: If x is between numbers r and s, then a^x is between a^r and a^s. Moreover, it can be shown that all the laws of exponents still hold.

By this discussion, for each real number x there corresponds a unique real number a^x if $a > 0$. Thus we can define the exponential function as follows:

> **(1) Definition** If $a > 0$, the **exponential function with base a** is the function f defined by
>
> $$f(x) = a^x$$
>
> where x is any real number.

For different values of the base a, the exponential function $f(x) = a^x$ (and its graph) have different characteristics.

EXAMPLE 1 Let $a = 2$ or $\frac{1}{2}$, and observe the symmetry with respect to $x = 0$ (the y-axis). See Table 1 and Fig. 1.

TABLE 1

x	2^x	$(\frac{1}{2})^x$
-3	$\frac{1}{8}$	8
-2	$\frac{1}{4}$	4
-1	$\frac{1}{2}$	2
0	1	1
1	2	$\frac{1}{2}$
2	4	$\frac{1}{4}$
3	8	$\frac{1}{8}$

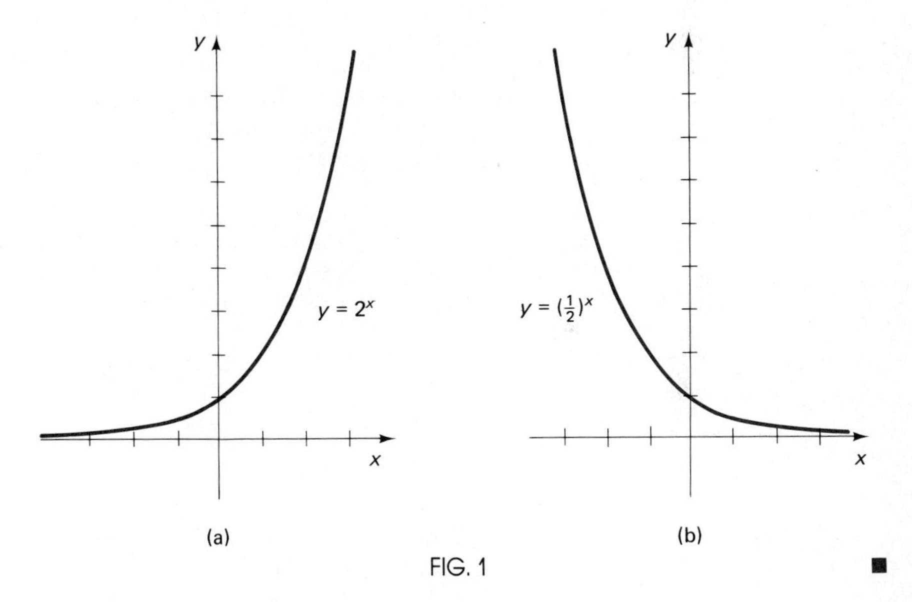

FIG. 1 ■

Example 1 illustrates the following general situation. If $a > 1$, the function $f(x) = a^x$ is strictly increasing; to the left the x-axis is a horizontal asymptote. If $0 < a < 1$, the function $f(x) = a^x$ is strictly decreasing; to the right the x-axis is a horizontal asymptote. For different values of a, these graphs differ in *steepness*, as indicated in Fig. 2.

Note that if $a > 1$, then $0 < a^{-1} < 1$, and the two graphs $y = a^x$ and $y = (a^{-1})^x$ are reflections of each other through the y-axis.

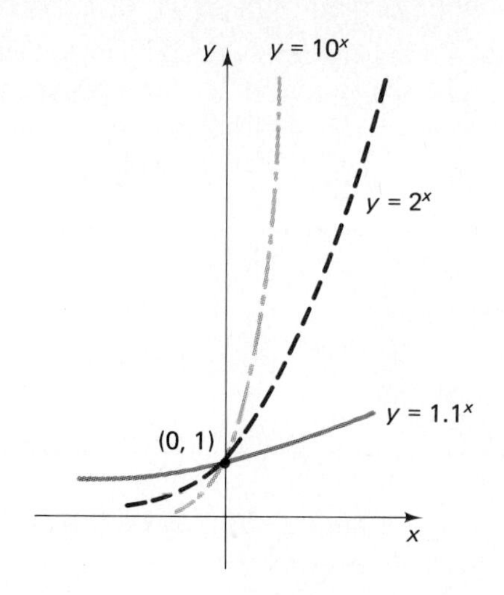

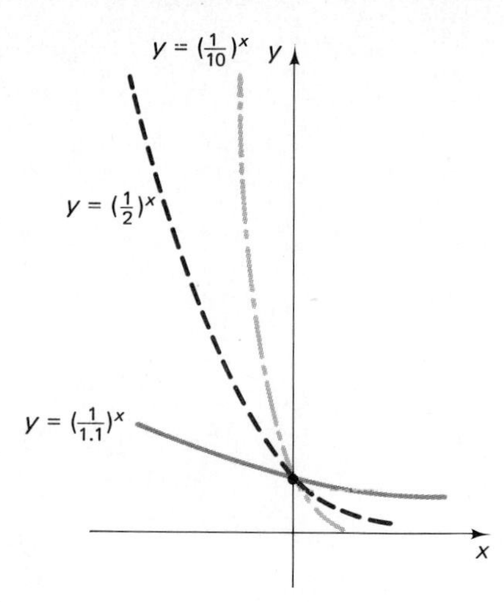

FIG. 2

GRAPHING TECHNIQUES

Graphing techniques can be used with exponential functions. In this situation, you should indicate the asymptote and one or two points. The following example involves translations only; the example after that involves a mixture of graphing techniques.

EXAMPLE 2 Sketch the graph of $y = 3^{x-2} - 3.$

Solution Start with the graph of $y = 3^x$. Translate it horizontally two units to the right. Then translate this down three units. Note that the asymptote is also translated.

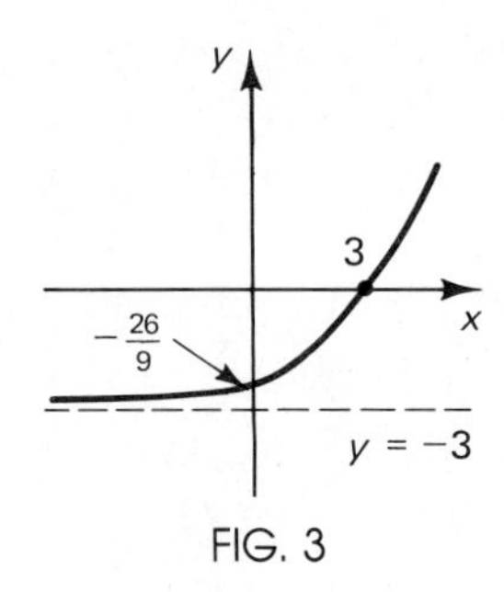

FIG. 3

■

EXAMPLE 3 Sketch the graph of $y = (-5)(\frac{1}{5})^x + 1.$

Solution Start with the graph of $y = (\frac{1}{5})^x$. Reflect it through the x-axis, and expand it vertically (i.e., multiply all the y-coordinates by -5), obtaining the graph of $y = (-5)(\frac{1}{5})^x$. Then translate this up one unit.

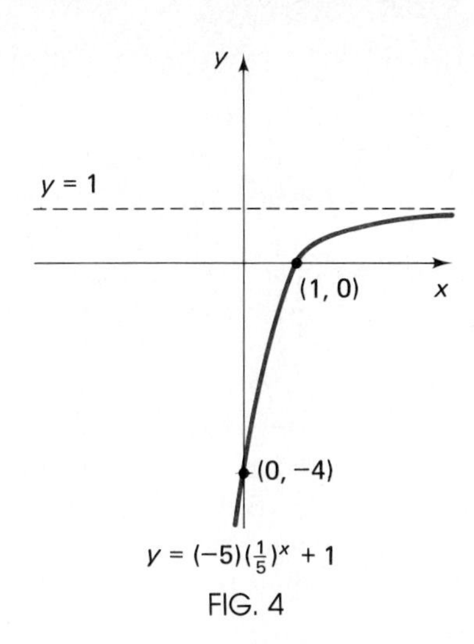

FIG. 4

■

There is a number denoted by e that arises frequently when dealing with exponentials and logarithms. This number plays a special role with these functions (just as π plays a special role with circles). It can be shown that e, like π, is irrational, $e = 2.71828 \ldots$. The exponential and logarithmic functions with base e are particularly useful in applications, as we shall see later.

The graph of $y = e^{-x^2}$ is a **normal distribution** curve. This is a very interesting and useful curve. To graph it, we first observe that it is symmetric to the y-axis. So to plot points, we need only fill in Table 2, sketch the curve for $x \geq 0$, and then use symmetry. Remember that e^u is always > 0. See Fig. 5(a). You might notice the graph is "bell-shaped." Of course, this graph can be expanded and translated, too. For example, the graph of $y = 2e^{-(x+1)^2} - 1$ is given in Fig. 5(b).

TABLE 2

x	$-x^2$	e^{-x^2}
0	0	1
0.5	−0.25	0.78
1	−1	0.37
1.5	−2.25	0.11
2	−4	0.02

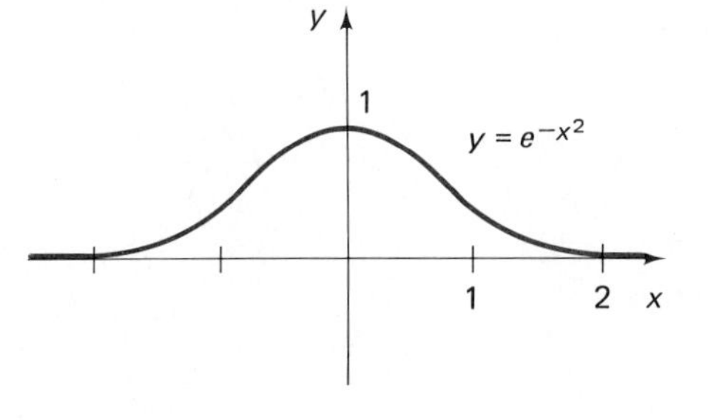

(a)

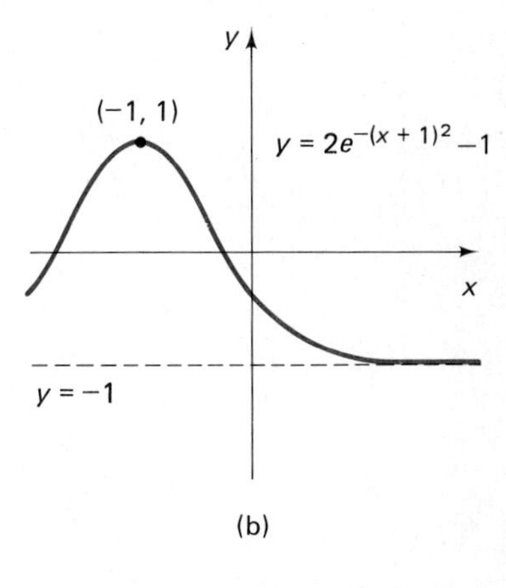

(b)

FIG. 5

ESTIMATING EXPONENTIALS

One of the themes of this text is that estimating is very important. When you are learning the material, it helps lead to a better understanding of the concepts. When doing a computation, it serves as a quick, useful check; even when using a calculator, it is extremely useful to have a rough estimate of the answer in order to check that you did not push the wrong button. (Usually if you push the wrong button during a calculator computation, the answer is so far off that *the error is obvious if you have a rough estimate* of the answer.)

The first method of estimation that we discuss we refer to as **rough interpolation.** This method is very useful, since, with a little practice, the estimate can be made in your head.

x	$f(x)$
a	$f(a)$
c	?
b	$f(b)$

Suppose you have a function f whose values at some points can easily be computed, and you wish to estimate a value elsewhere, say estimate $f(c)$. Let a and b be the closest points on either side of c such that you can easily compute their function values, and make the table on the left. You then determine if c is closer to one or the other of a or b, or about the middle, and then make a corresponding estimate for $f(c)$.

x	$y = \sqrt{x}$
1	1
π	?
4	2

For example, to estimate $\sqrt{\pi}$, we use $f(x) = \sqrt{x}$ and make the table shown in the margin. As π is between 1 and 4 but closer to 4, we see that $f(\pi) = \sqrt{\pi}$ is between 1 and 2 but closer to 2. Thus we might estimate $\sqrt{\pi} \approx 1.8$ or 1.9. In fact, $\sqrt{\pi} \approx 1.77245$, but either estimate would be close enough to tell us to check our work if the answer obtained using a calculator or tables was vastly different.

Another way to estimate the value of a function involves using its graph and is called **graphical estimation.** We now illustrate how to use both graphical estimation and rough interpolation to estimate exponential values.

EXAMPLE 4 Estimate $(\sqrt[3]{2})^7$.

Solution We are looking for an exponential function $f(x) = a^x$ which fits this situation. Since $(\sqrt[3]{2})^7 = 2^{7/3}$ (and 2 is an easy number to work with), we choose $f(x) = 2^x$. Then the problem may be restated: Estimate $f(\frac{7}{3})$.

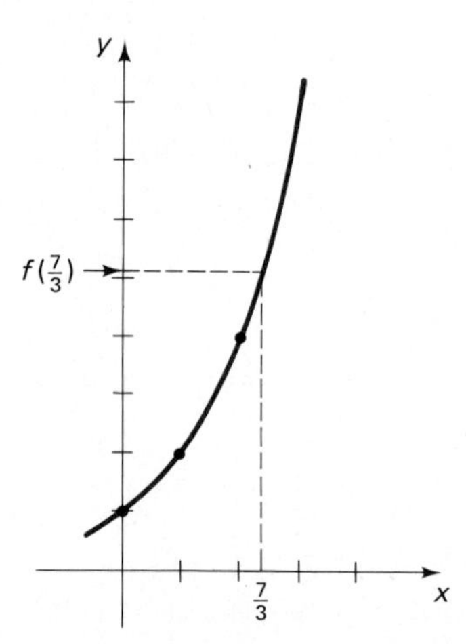

FIG. 6

TABLE 3

x	$y = 2^x$
2	4
$\frac{7}{3}$	?
3	8

To estimate graphically, plot $y = f(x)$ at $x = 0, 1, 2, 3$, and then sketch the graph, obtaining Fig. 6. From this, we guess that $f(\frac{7}{3}) \approx 5.1$.

To estimate $f(\frac{7}{3}) = 2^{7/3}$ by rough interpolation, we use Table 3. We pick 2 and 3 because $\frac{7}{3}$ is between them (and we can easily compute 2 raised to those powers). Since $\frac{7}{3}$ is a little closer to 2, we say $2^{7/3}$ is between 4 and 8 but a little closer to 4, say $2^{7/3} \approx 5$.

Using a calculator, $2^{7/3} \approx 5.03968$. ■

COMPUTING EXPONENTIALS

Computing by hand simple numbers raised to small integer powers, for example $3^4 = 81$ or $2^{-3} = 0.125$, is quite easy. Almost anything more complicated requires a calculator (or computer) or specialized tables. For this text, in Appendix C we give a table of the values of e raised to various numbers and explain how to use it. This table can be used to do some of the exercises at the end of this section. The tables in Appendix C could also be used to find other exponentials, such as $8.14^{1.112}$, and we shall explain how to do this in Sec. 5.3. However, we can explain how to do such computations now, if you have a calculator.

USING A CALCULATOR

Most calculators have three buttons related to exponential functions:

$$\boxed{10^x}, \quad \boxed{e^x}, \quad \boxed{y^x}$$

On some calculators, $\boxed{e^x} = \boxed{\text{INV}}\ \boxed{\text{LN } x}$ and $\boxed{10^x} = \boxed{\text{INV}}\ \boxed{\text{LOG } x}$.

The $\boxed{y^x}$ button computes the number y raised to the power x. To compute $23^{1.47}$, press

Alg.: $\boxed{23}\ \boxed{y^x}\ \boxed{1.47}\ \boxed{=}$

RPN: $\boxed{23}\ \boxed{\text{ENTER}}\ \boxed{1.47}\ \boxed{y^x}$

You should obtain 100.401.

Reminder: The preceding computation means $23^{1.47} \approx 100.401$. In this text, unless otherwise indicated, all intermediate calculations are carried to at least eight significant figures, and the answers are rounded to six.

The $\boxed{10^x}$ button computes 10 raised to the power x. To compute $10^{3.1}$, press (on any type of calculator)

$$\boxed{3.1}\ \boxed{10^x}$$

You should obtain 1258.93.

Note: A few calculators do not have a $\boxed{10^x}$ button. On such calculators, you can easily use the $\boxed{y^x}$ button.

The $\boxed{e^x}$ button computes e raised to the power x. To compute $e^{-4.17}$, press

$$\boxed{4.17}\ \boxed{+/-}\ \boxed{e^x}$$

You should get 0.0154523.

EXERCISES

In Exercises 1–12, use a calculator to compute the given number. Exercises 3–6 can be evaluated using Table C3 in Appendix C.

Ⓒ **1.** $10^{3.124}$

Ⓒ **2.** $\sqrt[5]{10}$

3. $e^{7.13}$

4. $e^{-4.12}$

5. $(e^{-1.2})^{3.2}$

6. $20e^{-0.08(3)}$

Ⓒ **7.** $\sqrt[20]{5}$

Ⓒ **8.** 53.8^{22}

Ⓒ **9.** $\left(1 + \dfrac{0.08}{12}\right)^{12(5)}$

Ⓒ **10.** $(2.31 + e^{15.2})^3 14.1^{-2}$

Ⓒ **11.** $381.7 + \sqrt{10^{4.31} - 22^3}$

Ⓒ **12.** $(3.21^{4.8} + 2.07^{16.2})^{0.07} 10^{30.65}$

13. Graph $y = 3^x$ and $y = (\frac{1}{3})^x$.

14. Use the graphs in Exercise 13 to estimate

a. $\sqrt{3}$ **b.** $3^{\sqrt{3}}$ **c.** $\sqrt[4]{\frac{1}{3}}$ **d.** $(\frac{1}{3})^{-2.1}$

15. Use rough interpolation to estimate the numbers in Exercise 14.

Ⓒ **16.** Use a calculator to compute approximations to the numbers in Exercise 14, and compare with Exercises 14 and 15.

In Exercises 17–24, estimate the given number first graphically and then by rough interpolation. Finally, compute the number with a calculator, and compare.

17. $\sqrt{5}$

18. $5^{0.2}$

19. $2^{-0.2}$

20. $2^{-0.9}$

21. $\sqrt[6]{2}$

22. $2^{-1/6}$

23. $\sqrt[3]{3^5}$

24. $\sqrt[5]{3^3}$

In Exercises 25–40, graph the equation, labeling at least two crucial things (a point and an asymptote or two points).

25. $y = 4^x + 1$

26. $y = 4^{-(x-2)}$

27. $y = 0.25^{x+1}$

28. $y = 0.25^{-x} - 1$

29. $y = 3^{-(x+1)} + 2$

30. $y = 3^{x-2} - 1$

31. $y = 0.1^{x-1} - 1$

32. $y = 0.1^{-(x-1)} - 2$

33. $y = e^{2x} - e$

34. $y = -3^{2x} - 3$

35. $y = -2e^{x-1} + e$

36. $y = (-2)4^{2x} + 2$

37. $y = 4^{-x^2} - 1$

38. $y = -2(3^{-x^2}) + 2$

39. $y = 3^{x^2} - 3$

40. $y = 2(2^{x^2}) + 1$

41. A certain factory has 3000 light bulbs. It was determined that if all the light bulbs were replaced at essentially the same time, the number of bulbs which had burned out after t hours was $N = 3000(1 - e^{-t/500})$. (Such functions can be used to show it is more economical to replace all light bulbs at once rather than just as they burn out.)

a. Graph this function (after appropriately labeling the axes).
b. How many bulbs have burned out after 100 hours? 500 hours? 900 hours?
c. If you want to replace all bulbs after 35% of them have burned out, when should you do this?

42. If a rock is dropped from a cliff or a building, then we can usually neglect air resistance, and its velocity is $V = -32t$ feet per second. However, if a sky diver jumps from an airplane with arms and legs spread out, then air resistance has an important effect, and the velocity is given by $V = K(1 - e^{-at})$, where K and a are constants. Suppose after jumping from an airplane at 10,000 feet, the velocity of a sky diver is given by $V = 220(1 - e^{-3t})$ feet per second, t in seconds, up until the time she opens her parachute.
a. Graph this function.
b. What is her velocity after 3 seconds?
c. How long does it take her to reach a velocity of 200 feet per second?
d. What is her *terminal velocity,* i.e., the velocity she approaches but never quite reaches? Use 30 miles per hour = 44 feet per second to express this in miles per hour.

5.2 Logarithms

LOGARITHMIC FUNCTIONS

If $a = 1$, the graph of $f(x) = a^x$ is a horizontal straight line. Hence it does not have an inverse function. However if $a > 1$, the graph of $f(x) = a^x$ is strictly increasing; if $0 < a < 1$, the graph is strictly decreasing. (See Fig. 1 for the cases $a = 2$, $a = \frac{1}{2}$.) Hence, by the horizontal line test, $f(x) = a^x$ has an inverse function in these cases.

> **(2) Definition** If $0 < a < 1$ or $a > 1$, the inverse of the exponential function $f(x) = a^x$ is called the **logarithm function (to the base *a*)** and is denoted by $\log_a x$.

Warning: Although $\log_a$ contains four letters, it should be thought of as one symbol (as is f^{-1}). The symbol $\log_a$ is read "log to the base a." A related difficulty is that $\log_a x$ should be written $\log_a(x)$, as in $f^{-1}(x)$. Unfortunately, the parentheses are traditionally left out.

From the basic relationship between a function and its inverse, we have the following, which is really just another form of the definition:

> **(3)** If $0 < a < 1$ or $a > 1$, then
> $$\log_a x = y \qquad \text{if and only if} \qquad a^y = x$$

The function $f(x) = \log_a x$ is not defined for all values of x. Since a is positive, a^y is positive. Thus with $x = a^y$, we see that

> **(4)** If $0 < a < 1$ or $a > 1$, $\log_a x$ is defined only for $x > 0$.

This restriction will be very important in some of the problems that follow.

Many logarithms can be calculated mentally using the definition in the form (3). To compute $\log_a x$, just ask "a to what power is x?"

EXAMPLE 1 Compute
(a) $\log_2 16$; (b) $\log_{10} 0.0001$; (c) $\log_{16} 2$; (d) $\log_3(-3)$
(e) $\log_a a$; if $a > 0, a \neq 1$; (f) $\log_a 1$, if $a > 0, a \neq 1$

Solution
(a) By (3), $\log_2 16 = y$ if and only if $2^y = 16$. Since $2^4 = 16$, $\log_2 16 = 4$. This demonstrates what we said above: To find $\log_2 16$ ask "2 to what power is 16?"
(b) To find $\log_{10} 0.0001$, ask "10 to what power is 0.0001?" Since $10^{-4} = 0.0001$, $\log_{10} 0.0001 = -4$.
(c) To find $\log_{16} 2$, ask "16 to what power is 2?" Since $2 = \sqrt[4]{16} = 16^{1/4}$, $\log_{16} 2 = \frac{1}{4} = 0.25$.
(d) By (4), $\log_a x$ is defined only for $x > 0$, so $\log_3(-3)$ does not exist.
(e) To find $\log_a a$, ask "a to what power is a?" Since $a^1 = a$, we have $\log_a a = 1$.
(f) To find $\log_a 1$, ask "a to what power is 1?" Since $a^0 = 1$, we have $\log_a 1 = 0$. ■

The last two computations are special enough to emphasize.

(5) Theorem If $0 < a < 1$ or $a > 1$, then

(i) $\log_a a = 1$ and (ii) $\log_a 1 = 0$

GRAPHS OF LOGARITHMIC FUNCTIONS

Recall that if you have the graph of $y = f(x)$, the graph of $y = f^{-1}(x)$ is obtained by reflection through the line $y = x$. Applying this to $f(x) = a^x$ so that $f^{-1}(x) = \log_a x$, we easily get the graph of $y = \log_a x$, shown in Fig. 7.

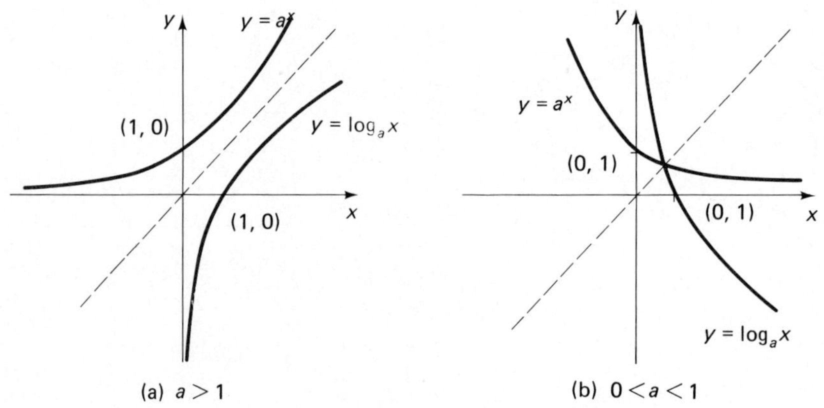

FIG. 7

From these graphs, we see that the graph of $y = \log_a x$ is strictly increasing if $a > 1$ and strictly decreasing if $0 < a < 1$.

Note that the graph of $y = \log_a x$ crosses the x-axis at $x = 1$ (since $\log_a 1 = 0$) and that the y-axis is a vertical asymptote. For different values of a, the graphs of $y = \log_a x$ differ in *steepness,* as indicated in Fig. 8. (The case $0 < a < 1$ is analogous. Compare with Fig. 2, Sec. 5.1.)

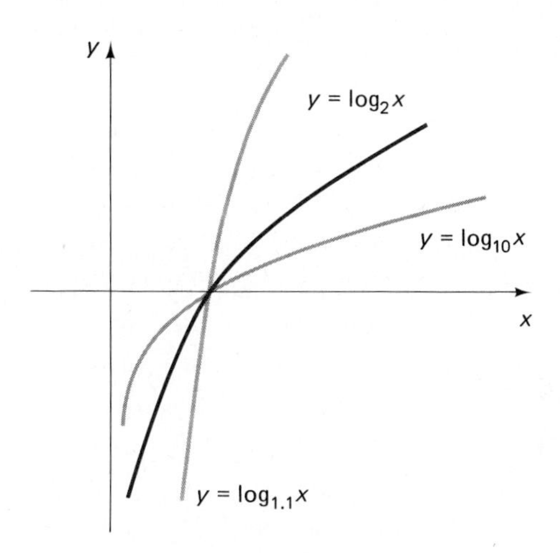

Graphs of $y = \log_a x$
various $a > 1$.
FIG. 8

We can also observe what the domains and ranges of $f(x) = a^x$ and $f^{-1}(x) = \log_a x$ are, namely,

(6)	$f\colon \mathbb{R} \to (0, \infty)$; $\quad f^{-1} = \log_a\colon (0, \infty) \to \mathbb{R}$

In other words, the domain of $f(x) = a^x$ is the set of all real numbers, and the range is the set of all positive real numbers. The domain of $f^{-1}(x) = \log_a x$ is the set of all positive real numbers, and the range is the set of all real numbers. This illustrates the earlier discussion of inverse functions where we observed that the domain of a function is the range of its inverse and vice versa.

Just as with exponential functions, the graphing techniques can be used with logarithmic functions. Again, you should indicate the asymptote and one or two points.

EXAMPLE 2 Sketch the graph of $\quad y = \log_3 (x - 2)$.

Solution Start with the graph of $y = \log_3 x$ and translate it two units to the right.

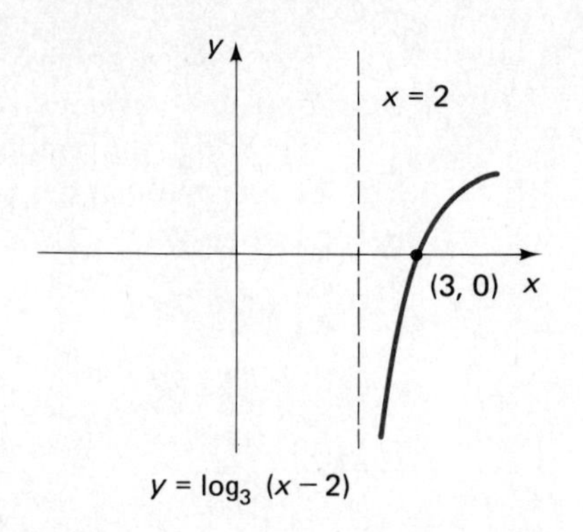

EXAMPLE 3 Sketch the graph of $y = 3 \log_{1/4}(x + 4)$.

Solution Start with the graph of $y = \log_{1/4} x$. Expand it vertically (by multiplying all y-coordinates by 3), obtaining the graph of $y = 3 \log_{1/4} x$. Then translate this four units to the left.

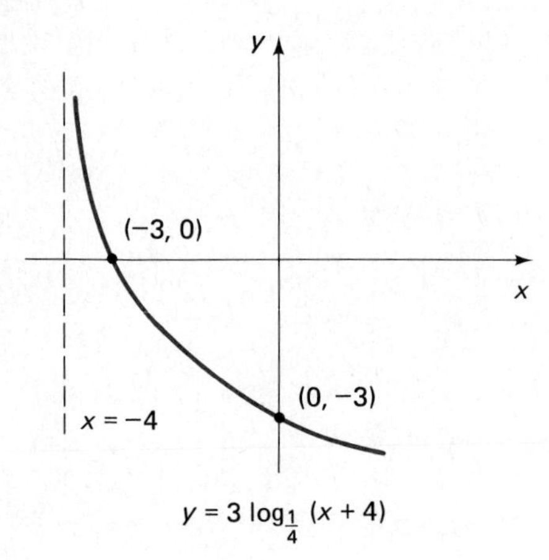

EXERCISES

In Exercises 1–14, find the given number without using a calculator or tables.

1. $\log_2 8$

2. $\log_8 2$

3. $\log_8 1$

4. $\log_8 8$

5. $\log_3(\log_3 27)$

6. $\log_2[\log_2(\log_2 4)]$

7. $\log_9 \frac{1}{81}$

8. $\log_9 3$

9. $\log_9 27$

10. $\log_{10} 10{,}000{,}000$

11. $\log_{10} 0.000000000001$

12. $\log_{0.01} 10$

13. $\log_{1/2} 4$

14. $\log_{0.2} 125$

In Exercises 15–38, graph the equation, labeling at least two crucial things (a point and an asymptote or two points).

15. $y = \log_4 x$
16. $y = \log_4(-x)$
17. $y = -\log_4 x$
18. $y = -\log_4(-x)$
19. $y = \log_3(x + 3)$
20. $y = \log_{1/3}(x - 1)$
21 $y = \log_e(x + e)$
22. $y = \log_e x - 1$
23. $y = \log_{0.5}(x - 1) + 2$
24. $y = \log_5(x + 5) + 1$
25. $y = -\log_{10}(x - 2)$
26. $y = 2 \log_{0.3}(x + 2)$
27. $y = \log_{0.2}(2x)$
28. $y = 3 \log_{10}(x + 3)$
29. $y = 3 \log_4(3 - x)$
30. $y = \log_{1/3}(2x - 2)$
31. $y = \log_{0.5}(1 - x) + 2$
32. $y = -2 \log_2(4 - 2x)$
33. $y = \log_7|x|$
34. $y = 2 \log_4(x + 2) + 2$
35. $y = e^{\log_e x}$
36. $y = |\log_7 x|$
37. $y = |\log_{0.25} x|$
38. $y = 10^{\log_{10}|x|}$

5.3 Estimating and Computing Logarithms

The purpose of this section is to present evaluation of logarithmic functions, first by estimation and then using tables or calculators. We also introduce two properties which are used in computations.

ESTIMATING LOGARITHMS

Logarithms may be estimated by rough interpolation in essentially the same way that exponentials are.

EXAMPLE 1 Estimate $\log_2 14$ by rough interpolation.

Solution Use Table 4. We pick 8 and 16 because we can easily compute their logarithms to the base 2, and 14 is between them. Since 14 is closer to 16, we say $\log_2 14$ is between 3 and 4 but closer to 4. We might guess $\log_2 14 \approx 3.8$. (In fact, $\log_2 14 \approx 3.80736$.) ■

TABLE 4

x	$y = \log_2 x$
$8(= 2^3)$	3
14	?
$16(= 2^4)$	4

TABLE 5

x	$y = \log_2 x$
$\frac{1}{4}(= 2^{-2})$	-2
$\frac{1}{5}$	?
$\frac{1}{8}(= 2^{-3})$	-3

EXAMPLE 2 Estimate $\log_2 \frac{1}{5}$ by rough interpolation.

Solution Use Table 5. Since $\frac{1}{5}$ is between $\frac{1}{4}$ and $\frac{1}{8}$ but closer to $\frac{1}{4}$, we say $\log_2 \frac{1}{5}$ is between -2 and -3 but closer to -2. We might guess $\log_2 \frac{1}{5} \approx -2.2$. (In fact, $\log_2 \frac{1}{5} \approx -2.32193$.) ■

The techniques of graphical estimation can be applied to logarithmic functions.

EXAMPLE 3 Graphically estimate $\log_2 3$ and $\log_2 \frac{1}{5}$.

Solution In Table 6, we list a few easy-to-find values near the numbers we are interested in. Then we plot them to get Fig. 9. From the graph, $\log_2 3 \approx 1.6$ and $\log_2 \frac{1}{5} \approx -2.3$.

TABLE 6

x	$y = \log_2 x$
$\frac{1}{8}$	-3
$\frac{1}{4}$	-2
$\frac{1}{2}$	-1
1	0
2	1
4	2

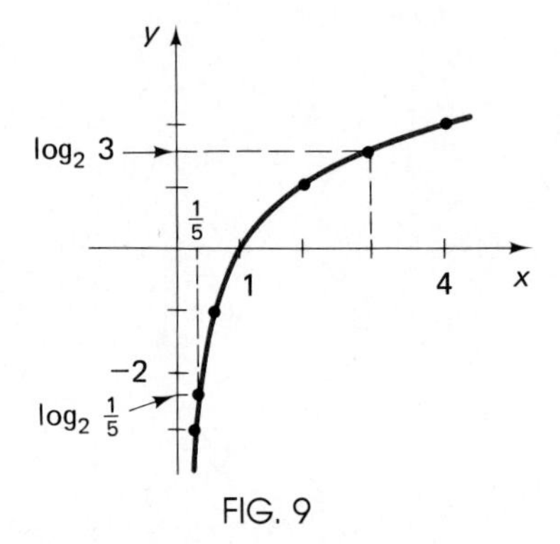

FIG. 9 ■

SPECIAL BASES

These are two bases that are very important for both historical and scientific reasons. Their logarithms have special names and usually have special notations.

> **Notation**
> **i.** If the base is 10, the logarithm $\log_{10} x$ is called the **common logarithm** of x. Often the 10 is not written, so that $\log x$ usually means $\log_{10} x$.
> **ii.** If the base is e, the logarithm $\log_e x$ is called the **natural logarithm** of x. Often $\ln x$ is written for $\log_e x$.

The symbol ln is usually read "ell-en" or "lin."

Hereafter in this text, we shall usually use $\log x$ for $\log_{10} x$ and $\ln x$ for $\log_e x$. Most calculators use this notation. However, some texts use $\log x$ for $\log_e x$, so you should be careful about this abbreviation.

USING TABLES

Appendix C contains tables of natural logarithms and common logarithms. There are also instructions as to how to use such tables, if you do not already know how. If you are using tables (rather than, or in addition to, calculators) and are inexperienced at using tables, this would be an appropriate time to read Appendix C. For example, you should be able to use these tables to determine

$$\ln 9.13 \approx 2.2116 \qquad \text{and} \qquad \log 358.4 \approx 2.5544$$

USING A CALCULATOR

The functions $\log x$ and $\ln x$ are so important that most scientific calculators have separate buttons for them:

$$\boxed{\text{LOG } x} \text{ (or just } \boxed{\text{LOG}}\text{)}, \qquad \boxed{\text{LN } x} \text{ (or just } \boxed{\ln}\text{)}$$

To compute $\log 100 = \log_{10} 100$, press (on any calculator)

$$\boxed{100}\ \boxed{\text{LOG } x}$$

Of course, $\log 100 = \log_{10} 100 = 2$.

Computing $\ln 100 = \log_e 100$ is similar. Press

$$\boxed{100}\ \boxed{\text{LN } x}$$

obtaining $\ln 100 = \log_e 100 \approx 4.60517$.

Since $\log_b x$ is undefined for $x \leq 0$, if you attempt to compute $\log x$ or $\ln x$ for $x \leq 0$, the calculator will give an error message.

COMPUTING $\log_b x$

Whether using tables or a calculator, the computation of logarithms to bases other than 10 or e usually requires a small calculation. (A few calculators have a $\boxed{\log_x y}$ button. However, most calculators just have $\boxed{\log}$ and $\boxed{\ln}$ buttons.) To compute $\log_b x$, $b \neq e$, 10, you must use the following:

(7) Theorem

$$\log_b x = \frac{\log_a x}{\log_a b}, \quad \text{where } a, b, x > 0, \quad a, b \neq 1$$

This will be proved in Sec. 5.4 [see (15iii)].

By Th. 7, to compute $\log_b x$, we can use any base a for which we can find $\log_a x$ and $\log_a b$. In particular, when using a calculator or tables, we can choose either $a = 10$ or $a = e$.

EXAMPLE 4 Compute $\log_2 3$ and $\log_2 \frac{1}{5}$.

Solution

$$\log_2 3 = \frac{\log_{10} 3}{\log_{10} 2} = \frac{\log 3}{\log 2} \approx \frac{0.4771213}{0.3010300} \approx 1.58496$$

or

$$\log_2 3 = \frac{\log_e 3}{\log_e 2} = \frac{\ln 3}{\ln 2} \approx \frac{1.0986123}{0.69314718} \approx 1.58496$$

$$\log_2 0.2 = \frac{\log_{10} 0.2}{\log_{10} 2} = \frac{\log 0.2}{\log 2} \approx \frac{-0.698970}{0.3010300} \approx -2.32193$$

or

$$\log_2 0.2 = \frac{\log_e 0.2}{\log_e 2} = \frac{\ln 0.2}{\ln 2} \approx \frac{-1.6094379}{0.69314718} \approx -2.32193 \quad \blacksquare$$

Although on most problems you have a choice of using either the log or ln function,

For certain later problems it will be easier to use natural logarithms,

so you should get used to using them now. To do the computation on a calculator is not difficult. For example, to compute $\frac{\ln 3}{\ln 2}$,

Alg.: [3] [LN x] [÷] [2] [LN x] [=]

RPN: [3] [LNx] [ENT] [2] [LN x] [÷]

To do the computation with tables, you simply look up the numbers and then divide.

COMPUTING y^x

We mentioned in Sec. 5.1 that it is possible to use the tables in Appendix C to compute exponentials such as $8.14^{1.112}$. The relationship used for such computations is the following:

> **(8) Theorem**
>
> $$y^x = e^{x \ln y}, \qquad y > 0, \text{ all } x$$

This will be proved in Sec. 5.4. [See (15ii).] In fact, the relationship (8) is used by calculators to compute y^x when the [y^x] button is pressed. Since ln y is defined only for $y > 0$, this is why calculators do not compute y^x for $y \le 0$. In fact, on some algebraic calculators if you want to compute $(3.1)^{5.2}$ and press [3.1] [y^x], the number 1.131402 appears in the display. This is just ln 3.1; the machine has just started the computation $e^{5.2 \ln 3.1}$.

EXAMPLE 5 Use (8) (and a calculator or tables) to compute (a) 2^3; (b) $8.14^{1.112}$

Solution (a) By (8), $2^3 = e^{3 \ln 2}$, so press

Alg.: [2] [LN] [×] [3] [=] [e^x]

RPN: [2] [LN] [3] [×] [e^x]

You should, of course, get 8, but you might get something like 7.9999998 due to calculator error.

Using tables, look up ln 2 in Table C1,

$$\ln 2 \approx 0.6931$$

Then multiply by 3, $3 \ln 2 \approx 2.0793$, and finally look up $e^{2.0793}$ in Table C3

$$e^{2.0793} \approx 8.0008$$

which we obtained by interpolation.

(b) By (8), $8.14^{1.112} = e^{1.112 \ln 8.14}$. We evaluate this as in (a), obtaining 10.2947 using a calculator. Of course, if you are doing this by hand using tables, the computation 1.112 times ln 8.14 is unpleasant but straightforward. The answer is 10.03; this is different because of the round-off error in the tables. ■

EXERCISES

In Exercises 1–6, use either a calculator or tables to evaluate.

1. ln 4.92 **2.** ln 1.137 **3.** log 5.122 **4.** log 7.124
5. log 21390 **6.** log 0.0006141
7. Graph $y = \log_3 x$ and $y = \log_{1/3} x$.

In Exercises 8–14, use Exercise 7 to estimate graphically the given numbers. Then use a calculator or tables to compute the numbers, and compare.

8. $\log_3 8$, $\log_{1/3} 8$
9. $\log_3 4$, $\log_{1/3} 4$
10. $\log_3 1.2$, $\log_{1/3} 1.2$
11. $\log_3 0.8$, $\log_{1/3} 0.8$
12. $\log_3 0.2$, $\log_{1/3} 0.2$
13. $\log_3 \frac{1}{20}$, $\log_{1/3} \frac{1}{20}$
14. $\log_3 12$, $\log_{1/3} 12$

In Exercises 15–28, estimate the given number by rough interpolation. Then use a calculator or tables to find that number, and compare.

15. $\log_3 4$
16. $\log_3 11$
17. $\log_3 21$
18. $\log_2 11$
19. $\log_2 39$
20. $\log_5 31$
21. $\log_7 14$
22. $\log_3 60$
23. $\log_3 0.1$
24. $\log_3 0.01$
25. $\log_5 0.3$
26. $\log_7 150$
27. $\log_7 0.2$
28. $\log_{12} 140$

In Exercises 29–32, compute y^x by $e^{x \ln y}$ (on your calculator or using tables) after first computing the answer by hand.

29. 2^4 **30.** 4^{-1} **31.** 10^6 **32.** $(0.1)^{-3}$

33. a. Compute $\ln \frac{3}{4}$, $\frac{\ln 3}{\ln 4}$, $\ln 2^3$, and $(\ln 2)^3$.

b. Does

$$\log_b \frac{x}{y} = \frac{\log_b x}{\log_b y} \quad \text{or} \quad \log_b x^y = (\log_b x)^y\,?$$

c. Prove:

$$\frac{\log_b x}{\log_b y} = \frac{\ln x}{\ln y}, \quad x, y, b > 0 \quad \text{and} \quad y, b \neq 1.$$

[c] **34.** This is very important and should be done before Sec. 5.4. Make a copy of the following table and fill it in using the [LOG] button on your calculator. Then

a. Find three relationships that you would guess would always be true (for $x, y > 0$).
b. Find at least two relationships that look like they might be true but that the table shows are not [for example, $\log(x - y) \neq \log x - \log y$].

x	y	$\log x$	$\log y$	$\log(xy)$	$\log(x + y)$	$\log x + \log y$	$(\log x)(\log y)$
2	3						
5.1	0.21						
0.01	0.3						
1	5						

x^y	$\log x^y$	$(\log x)^y$	$y \log x$	$\log \frac{x}{y}$	$\frac{\log x}{\log y}$	$\log x - \log y$

5.4 Properties of Logarithms

In this section we introduce and verify several properties of logarithms and show how to use these properties to solve equations that arise in applications. The most basic property of a logarithm is its definition. We shall mainly be using the form of the definition given in (3), which we restate here for convenience:

> **(9)** If $0 < a < 1$ or $a > 1$, then
> $$\log_a x = y \quad \text{if and only if} \quad a^y = x$$

The definition of $\log_a$ in this form is useful in solving certain equations.

EXAMPLE 1 Solve the following: (a) $\log_4 2 = x$; (b) $\log_4 x = 2$; (c) $\log_x 2 = 4$.

Solution In all problems of this form, the first step is to use (9). What you do from there depends on the form of the resulting equation.

(a) $\log_4 2 = x$ becomes $2 = 4^x$. Since $2 = \sqrt{4} = 4^{1/2}$, $x = \frac{1}{2}$.
(b) $\log_4 x = 2$ becomes $x = 4^2$, so $x = 16$.
(c) $\log_x 2 = 4$ becomes $2 = x^4$, so $x = 2^{1/4} \approx 1.18921$. (*Note:* x, as a base of a log, must be greater than zero.) ■

There are two very useful identities which follow from the definition in (9). To simplify $\log_a a^x$, we ask "a to what power is a^x?" Since $a^x = a^x$, we get

(10) Theorem If $0 < a < 1$ or $a > 1$, then $\log_a a^x = x$ for all real x.

To simplify $a^{\log_a x}$, we use (9) and substitute $y = \log_a x$ into the expression $a^y = x$, obtaining

(11) Theorem If $0 < a < 1$ or $a > 1$, then $a^{\log_a x} = x$ for all $x > 0$.

EXAMPLE 2 Solve (a) $3^{\log_3 x^2} = 25$; (b) $\log_{0.25} (\frac{1}{4})^{2x-1} = 5$.

Solution (a) By (11), $3^{\log_3 x^2} = x^2$. Thus the equation may be rewritten as $x^2 = 25$. The solutions are $x = \pm 5$.

(b) By (10), $\log_{0.25} (\frac{1}{4})^{2x-1} = 2x - 1$. Thus the equation may be rewritten as $2x - 1 = 5$. The solution is $x = 3$. ■

Note: Theorems (10) and (11) merely restate that $f^{-1}(f(x)) = x$ and $f(f^{-1}(x)) = x$ in the case $f(x) = a^x$ and $f^{-1}(x) = \log_a x$. Two additional fundamental properties are as follows:

(12) Theorem If $0 < a < 1$ or $a > 1$, then

i. $a^x = a^y$ if and only if $x = y$.

ii. $\log_a x = \log_a y$ if and only if $x = y$ with $x > 0$ and $y > 0$.

In the first paragraph of Sec. 5.2, we observed that the function $f(x) = a^x$ is strictly increasing (if $a > 1$) or strictly decreasing (if $0 < a < 1$). Thus two different values of x give two different values of a^x, yielding part i. (This allows a^x to have an inverse.) Part ii follows similarly.

EXAMPLE 3 Solve $3^x = 81$.

Solution $3^x = 81, \quad 3^x = 3^4, \quad x = 4$ by (12i). ■

EXAMPLE 4 Solve $\log_{0.1}(x^2 - 10) = \log_{0.1} 3x$.

Solution

$$\log_{0.1}(x^2 - 10) = \log_{0.1} 3x$$

$$x^2 - 10 = 3x \qquad \text{[by (12ii)]}$$

$$x^2 - 3x - 10 = 0$$

$$(x - 5)(x + 2) = 0, \qquad x = 5, -2$$

However, here we must be careful. By (4), $\log_a x$ is defined only for $x > 0$. If we substitute $x = -2$ back into the original equation, we get $\log_{0.1}(-6)$, which is undefined. However, substituting $x = 5$ yields $\log_{0.1}(25 - 10) = \log_{0.1} 15$, which is true. Thus, $x = 5$ is the only answer for this problem. Hence it is important to remember always to check any problem which has a logarithm in its original formulation. ■

LOGARITHMS AND THE ARITHMETIC OPERATIONS

We now turn to the relationships between logarithms and the operations of addition, subtraction, multiplication, division, and exponentiation.

> **(13) Theorem** If $0 < a < 1$ or $a > 1$, then
>
> **i.** $\log_a (uw) = \log_a u + \log_a w, \qquad u, w > 0.$
>
> **ii.** $\log_a \left(\dfrac{u}{w}\right) = \log_a u - \log_a w, \qquad u, w > 0.$
>
> **iii.** $\log_a (u^c) = c \log_a u, \qquad u > 0$, c any real number.

The properties of logarithms listed in (13) are the reasons logarithms were invented in the early seventeenth century, as the first two change a multiplication or division problem into an addition or subtraction problem, while the third changes an exponentiation problem into a multiplication problem.

Proof of (13i): Let $r = \log_a u$ and $s = \log_a w$. Then $a^r = u$ and $a^s = w$, by (9). Multiplying, we obtain $a^r a^s = uw$, so $a^{r+s} = uw$ by the laws of exponents. By (9), $\log_a (uw) = r + s$. Substituting $r = \log_a u$ and $s = \log_a w$ yields $\log_a (uw) = \log_a u + \log_a w$.

Since the proof of (13ii) uses (13iii), we prove (13iii) first.

Proof of (13iii): Let $r = \log_a u$. By (9), $a^r = u$. Raising both sides to the power c, we obtain $(a^r)^c = u^c$. By the laws of exponents, $a^{cr} = u^c$. By (9) again, $cr = \log_a u^c$. Substituting $r = \log_a u$, we obtain $c \log_a u = \log_a u^c$.

Proof of (13ii): This follows from (13i) and (13iii):

$$\begin{aligned} \log_a \left(\frac{u}{w}\right) &= \log_a (uw^{-1}) \\ &= \log_a u + \log_a w^{-1} && \text{[by (13i)]} \\ &= \log_a u + (-1) \log_a w && \text{[by (13iii)]} \\ &= \log_a u - \log_a w \end{aligned}$$ ■

Example 5 is very helpful in understanding the properties of logarithms given in (13).

EXAMPLE 5 Suppose $\log_7 2 \approx 0.36$ and $\log_7 3 \approx 0.56$. Using only this and fundamental properties, find (a) $\log_7 6$ and (b) $\log_7 \frac{72}{7}$.

Solution (a) $\log_7 6 = \log_7 (3 \cdot 2) = \log_7 3 + \log_7 2 \approx 0.56 + 0.36 = 0.92.$

(b) $\log_7 \frac{72}{7} = \log_7 (8 \cdot 9) - \log_7 7 = \log_7 2^3 + \log_7 3^2 - 1$
$= 3 \log_7 2 + 2 \log_7 3 - 1 \approx 3(0.36) + 2(0.56) - 1 = 1.2.$ ■

EXAMPLE 6 Simplify $9^{2 \log_9 7 + 3 \log_9 2}$.

Solution

$$9^{2 \log_9 7+3 \log_9 2} = 9^{\log_9 7^2+\log_9 2^3} \quad \text{[by (13iii)]}$$
$$= 9^{\log_9 (7^2 2^3)} \quad \text{[by (13i)]}$$
$$= 7^2 2^3 = 49(8) = 392 \quad \text{[by (11)]}$$ ■

EXAMPLE 7 Simplify $\log_3 (x^2 - 9) - \log_3 (x - 3)$.

Solution

$$\log_3 (x^2 - 9) - \log_3 (x - 3) = \log_3 \frac{x^2 - 9}{x - 3} \quad \text{[by (13ii)]}$$
$$= \log_3 (x + 3)$$ ■

Frequently it is necessary to use the properties of logarithms when solving equations involving exponentials or logarithms.

EXAMPLE 8 Solve $30e^{0.12x} = 6$.

Solution Divide by 30 first:

$$e^{0.12x} = \tfrac{6}{30} = 0.2$$

Next take ln's, using (12ii):

$$\ln e^{0.12x} = \ln 0.2$$

Now use $\ln e^u = u$, by (10):

$$0.12x = \ln 0.2$$

Finally, solve for x:

$$x = \frac{\ln 0.2}{0.12} \approx -13.4120$$ ■

There are problems that appear similar where the solutions require that you take either ln's or roots. The key to deciding what to do is the following:

> **(14)** If the unknown is in the exponent, take ln's.
> If the unknown is in the base, take roots.

EXAMPLE 9 Solve for x: (a) $12^x = 30$ and (b) $x^{12} = 30$.

Solution

(a) $12^x = 30$
$\ln 12^x = \ln 30$ [by (12ii)]
$x \ln 12 = \ln 30$ [by (13iii)]
$$x = \frac{\ln 30}{\ln 12} \approx 1.36874$$

(b) $x^{12} = 30$
$(x^{12})^{1/12} = \pm 30^{1/12}$
$x = \pm 30^{1/12}$
$x \approx \pm 1.32768$ ■

EXAMPLE 10 If $5000 = 2000(1 + r)^n$, find (a) r if $n = 13$ and (b) n if $r = 0.06$.

Solution (a)
$$5000 = 2000(1 + r)^{13}$$
$$2.5 = (1 + r)^{13}$$
$$(2.5)^{1/13} = 1 + r$$
$$r = (2.5)^{1/13} - 1$$
$$r \approx 0.0730273$$

(b)
$$5000 = 2000(1 + 0.06)^n$$
$$2.5 = 1.06^n$$
$$\ln 2.5 = \ln 1.06^n$$
$$n = \frac{\ln 2.5}{\ln 1.06}$$
$$n \approx 15.7252 \quad \blacksquare$$

There are several further properties of logarithms in addition to those given in (13). Some of the more important are the following [(13) is used in their proofs]:

> **(15) Theorem** If a, $b > 0$ and a, $b \neq 1$, then
>
> **i.** $\log_a \frac{1}{x} = -\log_a x, \quad x > 0.$
>
> **ii.** $y^x = e^{x \ln y}, \quad y > 0$, all x.
>
> **iii.** $\log_b x = \frac{\log_a x}{\log_a b}, \quad x > 0.$

Proof of (15i):

$$\log_a \frac{1}{x} = \log_a x^{-1}$$
$$= (-1) \log_a x \qquad \text{[by (13iii)]}$$
$$= -\log_a x$$

Proof of (15ii):

$$y^x = e^{\ln (y^x)} \qquad \text{[by (11)]}$$
$$= e^{x \ln y} \qquad \text{[by (13iii)]}$$

Proof of (15iii):

$$x = b^{\log_b x} \qquad \text{[by (11)]}$$
$$\log_a x = \log_a (b^{\log_b x}) \qquad \text{[by (12ii)]}$$
$$\log_a x = (\log_b x)(\log_a b) \qquad \text{[by (13iii)]}$$
$$\log_b x = \frac{\log_a x}{\log_a b} \qquad \text{[divide by } \log_a b\text{]} \quad \blacksquare$$

You have already been using (15iii) to compute with logarithms. [See Theorem (7).] It is most frequently used with either $a = e$ or $a = 10$. However, the most important of the relationships in Theorem (15) is probably (15ii). As we mentioned in the previous section, this is used by calculators to compute y^x when the $\boxed{y^x}$ button is pressed.

EXERCISES

In Exercises 1–20, use (9)–(12) to solve.

1. $x = \log_3 3^{15}$

2. $x = (\frac{1}{2})^{\log_{0.5} 23}$

3. $\log_2 2^x = 4$

4. $(\frac{1}{4})^{\log_{0.25} x^2} = 9$

5. $\log_5 5^{2x-3} = 4$

6. $(\frac{3}{5})^{\log_{0.6} (2x)} = x^2 - 8$

7. $\log_3 9 = x$

8. $\log_x 9 = 3$

9. $\log_3 x = 9$

10. $\log_2 2 = x$

11. $\log_x 2 = 2$

12. $\log_2 x = 2$

13. $\log_9 3 = x$

14. $\log_x 3 = 9$

15. $\log_9 x = 3$

16. $2^x = 16$

17. $3^{2x-1} = 3^{21}$

18. $\log_3 (x + 2) = \log_3 8$

19. $\log x = \log (x^2 - 12)$

20. $\ln x^2 = \ln (3x + 10)$

In Exercises 21–36, simplify. Do not use a calculator or tables.

21. $3^{\log_3 5}$

22. $4^{3 \log_4 x}$

23. $7^{\log_7 5 + \log_7 4}$

24. $23^{4 \log_{23} 2 - 2 \log_{23} 4}$

25. $\log_7 7^x$

26. $\log_{0.3} 0.3^x - \log_{0.3} 0.3^y$ (Do two ways.)

27. $(\log_3 9^{-1})^{-1}$ (Do two ways.)

28. $\log_2 (2 \cdot 8^x)$

29. $\{\log_{2/3} [(\frac{2}{3})^2]^3\}^2$

30. $\log_{0.1} x^4 - \log_{0.1} \sqrt{x}$

31. $3 \log_4 x^2 - \frac{1}{2} \log_4 \sqrt{x}$

32. $\log_9 (x^2 - 1) - \log_9 (x + 1)$

33. $\log_{12} (x^3 + 1) - \log_{12} (x + 1)$

34. $\log_a [a(a + 1)]$

35. $\log_a \dfrac{a}{\sqrt{x}} + \log_a \sqrt{\dfrac{x}{a}}$

36. $\log_a a\sqrt{x} - \log_a \sqrt{ax}$

In Exercises 37–46, assume that $\log_5 2 \approx 0.43$ and $\log_5 3 \approx 0.68$. Using only this and the fundamental properties, find the given numbers.

37. $\log_5 6$

38. $\log_5 10$

39. $\log_5 12$

40. $\log_5 \frac{3}{2}$

41. $\log_5 \frac{2}{5}$

42. $\log_5 \frac{4}{15}$

43. $\log_5 125$

44. $\log_5 \frac{9}{50}$

45. $\log_5 0.1$

46. $\log_5 0.01$

In Exercises 47–54, solve for x.

47. $2^x = 8$

48. $2^x = 9$

49. $x^{20} = 8$

50. $(1 + x)^5 = 4$

51. $5^{1+x} = 4$

52. $e^{0.423x} = 16.1$

53. $\log_5 (1 + x) = 2$

54. $\ln (51.7x) = 3.2$

55. Suppose $A = P(1 + r)^n$.
- **a.** If $P = 200$, $r = 0.02$, and $n = 12$, find A.
- **b.** If $A = 4000$, $r = 0.04$, and $n = 8$, find P.
- **c.** If $A = 5000$, $P = 2000$, and $n = 12$, find r.
- **d.** If $A = 3000$, $P = 1000$, and $r = 0.07$, find n.

[C] **56.** Compute in two ways:

$$x = \frac{14.38\sqrt{72.79}\,(3.91)^4}{37.91}$$

a. Using your calculator, the way it is.
b. First show that

$$\ln x = \ln 14.38 + \tfrac{1}{2}\ln 72.79 + 4\ln 3.91 - \ln 37.91$$

Then use your calculator to compute the right-hand side. Denote the result by y. Finally, compute $e^y = e^{\ln x} = x$, and compare with part **a.**

5.5 Applications*

In this section we shall discuss some applications of exponentials and logarithms. In doing so, we shall see one place where the number e arises naturally. We begin by examining compound interest.

Suppose you invest \$1000 for 2 years in a savings account that pays 6% per year, compounded quarterly. What this means is that each quarter, interest is earned at the rate of $\frac{1}{4}$ of 6% or 1.5% = 0.015, and this interest is accumulated with the principal to earn interest in subsequent quarters. Thus, during the first 3 months, the \$1000 earns \$1000(0.015) = \$15 interest, so you have \$1000 + \$15 = \$1015 after 3 months. During the second 3 months, the \$1015 earns \$1015(0.015) ≈ \$15.23 interest, so you have \$1015 + \$15.23 = \$1030.23 after 6 months. During the third 3 months, the \$1030.23 earns \$1030.23(0.015) ≈ \$15.45, so after 9 months you have \$1030.23 + \$15.45 = \$1045.68. It keeps going like this five more times until you have \$1126.49 after 2 years.

If it had been compounded monthly, then each month interest is earned at the rate of $\frac{1}{12}$ of 6% or 0.5% = 0.005, and it would take 24 computations like the above to compute that you would have \$1127.16 after 2 years. Fortunately, there is an easy formula.

First we observe how much money we have after one quarter or month, etc., if our money is invested in an account which compounds interest quarterly or monthly, etc.

> **(16)** Let Q be an amount of money invested for a period of time at a rate of interest r in decimal form. Then Q earns rQ for the time period. Thus at the end of the time period, the amount A that the investment is worth is
>
> $$A = Q + rQ = Q(1 + r)$$

* All the computations in this section have been done using a calculator; however, they can also be performed using tables. If tables are used, the answers may be slightly different because of round-off error.

The expression $Q(1 + r)$ is more useful, as will be seen below. We now derive the formula.

Suppose that we invest P dollars for n time periods and that r is the rate of interest for each period. In the first example $n = 8$; in the second, $n = 24$. Using (16), we then have $P(1 + r)$ after the first period. Now for the second period $Q = P(1 + r)$, so we have $P(1 + r)(1 + r) = P(1 + r)^2$ at the end of the second period. For the third period, $Q = P(1 + r)^2$, so we have $P(1 + r)^2(1 + r) = P(1 + r)^3$ at the end of the third period. After four periods, we have $P(1 + r)^4$, after five, $P(1 + r)^5$, and so on. In general,

(17) Theorem If P dollars is invested at a rate of interest of r per period, then the amount A that the investment is worth after n time periods is

$$A = P(1 + r)^n$$

EXAMPLE 1 If \$1000 is invested at 6% compounded quarterly for 2 years, then $r = \frac{0.06}{4} = 0.015$, $n = 2(4) = 8$. By (17),

$$A = 1000(1 + 0.015)^8 \approx 1126.49$$

If it is compounded monthly for 2 years, then

$$r = \frac{0.06}{12} = 0.005$$

$$n = 2(12) = 24$$

and

$$A = 1000(1 + 0.005)^{24} \approx 1127.16$$

If it is compounded daily, then

$$A = 1000\left(1 + \frac{0.06}{365}\right)^{730} \approx 1127.49 \quad \blacksquare$$

Comparing this answer with the previous one shows that very little is gained by increasing the number of times it was compounded. Monthly to daily increased the earnings only by 33 cents over 2 years.

This leads to a very interesting question: What exactly does happen as we shorten the time interval when the interest is compounded? As an example, we examine how much interest \$100,000 earns for 1 year at 8% compounded at different time intervals:

Compounded	Amount After 1 Year	Interest Earned for 1 Year
Annually	$100{,}000(1 + 0.08) = 108{,}000$	\$8000
Semiannually	$100{,}000\left(1 + \frac{0.08}{2}\right)^2 = 108{,}160$	8160
Quarterly	$100{,}000\left(1 + \frac{0.08}{4}\right)^4 \approx 108{,}243.22$	8243.22
Monthly	$100{,}000\left(1 + \frac{0.08}{12}\right)^{12} \approx 108{,}299.95$	8299.95
Daily	$100{,}000\left(1 + \frac{0.08}{365}\right)^{365} \approx 108{,}327.75$	8327.75
Hourly	$100{,}000\left(1 + \frac{0.08}{8760}\right)^{8760} \approx 108{,}328.69$	8328.69
Each minute	$100{,}000\left(1 + \frac{0.08}{525{,}600}\right)^{525{,}600} \approx 108{,}328.7067$	8328.7067

These numbers appear to be getting closer and closer to some particular number. This is in fact the case, and that number is very interesting: It is $100{,}000e^{0.08} \approx 108{,}328.7068$. Using calculus, one can prove the following:

(18) Theorem As k gets larger and larger, $\left(1 + \frac{R}{k}\right)^k$ approaches e^R.

This is one of the reasons the number e is so important and interesting (other reasons arise in calculus).

Notation: We shall use R for the stated annual interest and r for the interest per time period (month, quarter, etc.).

The preceding discussion shows how compounding more often leads to exponential growth. The limit is what is sometimes called **compounded continuously.**

(19) Definition If P dollars is invested at an annual rate of interest R compounded continuously, then after n years it is worth

$$A = P(e^R)^n = Pe^{Rn}$$

EXAMPLE 2 The \$100,000 compounded continuously for 1 year at 8% gives

$$\$100{,}000e^{0.08} = \$108{,}328.71$$

So the most interest that can be earned at 8% is \$8328.71, just 96 cents more than that earned when compounded daily. ■

EXAMPLE 3 Investing \$1000 for 5 years at 6% compounded continuously gives

$$A = \$1000(e^{0.06})^5 = \$1000e^{0.30} \approx \$1349.86 \quad ■$$

EXAMPLE 4 (a) If \$2000 is invested at 8% compounded quarterly, how long does it take to double?
(b) If it is compounded continuously, how long then?
(c) At what rate should we invest it to get it to double in $7\frac{1}{2}$ years if the interest is compounded monthly?

Before solving this example, it may be good to refer back to the "key" in Sec. 5.4 (14), and the examples that followed.

Solution (a) Let n = the number of years. After n years, the 2000 is worth $2000\left[1 + \frac{0.08}{4}\right]^{4n}$, so we want to solve

$$2000(1.02)^{4n} = 4000$$

Then

$$(1.02)^{4n} = 2$$

$$4n \ln (1.02) = \ln 2, \text{ so } n = \frac{\ln 2}{4 \ln 1.02} \approx 8.7507$$

or just over $8\frac{3}{4}$ years.

(b) We want to solve $2000e^{0.08n} = 4000$. Then

$$e^{0.08n} = 2$$

$$0.08n = \ln 2, \qquad n = \frac{\ln 2}{0.08} \approx 8.664 \text{ years}$$

(c) Let R = rate (per year). Since $7\frac{1}{2}$ years is 90 months, we want to solve $2000(1 + \frac{1}{12}R)^{90} = 4000$. Then

$$(1 + \tfrac{1}{12}R)^{90} = 2$$

$$1 + \tfrac{1}{12}R = 2^{1/90}$$

$$\tfrac{1}{12}R = 2^{1/90} - 1$$

$$R = 12(2^{1/90} - 1) \approx 0.09278, \qquad \text{so } 9.278\% \quad ■$$

There are many other situations in which exponential (and hence logarithmic) relations arise.

EXAMPLE 5 **Depreciation.** Suppose a new car depreciates at the rate of 30% a year. Thus each year the car loses in value 30% of what it was worth at the beginning of that year. Then the arithmetic works just like interest compounded annually except the rate is negative. For example, a \$12,000 car after 5 years is worth

$$\$12{,}000(1 - 0.30)^5 = \$12{,}000(0.7)^5 = \$2016.84 \quad ■$$

EXAMPLE 6 **Exponential growth.** Under favorable conditions many things such as populations grow exponentially. This means the following:

> **(20)** If P is the population after time t and P is growing exponentially, then
>
> $$P = P_0 e^{kt}$$
>
> where $k > 0$ is a proportionality constant and P_0 is the population at time $t = 0$.

For example, suppose that the population of a certain kind of rabbit is growing exponentially and that it doubles every 9 months (if you start with enough rabbits of both sexes). If you start with 100 rabbits (50 of each sex), how many do you have after 4 years?

Solution Since the population is growing exponentially, we use the formula $P = P_0 e^{kt}$. We know the initial population is $P_0 = 100$. To find k, we use the fact that the population doubles after 9 months ($= \frac{3}{4}$ year). So $P = 200$ when $t = \frac{3}{4}$ year. Solving for k, we obtain

$$200 = 100e^{3k/4}$$

$$\ln 2 = \frac{3k}{4}, \qquad k = \tfrac{4}{3}\ln 2 \approx 0.924196$$

Thus $P = 100e^{0.924196t}$. When $t = 4$, $P \approx 4032$ rabbits. ■

EXAMPLE 7 **Exponential decay.** Many things decay exponentially: radioactive substances, some chemicals during certain reactions, air pressure as a function of altitude. This means the following:

> **(21)** If A is the amount of a substance after time t and A is decaying exponentially, then
>
> $$A = A_0 e^{-kt}$$
>
> where $k > 0$ is a proportionality constant and A_0 is the amount at time $t = 0$.

For example, if you start with 10 grams of a certain radioactive isotope of iodine, you have 8 grams after 10 days. (a) What is its half-life? (b) How much do you have after a year?

Solution First we must find the constants in the formula $A = A_0 e^{-kt}$. We know $A_0 = 10$, so we only need to determine k. We are given that when $t = 10$ (days), $A = 8$, so

$$8 = 10e^{-10k}, \qquad k = -\tfrac{1}{10}\ln 0.8 \approx 0.0223144$$

(a) The **half-life** is the time that it takes for half of the isotope to decay. Thus we need to find the time t when we only have 5 grams left, i.e., when $A = 5$. Hence we solve the following for t.

$$5 = 10e^{-0.0223144t}$$

$$t = \frac{\ln 0.5}{-0.0223144} \approx 31.0628$$

so the half-life is about 31 days.

(b) $A \approx 10e^{-0.0223144(365)} \approx 0.00290253$ gram. ■

> **Note:** The only essential difference between the formulas for exponential growth and decay is the sign in this exponent. The sign is positive for growth and negative for decay.

EXERCISES

1. Compute $\left(1 + \frac{2}{n}\right)^n$ for $n = 1, 10, 100, 1000, 10^6$. Then compute e^2 and compare.

2. Compute $\left(1 + \frac{-2}{n}\right)^n$ for $n = 1, 10, 100, 1000, 10^6$. Then compute e^{-2} and compare.

3. Compute $\left(1 + \frac{-20}{n}\right)^n$ for $n = 1, 10, 100, 1000, 10^6$. Then compute e^{-20} and compare.

4. Compute $\left(1 + \frac{0.03}{n}\right)^n$ for $n = 1, 10, 100, 1000, 10^6$. Then compute $e^{0.03}$ and compare.

5. Suppose \$5000 is invested in an account which pays 8%. How much is this worth after 10 years if it is compounded
a. Annually? **b.** Quarterly? **c.** Daily? **d.** Continuously?

6. Suppose you had \$1000 to invest, and you wanted to double it in 8 years. At what rate of interest would you have to invest if it was compounded
a. Annually? **b.** Quarterly? **c.** Daily? **d.** Continuously?

7. Suppose you had \$10,000 to invest, and you wanted to triple it in 14 years. At what interest rate would you have to invest if it was compounded
a. Annually? **b.** Semiannually? **c.** Monthly? **d.** Continuously?

8. Suppose \$10,000 is invested in an account which pays 9.25%. How much is this worth after 5 years if it is compounded
a. Annually? **b.** Monthly? **c.** Daily? **d.** Continuously?

9. What is the best way to invest \$1000 for 3 years: **(a)** $9\frac{1}{2}\%$ compounded annually, **(b)** $9\frac{3}{8}\%$ compounded quarterly, or **(c)** $9\frac{1}{4}\%$ compounded continuously?

10. What is the best way to invest \$5000 for 2 years: **(a)** 10% compounded annually, **(b)** $9\frac{7}{8}\%$ compounded monthly, or **(c)** $9\frac{3}{4}\%$ compounded continuously?

11. Suppose you are going to need \$20,000 in 15 years. How much do you have to invest now in an account that pays 8.5% compounded
a. Annually? **b.** Quarterly? **c.** Weekly? **d.** Continuously?

12. A company has to pay off a loan of $2 million in 5 years. To cover this loan, how much would it have to set aside right now if it could earn 10.5% compounded

a. Annually? **b.** Monthly? **c.** Semimonthly? **d.** Continuously?

13. Suppose a $100,000 piece of machinery is depreciating at 10% a year.

a. How much is it worth after 3 years?

b. How long will it take to be worth only $20,000?

14. Suppose a Thunderbird depreciates at 20% a year and a Ford depreciates at 60% a year. How long does it take for each one to be worth half of its original price? How much of their original price are they worth after 3 years?

15. A woman buys an apartment house for $5 million as a tax shelter. She wants to depreciate the building at such a rate that it will be worth only $1 million after 7 years, when she will sell it. What rate of depreciation should she claim on her income tax form? If she wanted to claim 15% depreciation per year, how long would it take to depreciate to $1 million?

16. Suppose 100 mice (of both sexes) got free in a psychology lab and the population began to grow exponentially.

a. If there were 110 mice after 1 week, in how many weeks would the population double?

b. If the building could hold at most 2000 mice, how long would it take before they began falling out the windows?

17. Suppose a population of ants is growing exponentially. At the beginning of June there are 800 ants, and at the beginning of July 4 of the same year, there are 1000.

a. How many ants will there be at the end of September 15 of the same year?

b. On what day will the population have doubled?

18. If you start with 1 gram of a certain isotope, you have 0.9 gram after a week.

a. Find the formula that gives you how much is left after t days.

b. How much is left after a year (365 days)?

c. What is the half-life of this isotope?

d. How long will it be until there is $\frac{1}{10}$ gram left?

19. Suppose a certain radioactive substance has a half-life of 21 days. Suppose you start with 50 grams of this substance.

a. Find a formula that gives you how much is left after t days.

b. How long will it take until there are 5 grams left?

c. How much is left after a year (365 days)?

20. Suppose the atmospheric pressure at sea level is 15 pounds per square inch and the pressure is halved every 3 miles of vertical ascent.

a. Find the exponential equation that expresses this.

b. Find the pressure at 10 miles above sea level.

21. A certain satellite has a power supply whose output in watts is given by the equation $P = 40e^{-t/900}$, where t is the number of days the battery has operated.

a. If it is operated continuously after the satellite is placed into orbit, how many watts is the battery putting out after 1 year?

b. If it takes at least 10 watts to operate the satellite, how many days can the satellite be used?

22. After a person is about 25 years old, his basal metabolism slowly decreases. Let N be the number of calories per day that a person needs and $M = N - 1200$. For the average person, M decreases exponentially at the rate of 10% every 7 years.

a. Find a formula expressing N as a function of time t in years.

b. Suppose a certain person requires about 2500 calories per day when he is 25 years old. If he does not increase his physical activity as he gets older, how many calories will he require per day when he is 40 years old? 60 years old?

c. Graph the function obtained in part a as applied to the person in part b.

23. The simple circuit in Fig. 10 consists of a constant voltage E, an induction of L henries, and a resistance of R ohms. The current I at any time t is given by

$$I = \frac{E}{R}(1 - e^{-(R/L)t})$$

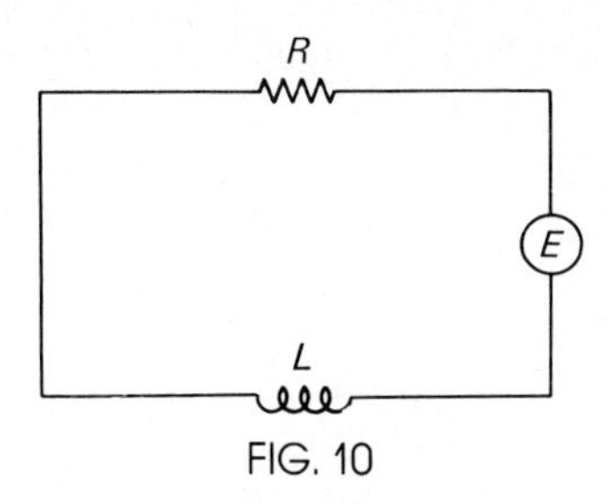

FIG. 10

Suppose for a given circuit that $E = 120$, $R = 50$, and $L = 40$.

a. Find the current I when $t = 2$, 10, and 100.
b. Graph this function.
c. For what time t will $I = 2$? $I = 4$?

24. The pH of a solution is defined by

$$\text{pH} = -\log[\text{H}^+]$$

where $[\text{H}^+]$ is the concentration of hydrogen ions in solution (in moles per liter).
a. Find the pH of a solution for which $[\text{H}^+] = 8.321 \times 10^{-7}$.
b. Find $[\text{H}^+]$ when the pH of a solution is 4.61.

25. The magnitude M of an earthquake is usually measured on the Richter scale. The intensity I of an earthquake (i.e., how strong the earthquake is) and M are related by the formula $M = \log\left(\frac{I}{I_0}\right)$, where I_0 was the intensity of an arbitrarily chosen earthquake. If one earthquake which measured 4.2 on the Richter scale hit San Francisco and a day later a second one hit which measured 5.7, how much stronger was the second than the first? [*Hint:* Translate $M = \log\left(\frac{I}{I_0}\right)$ to the equivalent exponential equation.]

26. The *logistic function*

$$P = \frac{aP_0e^{at}}{bP_0e^{at} + (a - bP_0)}$$

is another population growth function (which also occurs in the study of the spread of information).
a. Show that $P = P_0$ when $t = 0$.
b. If $a = 3$, $b = 0.002$, and $P_0 = 1000$, find P when $t = 0.1$.
c. Find t when $a = 3$, $b = 0.002$, $P_0 = 1000$, and $P = 1200$. (*Hint:* You might substitute $u = e^{at}$ and solve for u first.)

5.6 Two Important Formulas and Their Applications*

There are two very important formulas relating interest and making deposits or payments over a period of time. The derivations of these formulas are given in Appendix F.

Suppose that you invest P dollars at the beginning of each time period (each month, each year, etc.) and that r is the rate of interest per period. Then

*This section is optional, and calculators are required for almost all computations.

after n time periods the total amount A of money invested plus interest earned is

> **(22) Formula 1**
>
> $$A = P\left(1 + \frac{1}{r}\right)[(1 + r)^n - 1].$$

EXAMPLE 1 Ⓒ Suppose that your local savings bank pays 5% per year compounded monthly and that at the beginning of each month you deposit \$100. How much do you have after 5 years? How much did you deposit? How much interest did your money earn?

Solution The time period is a month. Since 0.05 (i.e., 5%) is the interest rate per year, the interest rate per month is $r = \frac{0.05}{12}$. The number of time periods is $n = 12(5) = 60$. By Formula 1, the amount you have after 5 years is

$$A = \$100\left(1 + \frac{12}{0.05}\right)\left[\left(1 + \frac{0.05}{12}\right)^{60} - 1\right] \approx \$6828.94$$

You deposited \$100 per month for 60 months, or \$6000, so your money earned \$828.94 in interest. ■

For the second formula, suppose that an amount A of money is borrowed and that it will be paid back by installment payments, one at the end of each time period for n periods. Let P = the amount of each payment per period, let r = the interest rate per period, and assume the interest is compounded at the end of each period.

> **(23) Formula 2**
>
> $$A = P\frac{1 - (1 + r)^{-n}}{r}$$

Note: The A in Formula 1 represents the amount of money *accumulated* at the *end*; the A in Formula 2 represents the amount of money at the *beginning*.

Note: With an installment plan payment, the interest due is taken out first, and the remainder reduces the amount borrowed. This has the effect of compounding the interest each period.

EXAMPLE 2 Ⓒ Suppose you borrow \$9800 for a new car at 11% interest for 36 months with monthly payments (hence compounded monthly). (a) How much are your payments? (b) How much do you pay altogether? (c) How much interest do you pay?

Solution (a) Here, $r = \frac{0.11}{12}$, $n = 36$, and you wish to solve

$$\$9800 = P\frac{1 - \left(1 + \frac{0.11}{12}\right)^{-36}}{\frac{0.11}{12}}$$

for P, so

$$P = \$9800\frac{\frac{0.11}{12}}{1 - \left(1 + \frac{0.11}{12}\right)^{-36}} \approx \$320.84$$

(b) You pay $\$320.84(36) = \$11,550.24$ altogether.
(c) You pay $\$11,550.24 - \$9800 = \$1750.24$ in interest.

Note: If you made payments for 48 months instead, the answers would be (a) \$253.29 (c) \$2357.92. *Be aware of the differences.* ■

EXAMPLE 3 [C] Suppose you buy a new washer and dryer for \$441.35. You can afford to pay \$20 a month. How many payments would you make if you (a) Charged it on a charge card at 18% per year = 1.5% per month? (b) Borrowed the money from a credit union at 12% per year = 1% per month?

Solution (a) Here $A = \$441.35$, $P = \$20$, $r = 1.5\%$, and you want to solve for n in

$$441.35 = 20\frac{1 - (1 + 0.015)^{-n}}{0.015}$$

So

$$\frac{441.35(0.015)}{20} = 1 - (1.015)^{-n}$$

$$(1.015)^{-n} = 1 - \frac{441.35(0.015)}{20}$$

$$n = \frac{-\ln\left[1 - \frac{441.35(0.015)}{20}\right]}{\ln 1.015} \approx 26.9998 \approx 27$$

Thus there are 27 payments.

(b) Just change all the 0.015s to 0.01, and you get

$$n = \frac{-\ln\left[1 - \frac{441.35(0.01)}{20}\right]}{\ln 1.01} \approx 25.057 \approx 25$$

So there are 25 payments (your twenty-fifth payment will be slightly more). ■

EXERCISES

[C] **1.** In Example 2, suppose you made payments semimonthly for 36 months. What are your answers to parts (a)–(c) then?

[C] **2.** Suppose you deposit \$10 each week in a savings account that pays 5% annually compounded weekly. How much do you have at the end of a year? Two years? How much interest did your money earn the first year?

[C] **3.** Suppose you buy a home and take out a mortgage for \$50,000 at 12% per year, with monthly payments (compounded monthly). Answer the following questions if it is a 20-year, 25-year, or 30-year mortgage.

a. How much is each monthly payment?
b. How much do you pay altogether?
c. How much interest do you pay?

[C] **4.** As in Example 3, suppose that you buy a color TV for \$300 and that you make payments of \$20 per month. How many payments would you make

a. If you charge it on a charge card at 18% per year?
b. If you borrow the money from your credit union at 12% per year?

[C] **5.** Suppose that you put an amount A into an account that compounds interest at a rate $= r$ each time period and that you wish this account to pay you P dollars at the end of each period for n periods (at which time A will be all used up).

a. Explain why the formula relating A, P, r, and n is exactly the same as Formula 2. (*Hint:* Change places with the bank.)
b. If you put \$10,000 into a 6% annual account, compounding monthly, how much could it pay you at the end of each month for the next 30 years (and be all used up then)?
c. Could this account pay out a fixed amount at the end of each month forever? (If that is too long, say 1000 years.)
d. If in part b you wanted \$100 a month, how long would it last?

[C] **6.** Suppose you win \$100,000 in the lottery, but the way the state pays you is \$10,000 now and \$10,000 at the end of each year for the next 9 years. Suppose that the way the state does this, after making that first payment, is to put a fixed sum into a savings account paying $7\frac{1}{2}$% compounded annually and that then that savings account pays you the \$10,000 at the end of each year (and that the account is empty after the 9 years.)

a. How much does the state have to set aside?
b. How much money does the state save (and you lose) doing it this way rather than all at once?

[C] **7.** I have just received a \$50,000 inheritance. I wish to use this to increase my income for the next 25 years by giving me a check at the end of each month. Suppose I can invest the money in an account which pays 7% per year, compounded monthly (and which will send me a monthly check).

a. If I want the maximum amount I can get each month and have the \$50,000 all used up at the end of the 25 years, how much will I get each month?
b. If instead I only want \$300 per month, how much is left at the end of the 25 years? (*Hint:* First compute how much of the \$50,000 I would have to set aside just to give me the \$300 per month.)

[C] **8.** A large corporation sells bonds worth \$20 million. To buy back these bonds in 30 years, it sets up an account (called a **sinking fund**) into which it will make a payment at the end of each quarter, starting 3 months from now. (Of course, it pays interest on the bonds for the 30 years, too.)

a. If the account pays 8.5% compounded quarterly, how much does the corporation have to deposit each quarter to have the \$20 million in 30 years?
b. How much will it actually have paid into the sinking fund?
c. How much interest will the fund have earned?

[C] **9.** A company wants to **float** (i.e., to sell) a 20-year bond issue. It estimates it can afford to pay into a sinking fund (see Exercise 8) \$20,000 at the end of each quarter. Suppose the sinking fund pays 9.2% compounded quarterly.

a. What is the largest amount of money the company can raise?
b. If the company wanted to raise \$8 million but still only pay \$20,000 per quarter into the sinking fund, how long do the bonds have to be issued for?

Ⓒ**10.** A town is about to put in a sewer system at a cost of $1.5 million. They will pay for it by selling 10-year bonds and then assessing all property owners a special semiannual sewer tax to be deposited into a sinking fund. Suppose the sinking fund will pay 8.5% compounded semiannually.

a. How much will the total semiannual assessment for the whole town have to be?

b. How much will the property owners have to pay over the 10-year period?

c. How much interest will the sinking fund earn altogether?

Ⓒ**11.** Suppose you buy a new bedroom suite for $600. You put $100 down and borrow the balance, paying off the loan in monthly installments for 3 years.

a. If you borrow the balance from the furniture store which charges 18% annual interest, compounded monthly, what are your payments?

b. If you borrow the balance from a credit union which charges 12% annual interest, compounded monthly, what are your payments?

Ⓒ**12.** Suppose in Exercise 11 that instead of paying off the loan in exactly 3 years, you decide to pay exactly $30 at the end of each month until the loan is paid off. How many months will it take to pay off the loan if you borrow the $500 from

a. The furniture store? **b.** The credit union?

Ⓒ**13.** Suppose that you and your spouse are living in an apartment and that you decide you want to save $8000 for a down payment on a house in 3 years. If you make monthly payments into a savings account, how much do you have to save each month if the account pays

a. 5% compounded monthly? **b.** 7% compounded monthly?

Ⓒ**14.** If in Exercise 13 you decide you can save $500 each month, how many months does it take you to save the $8000 if the account pays

a. 5% compounded monthly? **b.** 7% compounded monthly?

5.7 Exponential and Logarithmic Equations*

In the preceding sections, we have been solving equations involving exponentials and logarithms. In this section we shall expand on those techniques.

EXPONENTIAL EQUATIONS

When the unknown is in the exponent, we have already seen that taking logarithms of both sides often leads to a solution.

EXAMPLE 1 Solve for x: $\quad 4^{2x-1} = 7^{x+2}$.

Solution

$$\begin{aligned} 4^{2x-1} &= 7^{x+2} \\ \ln 4^{2x-1} &= \ln 7^{x+2} \\ (2x - 1) \ln 4 &= (x + 2) \ln 7 \\ 2(\ln 4)x - \ln 4 &= (\ln 7)x + 2 \ln 7 \\ (2 \ln 4 - \ln 7)x &= 2 \ln 7 + \ln 4 \end{aligned}$$

* As we commented in Sec. 5.5, we have performed all computations with a calculator, but they can be done with tables yielding slightly different answers.

$$x = \frac{2 \ln 7 + \ln 4}{2 \ln 4 - \ln 7} \approx 6.38472 \quad \blacksquare$$

Sometimes it is necessary to do some algebra first.

EXAMPLE 2 Solve for x: $\quad 2 \cdot 3^{2x} = 4 \cdot 5^{3x}$.

Solution

$$2 \cdot 3^{2x} = 4 \cdot 5^{3x}$$

$$2 \cdot 9^{x} = 4 \cdot 125^{x}$$

$$\frac{9^x}{125^x} = \frac{4}{2}$$

$$\left(\frac{9}{125}\right)^x = 2$$

$$\ln\left(\frac{9}{125}\right)^x = \ln 2$$

$$x = \frac{\ln 2}{\ln \frac{9}{125}} \approx -0.263445 \quad \blacksquare$$

Substitution is a basic mathematical tool. Its use is to simplify problems, making them more manageable.

EXAMPLE 3 Solve for x: $\quad \dfrac{3^x + 3^{-x}}{2} = 7$.

Solution Let $u = 3^x$. Then the equation becomes $\dfrac{u + u^{-1}}{2} = 7$. We now solve this equation for u and then use the values obtained for u to solve $3^x = u$ for x:

$$u + \frac{1}{u} = 14$$

$$u^2 + 1 = 14u$$

$$u^2 - 14u + 1 = 0$$

$$u = \frac{14 \pm \sqrt{14^2 - 4}}{2} = 7 \pm \tfrac{1}{2}\sqrt{192}$$

Then, since $3^x = u$, we have

$$3^x = 7 + \tfrac{1}{2}\sqrt{192}, \qquad 3^x = 7 - \tfrac{1}{2}\sqrt{192}$$

Solving these for x yields

$$x = \frac{\ln (7 + \frac{1}{2}\sqrt{192})}{\ln 3} \approx 2.39749$$

$$\text{or} \qquad x = \frac{\ln (7 - \frac{1}{2}\sqrt{192})}{\ln 3} \approx -2.39749 \quad \blacksquare$$

Substitution is also a powerful tool in logarithmic equations. The other tools used to solve logarithmic equations are the properties of Secs. 5.2, 5.3, and 5.4. Sometimes it is difficult to decide whether to use the properties or substitution in a particular problem. The key is the following:

(24)

1. If all the logarithms are functions of the same algebraic expression, try substitution [$u = \log_b(\ )$].
2. If the logarithms are functions of different expressions, try the properties.

EXAMPLE 4 Solve for x: $\quad [\log_3(x - 1)]^2 - 3\log_3(x - 1) = 4.$

Solution Here, both $\log_3$'s are functions of $x - 1$, so we try substituting $u = \log_3(x - 1)$. The equation becomes

$$u^2 - 3u = 4 \qquad \text{or} \qquad u^2 - 3u - 4 = 0$$

Factoring gives

$$(u - 4)(u + 1) = 0, \qquad \text{so } u = 4, -1$$

Substituting back $\log_3(x - 1) = u$, we obtain

$$\log_3(x - 1) = 4 \qquad \text{or} \qquad \log_3(x - 1) = -1$$

Solving for x, we obtain

$$x - 1 = 3^4 \qquad \text{or} \qquad x - 1 = 3^{-1}$$

$$x = 3^4 + 1 = 82 \qquad \text{or} \qquad x = 3^{-1} + 1 = \tfrac{1}{3} + 1 = \tfrac{4}{3}$$

Both answers $x = 82$ and $\frac{4}{3}$ check. ■

EXAMPLE 5 Solve for x: $\quad 2\log_3(x + 1) - \log_3(x + 4) = 2\log_3 2.$

Solution Here, the $\log_3$'s are functions of different algebraic expressions, so we use the properties:

$$2\log_3(x + 1) - \log_3(x + 4) = 2\log_3 2$$

$$\log_3(x + 1)^2 - \log_3(x + 4) = \log_3 2^2$$

$$\log_3 \frac{(x + 1)^2}{x + 4} = \log_3 4$$

$$\frac{(x + 1)^2}{x + 4} = 4$$

$$x^2 + 2x + 1 = 4(x + 4) = 4x + 16$$

$$x^2 - 2x - 15 = 0$$

$$(x - 5)(x + 3) = 0, \qquad x = 5, -3$$

But $x = -3$ does not work in the original equation because $\log_3(-2)$ is undefined. Since $x = 5$ does check, it is the only solution. ■

EXERCISES

In Exercises 1–37, solve for x.

1. $2^x = 16$
2. $2^{x-1} + 2 = 34$
3. $3^x = 16$
4. $3^{x-1} + 1 = 34$
5. $\log_4(2x - 1) = 2$
6. $[\log_5(x + 1)]^2 = 4$
7. $\log x^2 = 1$
8. $\log_2 x = \log_2(x^2 - 2)$
9. $3^{2x} \cdot 3^{x^2} = 27$
10. $7^{-x} = 5 \cdot 7^x$
11. $4^{x+2} = 7^{x-1}$
12. $(10^x)^2 = 6$
13. $\log x + \log(x - 9) = 1$
14. $\log x - \log(x - 9) = 1$
15. $(\log_4 x)^2 + 15 \log_4 x = 16$
16. $[\log_2(x - 1)]^2 = 4 \log_2(x - 1)$
17. $3 \cdot 4^{2x} + 5 \cdot 4^x - 2 = 0$
18. $3^{2x} - 4 \cdot 3^{x+1} + 27 = 0$
19. $[\log_4(x + 1)]^2 - 3 \log_4(x + 1) - 4 = 0$
20. $9(\log_3 x^2)^2 + 17 \log_3 x^2 = 2$
21. $\log(x + 1) - \log(x - 1) = 1$
22. $\log(2x + 1) - \log(x - 2) = 1$
23. $\log x^2 = (\log x)^2$
24. $(\log_2 x)^2 - \log_2 x^2 = 3$
25. $\log \sqrt{x} = \sqrt{\log x}$
26. $2[\log_4(x - 1)]^2 - 2 \log_4 \sqrt{x - 1} = 1$
27. $3^x + 8 \cdot 3^{-x} = 9$
28. $2 \cdot 4^x + 4^{-x} = \frac{9}{2}$
29. $\dfrac{4^x + 4^{-x}}{2} = 8$
30. $2^x + 2^{-x} = 0$
31. $4(2^x + 2^{-x}) = 17$
32. $4^{1+x} + 4^{1-x} = 10$
33. $\dfrac{2^x - 2^{-x}}{2^x + 2^{-x}} = \dfrac{1}{2}$
34. $\dfrac{3^x + 3^{-x}}{3^x - 3^{-x}} = 9$
35. $5^{2x} 7^{-3x} = 6$
36. $(2x)^5(7x)^{-3} = 6$
37. $\log x^3 = (\log x)^3$

REVIEW EXERCISES

1. Compute $3.68^{-4.1}$.
2. Compute $11.5\sqrt{5.9^3 - e^{2.3}}$
3. Use the graph of $y = 5^x$ to estimate $5^{0.3}$, and then compute the number on a calculator or use tables and compare.
4. Use rough interpolation to estimate $2^{4.6}$, and then compute the number on a calculator or use tables and compare.
5. Without using a calculator or tables, find $\log_{81} 3$ and $\log_3 81$.
6. Use rough interpolation to estimate $\log_5 87$, and then compute the number on a calculator or use tables and compare.

In Exercises 7–12, solve for x without using a calculator or tables.

7. $\log_2 32 = x$
8. $\log_3 x = -3$
9. $\log_x 8 = 3$
10. $\log_6 6^{4x-1} = 19$
11. $3.5^{\log_{3.5} 2x^2} = 98$
12. $\log_7(2x + 1) = \log_7(3x - 8)$

In Exercises 13–20, sketch the graph.

13. $y = 5^x$

14. $y = (\frac{1}{5})^x$

15. $y = 5^{x-2}$

16. $y = 5^x - 2$

17. $y = \log_5 x$

18. $y = \log_5(-x)$

19. $y = \log_5(x - 2)$

20. $y = (\log_5 x) - 2$

In Exercises 21–24, simplify without using a calculator or tables.

21. $\log_2 68 - \log_2 17$

22. $\log_9 \dfrac{3}{\sqrt{2}} + \log_9 \sqrt{2}$

23. $\log_4 4^{x^2-4x}$

24. $11^{2 \log_{11} 3 + \log_{11} 2}$

25. Using $\log_5 2 \approx 0.43$ and $\log_5 3 \approx 0.68$, compute $\log_5 \frac{2}{3}$, $\log_5 4$, and $\log_5 60$.

In Exercises 26–33, solve for x.

26. $\log_5(x - 2) + \log_5(x + 2) = 1$

27. $3^{2x+1} = 5.6$

28. $\ln(4.85x) = 5.981$

29. $(2x + 1)^3 = 5.6$

30. $[\log_3(x + 2)]^2 - 7 \log_3(x + 2) = -12$

31. $\dfrac{5^x + 5^{-x}}{5^x - 3(5^{-x})} = 3$

32. $3^{5x}4^{-2x} = 7$

33. $4^{2x} - 4^{x+1} - 21 = 0$

34. If $A = P(1 + r)^n$, $P = 1000$, $r = 0.01$, and $n = 24$, find A.

35. Suppose $5000 is invested in an account that pays 10.5%. How much is it worth after 4 years if interest is compounded

a. Annually? **b.** Monthly? **c.** Daily? **d.** Continuously?

36. If you have $10,000 to invest and you want to triple it in 16 years, at what interest rate would you have to invest if interest is compounded

a. Annually? **b.** Quarterly? **c.** Daily?

37. You have purchased a car for $9600. It depreciates 18% each year. How much is it worth after 4 years? How long does it take to be worth $4800?

38. Suppose a population of flies is growing exponentially. On May 1, there are 500 flies, and on May 18 there are 600 flies.

a. How many flies will there be on May 31?

b. On what day will the population have doubled?

39. Suppose you start with 100 grams of a radioactive substance which has a half-life of 25 days.

a. How many grams are left after 40 days?

b. How long will it take until there are 5 grams left?

[C] **40.** Suppose you deposit $60 into a savings account at the beginning of each month. The account earns interest at a rate of 7.5% annually compounded monthly. How much do you have after 6 years? How much did you de- posit? How much interest did your money earn?

[C] **41.** Suppose you purchase a lot for $12,000. You put $1000 down and borrow the rest at 13% for 5 years with monthly payments (compounded monthly).

a. How much is each monthly payment?

b. How much do you pay altogether?

c. How much interest do you pay?

APPENDIX A

Reference Material

A.1 Prefixes

The following prefixes, in combination with the basic unit names, provide multiples and submultiples. For example, 1 kilometer = 1000 meters, 1 kilogram = 1000 grams, etc.

tetra 10^{12}	kilo 10^3	deci 10^{-1}	micro 10^{-6}	femto 10^{-15}
giga 10^9	hecto 10^2	centi 10^{-2}	nano 10^{-9}	alto 10^{-18}
mega 10^6	deca 10	milli 10^{-3}	pico 10^{-12}	

A.2 Conversion Factors

Weight: 1 kilogram (kg) ≈ 2.2046226 pounds (lb) (avdp)
1 lb (avdp) = 0.45359237 kg (exactly)
= 16 ounces (oz)

Length: 1 inch (in.) = 2.54 centimeters (cm) (exactly)
1 foot (ft) = 0.3048 meters (m) (exactly)
1 mile (mi) = 1.609344 kilometers (km) (exactly)
= 5280 ft
= 8 furlongs (fur)

Volume: 1 liter (l) $\approx$ 1.0566882 quarts (qt)

1 qt = 0.946352946 l (exactly)

= 32 oz

= 2 pints (pt)

1 gallon (gal) = 3.785411784 l (exactly)

= 4 qt

Velocity of light $\approx$ 299,792.5 km/sec

$\approx$ 186,282 mi/sec

Velocity of sound $\approx$ 1088 ft/sec, in air at sea level at 32° F

A.3 The Greek Alphabet

Letters		*Names*	*Letters*		*Names*	*Letters*		*Names*	*Letters*		*Names*
A	α	Alpha	H	η	Eta	N	ν	Nu	T	τ	Tau
B	β	Beta	Θ	θ	Theta	Ξ	ξ	Xi	Υ	υ	Upsilon
Γ	γ	Gamma	I	ι	Iota	O	o	Omicron	Φ	ϕ	Phi
Δ	δ	Delta	K	κ	Kappa	Π	π	Pi	X	χ	Chi
E	ϵ	Epsilon	Λ	λ	Lambda	P	ρ	Rho	Ψ	ψ	Psi
Z	ζ	Zeta	M	μ	Mu	Σ	σ	Sigma	Ω	ω	Omega

APPENDIX B

Using a Calculator

B.1 Basic Information

When you turn on your calculator or start a new problem, always *clear the machine*. Although most machines clear automatically when turned on, you never know when that first surge will leave some unwanted numbers in the wrong place. It is best to get into the habit and be on the safe side.

When you are not using the machine, turn it off to save the battery.

Sometimes it takes a few moments for a calculator to perform an operation. The display is blank or lightly flashing when this is going on. **You must give your calculator the time it needs to complete a computation before pressing the next button,** or an error may occur.

ENTERING NUMBERS

To enter the number 231.72, press [2] [3] [1] [·] [7] [2]. To enter −231.72, you do the same thing, except you also press the *change sign* key, which usually looks like [+/−] but may be [SC] or [CHS]. This may be done *at any time* during the entry *after the first digit has been entered*. The entry of large and small numbers is discussed in Sec. 1.3.

THE BASIC OPERATIONS [+], [−], [×], [÷]

Your calculator has one of two kinds of logic: either *algebraic* or *reverse Polish notation* (RPN). If it has algebraic, it has an [=]; if it has RPN, it has an *enter* button, usually either [ENTER ↑], [ENT], or [EN]. To find $2 + 3$, you press

Alg.: [2] [+] [3] [=]; RPN: [2] [ENT] [3] [+]

and similarly for $-$, $\times$, and $\div$.

If you have an algebraic machine, you *must* know if it has standard or algebraic hierarchy. To illustrate what this means, consider the expression $2 + 3x$. You know that if you evaluate this expression at $x = 4$, you obtain 14. In particular you do *not* obtain 20 since $2 + 3 \times 4$ means $2 + (3 \times 4)$ and not $(2 + 3) \times 4$. This illustrates that when we write such expressions without parentheses, there is a **hierarchy** of operations; we perform all multiplications and divisions first and then additions and subtractions. An algebraic hierarchy machine has the electronics built into it to keep straight which operations take precedence over others; i.e., it keeps straight the hierarchy of the operations and will correctly compute the above as 14. On the algebraic hierarchy machine, if you want $(2 + 3) \times 4$, you press [2] [+] [3] [=] [×] [4] [=]. However, a standard algebraic machine just computes numbers as they go along. So if you want $(2 + 3) \times 4$, you need only press [2] [+] [3] [×] [4] [=] (you may but do not have to press [=] after the [3]). If you want $2 + 3 \times 4$, you must either use parentheses [2] [+] [(] [3] [×] [4] [)] [=] or change the order in which you enter the data, e.g., [3] [×] [4] [+] [2] [=]. Either way is not that hard to get used to, but *you must keep in mind what your calculator is doing as you are doing computations.*

On an RPN calculator, if you want $2 + 3 \times 4$, press [2] [ENT] [3] [ENT] [4] [×] [+]; if you want $(2 + 3) \times 4$, press [2] [ENT] [3] [+] [ENT] [4] [×]. An RPN calculator has a stack of (usually four) registers. In the first computation above, you are using the *bottom three* registers. When you then press [×], the 3 and 4 are multiplied, and 12 is put in the bottom, and the 2 is moved down one. When the [+] is then pressed, the bottom two are added. The thing you must keep track of with an RPN calculator is that when the stack is full and you enter another number, the top number is lost (out the top). Four registers are sufficient for most computations if you perform operations whenever possible and keep track of what is going on.

ERROR SIGN

If you try to compute something that is not allowable, for example, divide by zero or take the square root of a negative number or compute a number that is outside of the calculator's range, the machine will start to compute it and then given an error sign. These signs vary from calculator to calculator. The sign may be a flashing "E," a symbol such as "I" or "F" on the left-hand side of the display, the word "ERROR," a row of dots, etc. Try to compute $1 \div 0$ or $\sqrt{-1}$ on your machine to see what its error sign is, so you will recognize what happened the next time you see it.

EXERCISES

Use your calculator to compute the following after estimating first.

	Estimate	Answer
1. $571.81 + 219.96$	$\approx 600 + 200 = 800$	791.77
2. $0.08412 - 0.10908$	$\approx 0.08 - 0.11 = -0.03$	-0.02496
3. $4.821 \times (-621.5)$	$\approx 5 \times (-600) = -3000$	-2996.25
4. $0.9198 \div (0.0185)$	$\approx 1 \div 0.02 = 100 \div 2 = 50$	49.7189
5. $8.11 + 2.81 \times 3.10$	$\approx 8 + 3(3) = 17$	16.821
6. $42.1 - 8.11 \div (-4.92)$	$\approx 42 - 8 \div (-4) = 44$	43.7484
7. $121.9 \times (-8.21) \div (0.00212)$	$\approx \dfrac{100(-8)}{0.002} = \dfrac{100{,}000(-8)}{2}$ $= -400{,}000$	$-472{,}075$
8. $48.11 - 19.19 - 14.21$	$\approx 50 - 20 - 10 = 20$	14.71
9. $48.11 - (19.19 - 14.21)$	$\approx 50 - (20 - 10) = 40$	43.13

B.2 Memory and Parentheses

Calculators vary as to the number of memories and how you may use them. Consult your owner's manual to see how to use the memories on your calculator. Typically, calculators have an *add into memory* and/or *store in memory* and *recall from memory* buttons. The add into memory button, commonly labeled [M +] or [SUM], adds whatever is in the display to the contents of the memory. Before using this the first time in a computation, it is essential that the memory contain zero. Check your instructions as to how to clear your memory. The store into memory button, commonly labeled [M in] or [STO], replaces the contents of memory by whatever is in the display. Whatever was in the memory before is lost. The recall from memory button, commonly labeled [MR] or [RCL], puts the contents of the memory into the display.

Warning: The [MR] button will usually *not* set the memory to zero.

Calculators also vary as to how many levels of parentheses they have and even whether or not they have parentheses. Parentheses are used in a very natural way. The following examples will illustrate how parentheses and memory are used to do more complicated problems. Since we wish to illustrate only the capabilities of various calculators, all the examples will use very simple numbers. However, first consult your owner's manual to determine the capabilities of your calculator.

EXAMPLE 1 Use a calculator to compute $2 \cdot 8 + 4 \cdot 5 + 3 \cdot 3 = 45.$

Solution (a) Algebraic calculator with hierarchy: The hierarchy takes care of everything, so simply press

[2] [×] [8] [+] [4] [×] [5] [+] [3] [×] [3] [=]

(b) Algebraic, with parentheses: Press

[2] [×] [8] [+] [(] [4] [×] [5] [)]
[+] [(] [3] [×] [3] [)] [=]

(c) Algebraic, with [M +]: Press

[2] [×] [8] [=] [M +] [4] [×] [5]
[=] [M +] [3] [×] [3] [=] [M +] [MR]

(d) RPN: Press

[2] [ENT] [8] [×] [4] [ENT] [5] [×]
[+] [3] [ENT] [3] [×] [+] ■

EXAMPLE 2 Use a calculator to compute

$$(2 + 8)(4 + 5)(3 + 3) = 540$$

Solution (a) Algebraic, with parentheses: Press

[2] [+] [8] [=] [×] [(] [4] [+] [5] [)]
[×] [(] [3] [+] [3] [)] [=]

(b) Algebraic, with [STO] (*do not* use the [M +] button): Press

[2] [+] [8] [=] [STO] [4] [+] [5] [=] [×] [MR]
[=] [STO] [3] [+] [3] [=] [×] [MR] [=]

(c) RPN: Similar to part (d) of the solution to Example 1, except interchange [+] and [×]. ■

EXAMPLE 3 Use a calculator to compute $\frac{2 + 3}{3 + 5} = 0.625.$

Solution (a) Algebraic, with parentheses: Press

[2] [+] [3] [=] [÷] [(] [3] [+] [5] [)] [=]

(b) Algebraic, with memory (either [STO] or [M +]: Press

[3] [+] [5] [=] [STO] [2] [+] [3] [=] [÷] [MR] [=]

(c) RPN: Press

[2] [ENT] [3] [+] [3] [ENT] [5] [+] [÷] ■

EXERCISES

Use your calculator to compute the following after estimating first.

1. $23.41 \times 48.19 + 213.1 \times 9.181 + 7.191 \times 417.8$

Est.: $\approx 20 \times 50 + 200 \times 9 + 7 \times 400 \approx 1000 + 1800 + 2800 \approx 6000$
Ans.: 6089.00

2. $0.2133 \times 0.987 + 0.581 \times 0.612 + 0.018 \div 0.2113$

Est.: $\approx 0.2(1) + 0.6(0.6) + \frac{0.02}{0.2} = 0.2 + 0.36 + 0.1 \approx 0.7$

Ans.: 0.651286

3. $(54.11 + 23.1)(76.001 - 18.91)(52.13 + 4.009)$
Est.: $\approx 70(60)(60) \approx 4000(60) = 240{,}000$
Ans.: 247,460

4. $\frac{(0.1181 + 0.0932)(0.7919 - 0.5818)}{0.6912 + 0.1123}$

Est.: $\frac{0.2(0.2)}{0.8} = \frac{0.1}{2} = 0.05$

Ans.: 0.0552509

5. $\frac{8.127 + 2.916}{3.911 - 1.121}$

Est.: $\approx \frac{8 + 3}{4 - 1} = \frac{11}{3} \approx 3.7$

Ans.: 3.95806

6. $\frac{88.91 + 15.01}{23.04 - 8.12}$

Est.: $\approx \frac{100}{15} = \frac{20}{3} \approx 7$

Ans.: 6.96515

7. $\frac{0.0129 - 0.1281}{0.918 + 0.2001}$

Est.: $\approx \frac{-0.1}{1.1} \approx -0.1$

Ans.: -0.103032

B.3 The Square, Square Root, and Reciprocal Buttons

Most scientific calculators have the square, square root, and reciprocal buttons as separate buttons. The square button is usually labeled x^2. The square root button is labeled $\sqrt{\ }$ or $\sqrt{x}$; the reciprocal button is usually labeled $\frac{1}{x}$ or x^{-1} $\left(\text{remember that } x^{-1} \text{ means } \frac{1}{x}\right)$. Their use is the same for both algebraic and RPN machines. For example, to find 3^2, $\sqrt{2}$, or $\frac{1}{2}$, press [3] [x^2] (getting 9), [2] [$\sqrt{\ }$] (getting 1.414 . . .), or [2] [$1/x$] (getting 0.5).

It is important to understand how to use one of these buttons when the operation is part of a larger expression. For example, consider the two expressions

1. $7 + \sqrt{9}$ **2.** $\sqrt{7 + 9}$

You can easily do these in your head and see that expression 1 is equal to 10 and that expression 2 is equal to 4. The question is, How do you use your calculator to compute these numbers (so you can compute similar expressions with numbers which are not so simple)? You press

1. Alg.: [7] [+] [9] [$\sqrt{\ }$] [=]
RPN: [7] [ENT] [9] [$\sqrt{\ }$] [+]

2. Alg.: [7] [+] [9] [=] [$\sqrt{\ }$]

RPN: [7] [ENT] [9] [+] [$\sqrt{\ }$]

The point to understand is that the [$\sqrt{\ }$] button works *only* on the number that is in the display, and it leaves alone anything else that may be in the machine. So to compute the square root of an expression, make sure the value of that expression *only* is sitting in the display.

Thus the difference between computing expressions 1 and 2 is very slight: simply interchanging the last two operations. However, the consequences of such small differences can be extreme. This simple example illustrates the following point, which cannot be overemphasized: *Your calculator will not do any thinking for you*. It will simply compute exactly whatever corresponds to the button you push. Consequently, *you must keep track of what you are doing when using a calculator*. That is, you must know the algebra that is going on and the capabilities of the machine.

The preceding discussion for the [$\sqrt{\ }$] button applies to the [$1/x$] and [x^2] buttons in the corresponding way.

To compute $8 + \frac{1}{2}$ (= 8.5), press

Alg.: [8] [+] [2] [$1/x$] [=]

RPN: [8] [ENT] [2] [$1/x$] [+]

To compute $\dfrac{1}{8 + 2}$ $(= \frac{1}{10} = 0.1)$, press

Alg.: [8] [+] [2] [=] [$1/x$]

RPN: [8] [ENT] [2] [+] [$1/x$]

To compute $2(3^2)(= 18)$, press

Alg.: [2] [×] [3] [x^2] [=]

RPN: [2] [ENT] [3] [x^2] [×]

To compute $(2 \cdot 3)^2(= 36)$, press

Alg.: [2] [×] [3] [=] [x^2]

RPN: [2] [ENT] [3] [×] [x^2]

We now give an example that combines these properties.

EXAMPLE 1 Compute

$$\left(4.3\sqrt{5.128} - \frac{1}{0.1249}\right)^2$$

Solution Press

Alg.: [4.3] [×] [5.128] [$\sqrt{\ }$] [=] [−] [0.1249] [$1/x$] [=] [x^2]

RPN: [4.3] [ENT] [5.128] [$\sqrt{\ }$] [×] [0.1249] [$1/x$] [−] [x^2]

The answer is ≈ 2.99630. ■

EXERCISES

Compute the given number.

1. $2 + \sqrt{3}$
Ans.: ≈ 3.73205

2. $2 - \sqrt{3}$
Ans.: ≈ 0.267949

3. $3.1 + 4.2^2$
Ans.: 20.74

4. $8.1 - \dfrac{1}{7.3}$
Ans.: ≈ 7.96301

5. $(3.1 - \sqrt{4.2})^2$
Ans.: ≈ 1.10378

6. $\sqrt{4.81 - \dfrac{1}{3.41}}$
Ans.: ≈ 2.12526

7. $\left(5.7\sqrt{19.81} - \dfrac{1}{3.1}\right)^2$
Ans.: ≈ 627.363

8. $\left(1 + \dfrac{0.04}{6}\right)^2$
Ans.: ≈ 1.01338

9. $\sqrt{1 + \dfrac{0.08}{365}}$
Ans.: ≈ 1.00011

10. $4.2 + \dfrac{1}{2 - \dfrac{1}{1.8}}$
Ans.: ≈ 4.89231

11. $\sqrt{8.91 + \sqrt{3.41 + (2.81)^2}}$
Ans.: ≈ 3.50321

12. $\left(2.9 - 9.1 \div \sqrt{89.7} - \dfrac{1}{3.4}\right)^2$
Ans.: 2.70621

13. $\sqrt{\left(\dfrac{1}{14.97} - 3.1\sqrt{8.24}\right)^2 - 5.8^2}$
Ans.: ≈ 6.66048

B.4 Scientific Notation, Significant Figures, and the $\boxed{y^x}$ Button

Recall that a number is written in **scientific notation** if it is expressed in the form

$$\pm c \times 10^n$$

where n is an integer, c is in decimal form, and $1 \le c < 10$.

EXAMPLE 1 $38{,}760{,}000{,}000 = 3.876 \times 10^{10}$ and $-0.000000009919 = -9.919 \times 10^{-9}$. ■

SCIENTIFIC NOTATION ON A CALCULATOR

Most scientific calculators are equipped to handle scientific notation, within limits. They have a button that allows you to enter the exponent of a power of 10 up to two digits. (The two digits are plenty for virtually any practical application.) On most machines, the button looks like [EXP], [EEX], or [EE]. To enter 3.29×10^7, press [3] [·] [2] [9] [EXP] [7]. The display will read [3.29 07]. Of course, you could have entered [3] [2] [9] [0] [0] [0] [0] [0], but using scientific notation to enter this number is quicker and easier. To enter 0.000000000009123, you must use scientific notation and enter 9.123×10^{-12}. Press [9] [·] [1] [2] [3] [EXP] [1] [2] [+/−]. The [+/−] may be pressed any time after pressing [EXP]. The display will read [9.123 − 12]. Alternatively, the same number could be entered as 0.09123×10^{-10}, 0.9123×10^{-11}, 91.23×10^{-13}, etc.

EXAMPLE 2 Enter the following numbers into your calculator:
(a) 7.654×10^{12}; (b) 7.654×10^{-12}; (c) -7.654×10^{12};
(d) -7.654×10^{-12}.

Solution (a) [7.654] [EXP] [12]; (b) [7.654] [EXP] [12] [+/−]
(c) [7.654] [+/−] [EXP] [12]; (d) [7.654] [+/−] [EXP] [12] [+/−] ■

Be careful with the minus signs; on many machines they will not register if you press [+/−] *before* you enter the number.

Scientific notation can help make approximating easier.

EXAMPLE 3 For each of the following, approximate first and then use your calculator to compute.
(a) 4239(78,411)(0.00002991)
(b) $1.892 \times 10^{13}(7.196 \times 10^{12})(8.719 \times 10^{-9})$
(c) 894,100 + 912,812 − 42,371.

Solution Estimations:
(a) $4 \times 10^3 \times 8 \times 10^4 \times 3 \times 10^{-5} \approx 90 \times 10^2 = 9000$
(b) $2 \times 10^{13} \times 7 \times 10^{12} \times 9 \times 10^{-9} \approx 140 \times 10^{16} \approx 1 \times 10^{18}$
(c) $9 \times 10^5 + 9 \times 10^5 - 4 \times 10^4 \approx 18 \times 10^5$.

With the calculator: (a) 9941.61; (b) 1.18708×10^{18}; (c) 1,764,541. ■

SIGNIFICANT FIGURES

A digit of a number is **significant** unless it is used *only* to place the decimal point.

EXAMPLE 4 How many significant figures do the following numbers have?
(a) 15.89; (b) 0.001589; (c) 1.589×10^{-3}; (d) 0.003600;
(e) 1.589×10^6; (f) 3.600×10^6; (g) 1,589,217;
(h) 1.589000×10^6; (i) 1,589,000.

Solution (a)–(f) all have four significant figures. The first two zeros in (b) and (d) are used only to place the decimal point and hence are not significant. (g) and (h) have 7. (i) is ambiguous. It probably is a number like (g) rounded off, but it might be like (h). When there is a real question concerning significant figures, (i) would be written using scientific notation, like (e) or (h). ■

If you state that all numbers are accurate to three significant figures, then you may write 9 instead of 9.00, etc. We are using this convention in this text and are assuming all numbers to be accurate to eight significant figures unless stated otherwise (see Ch. 0).

EXPONENTS ON A CALCULATOR

Your calculator has a button for calculating general exponents that is commonly labeled y^x (but other variations are a^x and x^y). This button computes any *positive* number y raised to the power x *within the limits of the calculator*. For example, to compute 1.51^7, press

Alg: [1.51] [y^x] [7] [=]; RPN: [1.51] [ENT] [7] [y^x]

You should obtain 17.899406.

It is important to remember that your calculator *cannot* compute *negative* numbers raised to a power x. (It may or may not be able to compute 0^x, but of course this should cause no trouble either way.) When you want to compute negative numbers raised to an integer power, you should compute the signs yourself and use your calculator for the positive part.

RATIONAL EXPONENTS ON A CALCULATOR

The [y^x] button on your calculator works perfectly well for x a rational number. As before, you must remember that this button may work only for $y > 0$. So if you wanted to use your calculator to compute $(-8)^{5/3} = -32$, you should use algebra to determine the sign, compute $8^{5/3}$ on the calculator, and put in the signs yourself.

We first describe how to compute $y^{1/n}$, which is a special case. It can be computed using the [y^x] and [$1/x$] buttons. For example, to compute $2^{1/20} = \sqrt[20]{2}$, press

Alg.: [2] [y^x] [20] [$1/x$] [=]; RPN: [2] [ENT] [20] [$1/x$] [y^x]

Some calculators have a [$y^{1/x}$] button. This combines the [y^x] and [$1/x$] buttons, and its use saves one push.

The general case of $y^{m/n}$ is computed one of two ways:

A. Using the definition, $y^{m/n} = (y^{1/n})^m$. For example, to compute $9^{11/13} = (9^{1/13})^{11}$, press

Alg.: [9] [y^x] [13] [$1/x$] [=] [y^x] [11] [=]

(the first [=] may not be necessary)

RPN: [9] [ENT] [13] [$1/x$] [y^x] [11] [y^x]

B. Computing $\frac{m}{n}$ and raising y to this number.

On an RPN calculator, there is no problem. For example, to compute $9^{11/13}$, press [9] [ENT] [11] [ENT] [13] [÷] [y^x].

Warning: On an algebraic calculator there is a trap. To see it, suppose you wanted to compute $9^{3/2}$ (which, in fact, = 27). If you press

[9] [y^x] [3] [÷] [2] [=]

you get the wrong answer of 364.5. Here is what happens: When [÷] is pressed, the calculator computes what is in the machine so far, namely, $9^3 = 729$. Then you get $(9^3) \div 2 = 729 \div 2 = 364.5$. (*Note:* The [$1/x$] button does not cause pending operations to complete. The same is usually true of the [x^2], [$\sqrt{\ }$], and any other button that operates on *one* number at a time.)

In the preceding computation, you want $9^{(3 \div 2)}$, and there are several ways of getting it:

1. ***Hardest way:*** If your calculator has parentheses, press

   [9] [y^x] [(] [3] [÷] [2] [)] [=]

   If your calculator has no parentheses but has a memory, press

   [3] [÷] [2] [=] [M +] [9] [y^x] [MR] [=]

2. ***Easier way:*** Compute $\frac{3}{2} = 1.5$ in your head. As $9^{3/2} = 9^{1.5}$, simply press

   [9] [y^x] [1.5] [=]

   (Of course, if the problem had been $9^{11/13}$, you would have to do it as in method 1, or, which for this case is easier, as in method A.)

3. ***Easiest way:*** Do the whole problem in your head:

   $$9^{3/2} = (9^{1/2})^3 = (\sqrt{9})^3 = 3^3 = 27$$

One of the important things a calculator does is to relieve you of the drudgery of arithmetic so that you can focus your attention on the important aspects of the problem. Unfortunately, though, it is tempting just to push buttons without really thinking about the calculations. The preceding example demonstrates how this often results in doing things in a more complicated way than is necessary. *You should think about two things as you do calculations:*

1. *Can they be simplified* (which consequently decreases the likelihood of mistakes)?
2. *Are you pushing the right buttons* to do the calculations? (For example, on an algebraic calculator, does [9] [y^x] [3] [÷] [2] [=] give $9^{3/2}$?)

While the calculator is a great computational aid, it cannot do the thinking for you.

EXAMPLE 5 Use your calculator to compute (a) $15^{7/11}$; (b) $3^{3/20}$; (c) $8^{1/15}$

Solution (a) Use the definition $15^{7/11} = (15^{1/11})^7$ as in method A or as in method 1. $15^{7/11} \approx 5.60301$; (b) As $\frac{3}{20} = \frac{15}{100} = 0.15$, $3^{3/20} = 3^{0.15}$, so you can do this as in method 2: $3^{0.15} \approx 1.17915$; (c) Use the $\boxed{1/x}$ button. $8^{1/15} \approx 1.14870$. ■

EXERCISES

In Exercises 1–4, using scientific notation, enter the given number into your calculator.

1. 4.0135×10^{-13} (Ans., [4.0135 −13])

2. -8.91937×10^{52} (Ans., [−8.91937 52])

3. 0.0000000000008512 (Ans., [8.512 −13])

4. 5,121,870,000,000,000 (Ans., [5.12187 15])

In Exercises 5 and 6, how many significant figures does the given number have?

5. 15000.0020 (Ans., 9)

6. 0.00081200 (Ans., 5)

In Exercises 7–20, compute the given number.

7. $(4.12 \times 10^{13})(-5.121 \times 10^{-4})$ (Ans., $\approx -2.10985 \times 10^{10}$)

8. $(5.013 \times 10^{-4}) \div (7.011 \times 10^{-5})$ (Ans., ≈ 7.15019)

9. $(-5.12)^3$ (Ans., ≈ -134.218)

10. $(-3.121)^{-4}$ (Ans., ≈ 0.0105396)

11. $(-4.917 \times 10^5)^3$ (Ans., $\approx 1.18878 \times 10^{17}$)

12. $(-4.791 \times 10^{-4})^{-11}$ (Ans., $\approx -3.27576 \times 10^{36}$)

13. $8^{5/3}$ (Ans., $= 32$)

14. $9^{7/5}$ (Ans., ≈ 21.6740)

15. $11^{5/20}$ (Ans., ≈ 1.82116)

16. $2.14^{1/5}$ (Ans., ≈ 1.16435)

17. $8.1^{23.12}$ (Ans., $\approx 1.00966 \times 10^{21}$)

18. $\sqrt[5]{45}$ (Ans., ≈ 2.14113)

19. $\sqrt[4]{3^9}$ (Ans., ≈ 11.8447)

20. $\sqrt[4]{2\sqrt[3]{2}}$ (Ans., ≈ 1.25992)

APPENDIX C

Logarithmic and Exponential Tables

In this appendix, we give tables of logarithmic and exponential functions and explain how to use them.

NATURAL LOGARITHMS

Table C1 gives values of $\ln x$ rounded to four decimal places corresponding to values of x between 1.00 and 9.99 in steps of 0.01. Example 1 explains how to use Table C1.

EXAMPLE 1 Use Table C1 to find the value of $\ln 5.78$. (By this we mean the approximate value, rounded to four decimal places.)

Solution First, think of 5.78 as $5.7 + 0.08$. Locate the number 5.7 in the column on the left side of Table C1. Next locate the 0.08 on the top row of Table C1. Finally, find the number in the same row as the 5.7 and the same column as the 0.08. See Fig. 1, on the next page.

x	0.00	. . .	0.07	0.08	0.09
1.0	0.0000		0.0677	0.0770	0.0862
⋮	⋮		⋮	⋮	⋮
5.6	1.7228				
5.7	1.7405	. . .	1.7527	1.7544	1.7561
5.8	1.7579				

FIG. 1

This value is 1.7544, so we now know

$$\ln 5.78 \approx 1.7544 \quad \blacksquare$$

When using Table C1, you may want to find the value of the natural logarithm function at some number *between* those listed in the table. The method used to do this is **linear interpolation.** This is explained in Example 2.

EXAMPLE 2 Use Table 1 and linear interpolation to find the approximate value of ln 7.216.

Solution We first arrange our work in Table 1 (virtually the same way we do for estimating).

TABLE 1

x	$\ln x$
7.21	1.9755
7.216	?
7.22	1.9769

Table C1 was used to determine $\ln 7.21 \approx 1.9755$ and $\ln 7.22 \approx 1.9769$. We next compute various differences in Table 1, which are indicated in Table 2.

TABLE 2

		x	$\ln x$		
0.01	0.006	7.21	1.9755	d	0.0014
		7.216	?		
		7.22	1.9769		

(In Table 2, 0.006 brackets 7.21–7.216, 0.01 brackets 7.21–7.22, d brackets 1.9755–?, and 0.0014 brackets 1.9755–1.9769.)

Then we form ratios of differences from each side of the table and set the ratios equal:

$$\frac{\text{Smaller difference}}{\text{Larger difference}} = \frac{\text{Smaller difference}}{\text{Larger difference}} \quad \text{or} \quad \frac{0.006}{0.01} = \frac{d}{0.0014}$$

Solving for d, we obtain

$$d = 0.0014\left(\frac{0.006}{0.01}\right) = 0.0014(0.6) = 0.00084$$

which we round off to $d \approx 0.0008$. Thus the value is approximately

$$\ln 7.216 \approx \ln 7.21 + d \approx 1.9755 + 0.0008 = 1.9763 \quad ■$$

COMMON LOGARITHMS

We now turn to Table C2, which gives common logarithms in the same way Table C1 gives natural logarithms.

EXAMPLE 3 Use Table C2 to find (a) $\log 3.71$ and (b) $\log 2.198$.

Solution (a) Using the same procedure as we did in Example 1, we look up $\log 3.71$ in Table C2 to obtain $\log 3.71 \approx 0.5694$.

(b) For this we must use interpolation, as we did in Example 2. See Table 3.

TABLE 3

		x	$\log x$		
0.01	0.008	2.19	0.3404	d	0.0020
		2.198	?		
		2.20	0.3424		

We now form the ratios of differences

$$\frac{0.008}{0.01} = \frac{d}{0.0020}$$

and solve for d

$$d = 0.0020\,\frac{0.008}{0.01} = 0.002(0.8) = 0.0016$$

Then

$$\log 2.198 \approx 0.3404 + 0.0016 = 0.3420 \quad ■$$

The advantage of common logarithms is that it is easy to find logarithms outside of the range of the table, $1.00 \le x \le 9.99$. This is illustrated in Example 4.

EXAMPLE 4 Find (a) $\log 475000$ and (b) $\log 0.0000000391$.

Solution We rewrite the number in scientific notation and then use the properties $\log ab = \log a + \log b$ and $\log 10^n = n$.

(a)

$$475000 = 4.75 \times 10^5$$

$$\log 475000 = \log(4.75 \times 10^5)$$

$$= \log 4.75 + \log 10^5$$
$$= 0.6767 + 5 = 5.6767$$

(b)
$$0.0000000391 = 3.91 \times 10^{-8} \quad \text{so}$$
$$\log 0.0000000391 = \log(3.91 \times 10^{-8})$$
$$= \log 3.91 + \log 10^{-8}$$
$$= 0.5922 + (-8) = -7.4078 \quad \blacksquare$$

In solving each part of Example 4, it was natural to compute the logarithm as a *positive* decimal plus an integer. The decimal part between 0 and 1 is called the **mantissa** of the logarithm and the integer is called the **characteristic.** Thus, in Example 4(a), the mantissa of log 475000 is 0.6767 and the characteristic is 5. In Example 4(b), the mantissa of log 0.0000000391 is 0.5922 (*careful!*), and the characteristic is −8.

If we are given the number b, by reading Table C2 "backward" we can solve the equation $\log x = b$ for x (which is sometimes called the **antilogarithm** of b).

EXAMPLE 5 Use Table C2 to solve (a) $\log x = 0.6085$ (b) $\log x = 0.3915$ (c) $\log x = 3.6085$ (d) $\log x = -3.6085$

Solution (a) We look in the body of Table C2 and find 0.6085 in the row of 4.0 and the column of 0.06. Thus we see (by reading the table "backward")

$$\log 4.06 \approx 0.6085$$

and conclude $x \approx 4.06$.

(b) We again look in the body of Table C2 but we do not find 0.3915. However, it is between 0.3909 and 0.3927. Hence we must use linear interpolation. As in part (a), we recognize $\log 2.46 \approx 0.3909$ and $\log 2.47 \approx 0.3927$ and we form Table 4

TABLE 4

		$\log x$	x		
0.0018	0.006	0.3909	2.46	d	0.01
		0.3915	?		
		0.3927	2.47		

We next form the ratios of differences

$$\frac{0.006}{0.0018} = \frac{d}{0.01}$$

and solve for d, $d = \frac{6}{18}(0.01) \approx 0.003$. Thus if $\log x = 0.3915$, then

$$x \approx 2.46 + d \approx 2.463$$

(c) Since the numbers in the body of Table C2 are between 0 and 1, to solve $\log x = 3.6085$ we must break the problem up into two problems,

$$\log a = 0.6085 \qquad \text{and} \qquad \log b = 3$$

Since $\log 10^n = n$, we see $b = 10^3$. We know $a \approx 4.06$ by part (a). Since

$$\log ab = \log a + \log b = 0.6085 + 3 = 3.6085$$

we see the answer is

$$x = ab \approx (4.06)10^3 = 4060$$

(d) To solve $\log x = -3.6085$, we cannot rewrite the problem as we did in part (c) since

$$-3.6085 = -0.6085 - 3$$

and negative numbers are not in Table C2. Hence we must rewrite this as

$$-3.6085 = 4 - 3.6085 - 4 = 0.3915 - 4$$

Note that we have added and subtracted the next integer larger than $|-3.6065|$; this is the smallest integer we can add to give us a decimal number between 0 and 1 (the mantissa) minus an integer (the characteristic).

By part (b), $0.3915 \approx \log 2.463$, and we know $-4 = \log 10^{-4}$. Altogether we have

$$0.3915 - 4 \approx \log 2.463 + \log 10^{-4} = \log (2.463 \times 10^{-4})$$

so that the answer is $x \approx 2.463 \times 10^{-4} = 0.0002463$. ■

It is important to note that the computations in Example 5 also arise in another context. We sometimes need to compute numbers of the form $10^{6.8124}$. If we let $x = 10^{6.8124}$, then this is equivalent to solving $\log x = 6.8124$, which is exactly the type of problem we just solved in Example 5.

EXPONENTIAL FUNCTIONS

Table C3 gives values of e^x and e^{-x}, and we use it in much the same way as Tables C1 and C2.

EXAMPLE 6 Use Table C3 to find (a) $e^{1.9}$ (b) $e^{-5.27}$.

Solution (a) We look in Table C3 and find $e^{1.9} \approx 6.6859$.
(b) Here we must interpolate, but be careful with negative signs

TABLE 5

		x	e^{-x}		
0.1	0.07	5.2	0.0055	d	−0.0005
		5.27	?		
		5.3	0.0050		

We set up ratios and solve

$$\frac{0.07}{0.1} = \frac{d}{-0.0005} \quad \text{so}$$

$$d = 0.7(-0.0005) = -0.00035 \approx -0.0004$$

Therefore

$$e^{-5.27} \approx e^{-5.2} + d \approx 0.0055 - 0.0004 = 0.0051 \quad \blacksquare$$

When we wish to find e^b for some number b outside the range of Table C3, there is no easy analog for natural logarithms to the method demonstrated for common logarithms in Example 5, parts (c) and (d). Instead, we use the general rule

$$a^b = 10^{\log a^b} = 10^{b \log a}$$

with $a = e$. We noted following Example 5 that solving the problem $x = 10^{b \log e}$ was equivalent to solving the problem $\log x = b \log e$, and we can solve the latter using the method of Example 5. Note that $\log e \approx 0.43429$ is given at the top of Table C2.

EXAMPLE 7 Find e^{20}

Solution By the above discussion,

$$e^{20} = 10^{20 \log e} \approx 10^{20(0.43429)} = 10^{8.6858}$$

We solve

$$\log x = 8.6858 = 0.6858 + 8$$

as in Example 5(c). If $\log y = 0.6858$, then $y \approx 4.851$ by going backward in Table C2 and interpolating. Thus $e^{20} \approx 4.851 \times 10^8$. $\blacksquare$

TABLE C1 Natural Logarithms

x	0.00	0.01	0.02	0.03	0.04	0.05	0.06	0.07	0.08	0.09
1.0	0.0000	0.0100	0.0198	0.0296	0.0392	0.0488	0.0583	0.0677	0.0770	0.0862
1.1	0.0953	0.1044	0.1133	0.1222	0.1310	0.1398	0.1484	0.1570	0.1655	0.1740
1.2	0.1823	0.1906	0.1989	0.2070	0.2151	0.2231	0.2311	0.2390	0.2469	0.2546
1.3	0.2624	0.2700	0.2776	0.2852	0.2927	0.3001	0.3075	0.3148	0.3221	0.3293
1.4	0.3365	0.3436	0.3507	0.3577	0.3646	0.3716	0.3784	0.3853	0.3920	0.3988
1.5	0.4055	0.4121	0.4187	0.4253	0.4318	0.4383	0.4447	0.4511	0.4574	0.4637
1.6	0.4700	0.4762	0.4824	0.4886	0.4947	0.5008	0.5068	0.5128	0.5188	0.5247
1.7	0.5306	0.5365	0.5423	0.5481	0.5539	0.5596	0.5653	0.5710	0.5766	0.5822
1.8	0.5878	0.5933	0.5988	0.6043	0.6098	0.6152	0.6206	0.6259	0.6313	0.6366
1.9	0.6419	0.6471	0.6523	0.6575	0.6627	0.6678	0.6729	0.6780	0.6831	0.6881
2.0	0.6931	0.6981	0.7031	0.7080	0.7130	0.7178	0.7227	0.7275	0.7324	0.7372
2.1	0.7419	0.7467	0.7514	0.7561	0.7608	0.7655	0.7701	0.7747	0.7793	0.7839
2.2	0.7885	0.7930	0.7975	0.8020	0.8065	0.8109	0.8154	0.8198	0.8242	0.8286
2.3	0.8329	0.8372	0.8416	0.8459	0.8502	0.8544	0.8587	0.8629	0.8671	0.8713
2.4	0.8755	0.8796	0.8838	0.8879	0.8920	0.8961	0.9002	0.9042	0.9083	0.9123
2.5	0.9163	0.9203	0.9243	0.9282	0.9322	0.9361	0.9400	0.9439	0.9478	0.9517
2.6	0.9555	0.9594	0.9632	0.9670	0.9708	0.9746	0.9783	0.9821	0.9858	0.9895
2.7	0.9933	0.9969	1.0006	1.0043	1.0080	1.0116	1.0152	1.0188	1.0225	1.0260
2.8	1.0296	1.0332	1.0367	1.0403	1.0438	1.0473	1.0508	1.0543	1.0578	1.0613
2.9	1.0647	1.0682	1.0716	1.0750	1.0784	1.0818	1.0852	1.0886	1.0919	1.0953
3.0	1.0986	1.1019	1.1053	1.1086	1.1119	1.1151	1.1184	1.1217	1.1249	1.1282
3.1	1.1314	1.1346	1.1378	1.1410	1.1442	1.1474	1.1506	1.1537	1.1569	1.1600
3.2	1.1632	1.1663	1.1694	1.1725	1.1756	1.1787	1.1817	1.1848	1.1878	1.1909
3.3	1.1939	1.1970	1.2000	1.2030	1.2060	1.2090	1.2119	1.2149	1.2179	1.2208
3.4	1.2238	1.2267	1.2296	1.2326	1.2355	1.2384	1.2413	1.2442	1.2470	1.2499
3.5	1.2528	1.2556	1.2585	1.2613	1.2641	1.2669	1.2698	1.2726	1.2754	1.2782
3.6	1.2809	1.2837	1.2865	1.2892	1.2920	1.2947	1.2975	1.3002	1.3029	1.3056
3.7	1.3083	1.3110	1.3137	1.3164	1.3191	1.3218	1.3244	1.3271	1.3297	1.3324
3.8	1.3350	1.3376	1.3403	1.3429	1.3455	1.3481	1.3507	1.3533	1.3558	1.3584
3.9	1.3610	1.3635	1.3661	1.3686	1.3712	1.3737	1.3762	1.3788	1.3813	1.3838
4.0	1.3863	1.3888	1.3913	1.3938	1.3962	1.3987	1.4012	1.4036	1.4061	1.4085
4.1	1.4110	1.4134	1.4159	1.4183	1.4207	1.4231	1.4255	1.4279	1.4303	1.4327
4.2	1.4351	1.4375	1.4398	1.4422	1.4446	1.4469	1.4493	1.4516	1.4540	1.4563
4.3	1.4586	1.4609	1.4633	1.4656	1.4679	1.4702	1.4725	1.4748	1.4770	1.4793
4.4	1.4816	1.4839	1.4861	1.4884	1.4907	1.4929	1.4952	1.4974	1.4996	1.5019
4.5	1.5041	1.5063	1.5085	1.5107	1.5129	1.5151	1.5173	1.5195	1.5217	1.5239
4.6	1.5261	1.5282	1.5304	1.5326	1.5347	1.5369	1.5390	1.5412	1.5433	1.5454
4.7	1.5476	1.5497	1.5518	1.5539	1.5560	1.5581	1.5602	1.5623	1.5644	1.5665
4.8	1.5686	1.5707	1.5728	1.5748	1.5769	1.5790	1.5810	1.5831	1.5851	1.5872
4.9	1.5892	1.5913	1.5933	1.5953	1.5974	1.5994	1.6014	1.6034	1.6054	1.6074
5.0	1.6094	1.6114	1.6134	1.6154	1.6174	1.6194	1.6214	1.6233	1.6253	1.6273
5.1	1.6292	1.6312	1.6332	1.6351	1.6371	1.6390	1.6409	1.6429	1.6448	1.6467
5.2	1.6487	1.6506	1.6525	1.6544	1.6563	1.6582	1.6601	1.6620	1.6639	1.6658
5.3	1.6677	1.6696	1.6715	1.6734	1.6752	1.6771	1.6790	1.6808	1.6827	1.6845
5.4	1.6864	1.6882	1.6901	1.6919	1.6938	1.6956	1.6974	1.6993	1.7011	1.7029
5.5	1.7047	1.7066	1.7084	1.7102	1.7120	1.7138	1.7156	1.7174	1.7192	1.7210
5.6	1.7228	1.7246	1.7263	1.7281	1.7299	1.7317	1.7334	1.7352	1.7370	1.7387
5.7	1.7405	1.7422	1.7440	1.7457	1.7475	1.7492	1.7509	1.7527	1.7544	1.7561
5.8	1.7579	1.7596	1.7613	1.7630	1.7647	1.7664	1.7682	1.7699	1.7716	1.7733
5.9	1.7750	1.7766	1.7783	1.7800	1.7817	1.7834	1.7851	1.7867	1.7884	1.7901

TABLE C1 Natural Logarithms (*cont.*)

x	0.00	0.01	0.02	0.03	0.04	0.05	0.06	0.07	0.08	0.09
6.0	1.7918	1.7934	1.7951	1.7967	1.7984	1.8001	1.8017	1.8034	1.8050	1.8066
6.1	1.8083	1.8099	1.8116	1.8132	1.8148	1.8165	1.8181	1.8197	1.8213	1.8229
6.2	1.8245	1.8262	1.8278	1.8294	1.8310	1.8326	1.8342	1.8358	1.8374	1.8390
6.3	1.8406	1.8421	1.8437	1.8453	1.8469	1.8485	1.8500	1.8516	1.8532	1.8547
6.4	1.8563	1.8579	1.8594	1.8610	1.8625	1.8641	1.8656	1.8672	1.8687	1.8703
6.5	1.8718	1.8733	1.8749	1.8764	1.8779	1.8795	1.8810	1.8825	1.8840	1.8856
6.6	1.8871	1.8886	1.8901	1.8916	1.8931	1.8946	1.8961	1.8976	1.8991	1.9006
6.7	1.9021	1.9036	1.9051	1.9066	1.9081	1.9095	1.9110	1.9125	1.9140	1.9155
6.8	1.9169	1.9184	1.9199	1.9213	1.9228	1.9242	1.9257	1.9272	1.9286	1.9301
6.9	1.9315	1.9330	1.9344	1.9359	1.9373	1.9387	1.9402	1.9416	1.9430	1.9445
7.0	1.9459	1.9473	1.9488	1.9502	1.9516	1.9530	1.9544	1.9559	1.9573	1.9587
7.1	1.9601	1.9615	1.9629	1.9643	1.9657	1.9671	1.9685	1.9699	1.9713	1.9727
7.2	1.9741	1.9755	1.9769	1.9782	1.9796	1.9810	1.9824	1.9838	1.9851	1.9865
7.3	1.9879	1.9892	1.9906	1.9920	1.9933	1.9947	1.9961	1.9974	1.9988	2.0001
7.4	2.0015	2.0028	2.0042	2.0055	2.0069	2.0082	2.0096	2.0109	2.0122	2.0136
7.5	2.0149	2.0162	2.0176	2.0189	2.0202	2.0215	2.0229	2.0242	2.0255	2.0268
7.6	2.0282	2.0295	2.0308	2.0321	2.0334	2.0347	2.0360	2.0373	2.0386	2.0399
7.7	2.0412	2.0425	2.0438	2.0451	2.0464	2.0477	2.0490	2.0503	2.0516	2.0528
7.8	2.0541	2.0554	2.0567	2.0580	2.0592	2.0605	2.0618	2.0631	2.0643	2.0665
7.9	2.0669	2.0681	2.0694	2.0707	2.0719	2.0732	2.0744	2.0757	2.0769	2.0782
8.0	2.0794	2.0807	2.0819	2.0832	2.0844	2.0857	2.0869	2.0882	2.0894	2.0906
8.1	2.0919	2.0931	2.0943	2.0956	2.0968	2.0980	2.0992	2.1005	2.1017	2.1029
8.2	2.1041	2.1054	2.1066	2.1078	2.1090	2.1102	2.1114	2.1126	2.1138	2.1150
8.3	2.1163	2.1175	2.1187	2.1199	2.1211	2.1223	2.1235	2.1247	2.1258	2.1270
8.4	2.1282	2.1294	2.1306	2.1318	2.1330	2.1342	2.1353	2.1365	2.1377	2.1389
8.5	2.1401	2.1412	2.1424	2.1436	2.1448	2.1459	2.1471	2.1483	2.1494	2.1506
8.6	2.1518	2.1529	2.1541	2.1552	2.1564	2.1576	2.1587	2.1599	2.1610	2.1622
8.7	2.1633	2.1645	2.1656	2.1668	2.1679	2.1691	2.1702	2.1713	2.1725	2.1736
8.8	2.1748	2.1759	2.1770	2.1782	2.1793	2.1804	2.1815	2.1827	2.1838	2.1849
8.9	2.1861	2.1872	2.1883	2.1894	2.1905	2.1917	2.1928	2.1939	2.1950	2.1961
9.0	2.1972	2.1983	2.1994	2.2006	2.2017	2.2028	2.2039	2.2050	2.2061	2.2072
9.1	2.2083	2.2094	2.2105	2.2116	2.2127	2.2138	2.2148	2.2159	2.2170	2.2181
9.2	2.2192	2.2203	2.2214	2.2225	2.2235	2.2246	2.2257	2.2268	2.2279	2.2289
9.3	2.2300	2.2311	2.2322	2.2332	2.2343	2.2354	2.2364	2.2375	2.2386	2.2396
9.4	2.2407	2.2418	2.2428	2.2439	2.2450	2.2460	2.2471	2.2481	2.2492	2.2502
9.5	2.2513	2.2523	2.2534	2.2544	2.2555	2.2565	2.2576	2.2586	2.2597	2.2607
9.6	2.2618	2.2628	2.2638	2.2649	2.2659	2.2670	2.2680	2.2690	2.2701	2.2711
9.7	2.2721	2.2732	2.2742	2.2752	2.2762	2.2773	2.2783	2.2793	2.2803	2.2814
9.8	2.2824	2.2834	2.2844	2.2854	2.2865	2.2875	2.2885	2.2895	2.2905	2.2915
9.9	2.2925	2.2935	2.2946	2.2956	2.2966	2.2976	2.2986	2.2996	2.3006	2.3016

TABLE C2 Common Logarithms—Note: log $e \approx 0.43429$

x	0.00	0.01	0.02	0.03	0.04	0.05	0.06	0.07	0.08	0.09
1.0	0.0000	0.0043	0.0086	0.0128	0.0170	0.0212	0.0253	0.0294	0.0334	0.0374
1.1	0.0414	0.0453	0.0492	0.0531	0.0569	0.0607	0.0645	0.0682	0.0719	0.0755
1.2	0.0792	0.0828	0.0864	0.0899	0.0934	0.0969	0.1004	0.1038	0.1072	0.1106
1.3	0.1139	0.1173	0.1206	0.1239	0.1271	0.1303	0.1335	0.1367	0.1399	0.1430
1.4	0.1461	0.1492	0.1523	0.1553	0.1584	0.1614	0.1644	0.1673	0.1703	0.1732
1.5	0.1761	0.1790	0.1818	0.1847	0.1875	0.1903	0.1931	0.1959	0.1987	0.2014
1.6	0.2041	0.2068	0.2095	0.2122	0.2148	0.2175	0.2201	0.2227	0.2253	0.2279
1.7	0.2304	0.2330	0.2355	0.2380	0.2405	0.2430	0.2455	0.2480	0.2504	0.2529
1.8	0.2553	0.2577	0.2601	0.2625	0.2648	0.2672	0.2695	0.2718	0.2742	0.2765
1.9	0.2788	0.2810	0.2833	0.2856	0.2878	0.2900	0.2923	0.2945	0.2967	0.2989
2.0	0.3010	0.3032	0.3054	0.3075	0.3096	0.3118	0.3139	0.3160	0.3181	0.3201
2.1	0.3222	0.3243	0.3263	0.3284	0.3304	0.3324	0.3345	0.3365	0.3385	0.3404
2.2	0.3424	0.3444	0.3464	0.3483	0.3502	0.3522	0.3541	0.3560	0.3579	0.3598
2.3	0.3617	0.3636	0.3655	0.3674	0.3692	0.3711	0.3729	0.3747	0.3766	0.3784
2.4	0.3802	0.3820	0.3838	0.3856	0.3874	0.3892	0.3909	0.3927	0.3945	0.3962
2.5	0.3979	0.3997	0.4014	0.4031	0.4048	0.4065	0.4082	0.4099	0.4116	0.4133
2.6	0.4150	0.4166	0.4183	0.4200	0.4216	0.4232	0.4249	0.4265	0.4281	0.4298
2.7	0.4314	0.4330	0.4346	0.4362	0.4378	0.4393	0.4409	0.4425	0.4440	0.4456
2.8	0.4472	0.4487	0.4502	0.4518	0.4533	0.4548	0.4564	0.4579	0.4594	0.4609
2.9	0.4624	0.4639	0.4654	0.4669	0.4683	0.4698	0.4713	0.4728	0.4742	0.4757
3.0	0.4771	0.4786	0.4800	0.4814	0.4829	0.4843	0.4857	0.4871	0.4886	0.4900
3.1	0.4914	0.4928	0.4942	0.4955	0.4969	0.4983	0.4997	0.5011	0.5024	0.5038
3.2	0.5051	0.5065	0.5079	0.5092	0.5105	0.5119	0.5132	0.5145	0.5159	0.5172
3.3	0.5185	0.5198	0.5211	0.5224	0.5237	0.5250	0.5263	0.5276	0.5289	0.5302
3.4	0.5315	0.5328	0.5340	0.5353	0.5366	0.5378	0.5391	0.5403	0.5416	0.5428
3.5	0.5441	0.5453	0.5465	0.5478	0.5490	0.5502	0.5514	0.5527	0.5539	0.5551
3.6	0.5563	0.5575	0.5587	0.5599	0.5611	0.5623	0.5635	0.5647	0.5658	0.5670
3.7	0.5682	0.5694	0.5705	0.5717	0.5729	0.5740	0.5752	0.5763	0.5775	0.5786
3.8	0.5798	0.5809	0.5821	0.5832	0.5843	0.5855	0.5866	0.5877	0.5888	0.5899
3.9	0.5911	0.5922	0.5933	0.5944	0.5955	0.5966	0.5977	0.5988	0.5999	0.6010
4.0	0.6021	0.6031	0.6042	0.6053	0.6064	0.6075	0.6085	0.6096	0.6107	0.6117
4.1	0.6128	0.6138	0.6149	0.6160	0.6170	0.6180	0.6191	0.6201	0.6212	0.6222
4.2	0.6232	0.6243	0.6253	0.6263	0.6274	0.6284	0.6294	0.6304	0.6314	0.6325
4.3	0.6335	0.6345	0.6355	0.6365	0.6375	0.6385	0.6395	0.6405	0.6415	0.6425
4.4	0.6435	0.6444	0.6454	0.6464	0.6474	0.6484	0.6493	0.6503	0.6513	0.6522
4.5	0.6532	0.6542	0.6551	0.6561	0.6571	0.6580	0.6590	0.6599	0.6609	0.6618
4.6	0.6628	0.6637	0.6646	0.6656	0.6665	0.6675	0.6684	0.6693	0.6702	0.6712
4.7	0.6721	0.6730	0.6739	0.6749	0.6758	0.6767	0.6776	0.6785	0.6794	0.6803
4.8	0.6812	0.6821	0.6830	0.6839	0.6848	0.6857	0.6866	0.6875	0.6884	0.6893
4.9	0.6902	0.6911	0.6920	0.6928	0.6937	0.6946	0.6955	0.6964	0.6972	0.6981
5.0	0.6990	0.6998	0.7007	0.7016	0.7024	0.7033	0.7042	0.7050	0.7059	0.7067
5.1	0.7076	0.7084	0.7093	0.7101	0.7110	0.7118	0.7126	0.7135	0.7143	0.7152
5.2	0.7160	0.7168	0.7177	0.7185	0.7193	0.7202	0.7210	0.7218	0.7226	0.7235
5.3	0.7243	0.7251	0.7259	0.7267	0.7275	0.7284	0.7292	0.7300	0.7308	0.7316
5.4	0.7324	0.7332	0.7340	0.7348	0.7356	0.7364	0.7372	0.7380	0.7388	0.7396
5.5	0.7404	0.7412	0.7419	0.7427	0.7435	0.7443	0.7451	0.7459	0.7466	0.7474
5.6	0.7482	0.7490	0.7497	0.7505	0.7513	0.7520	0.7528	0.7536	0.7543	0.7551
5.7	0.7559	0.7566	0.7574	0.7582	0.7589	0.7597	0.7604	0.7612	0.7619	0.7627
5.8	0.7634	0.7642	0.7649	0.7657	0.7664	0.7672	0.7679	0.7686	0.7694	0.7701
5.9	0.7709	0.7716	0.7723	0.7731	0.7738	0.7745	0.7752	0.7760	0.7767	0.7774

TABLE C2 Common Logarithms (*cont.*)

x	0.00	0.01	0.02	0.03	0.04	0.05	0.06	0.07	0.08	0.09
6.0	0.7782	0.7789	0.7796	0.7803	0.7810	0.7818	0.7825	0.7832	0.7839	0.7846
6.1	0.7853	0.7860	0.7868	0.7875	0.7882	0.7889	0.7896	0.7903	0.7910	0.7917
6.2	0.7924	0.7931	0.7938	0.7945	0.7952	0.7959	0.7966	0.7973	0.7980	0.7987
6.3	0.7993	0.8000	0.8007	0.8014	0.8021	0.8028	0.8035	0.8041	0.8048	0.8055
6.4	0.8062	0.8069	0.8075	0.8082	0.8089	0.8096	0.8102	0.8109	0.8116	0.8122
6.5	0.8129	0.8136	0.8142	0.8149	0.8156	0.8162	0.8169	0.8176	0.8182	0.8189
6.6	0.8195	0.8202	0.8209	0.8215	0.8222	0.8228	0.8235	0.8241	0.8248	0.8254
6.7	0.8261	0.8267	0.8274	0.8280	0.8287	0.8293	0.8299	0.8306	0.8312	0.8319
6.8	0.8325	0.8331	0.8338	0.8344	0.8351	0.8357	0.8363	0.8370	0.8376	0.8382
6.9	0.8388	0.8395	0.8401	0.8407	0.8414	0.8420	0.8426	0.8432	0.8439	0.8445
7.0	0.8451	0.8457	0.8463	0.8470	0.8476	0.8482	0.8488	0.8494	0.8500	0.8506
7.1	0.8513	0.8519	0.8525	0.8531	0.8537	0.8543	0.8549	0.8555	0.8561	0.8567
7.2	0.8573	0.8579	0.8585	0.8591	0.8597	0.8603	0.8609	0.8615	0.8621	0.8627
7.3	0.8633	0.8639	0.8645	0.8651	0.8657	0.8663	0.8669	0.8675	0.8681	0.8686
7.4	0.8692	0.8698	0.8704	0.8710	0.8716	0.8722	0.8727	0.8733	0.8739	0.8745
7.5	0.8751	0.8756	0.8762	0.8768	0.8774	0.8779	0.8785	0.8791	0.8797	0.8802
7.6	0.8808	0.8814	0.8820	0.8825	0.8831	0.8837	0.8842	0.8848	0.8854	0.8859
7.7	0.8865	0.8871	0.8876	0.8882	0.8887	0.8893	0.8899	0.8904	0.8910	0.8915
7.8	0.8921	0.8927	0.8932	0.8938	0.8943	0.8949	0.8954	0.8960	0.8965	0.8971
7.9	0.8976	0.8982	0.8987	0.8993	0.8998	0.9004	0.9009	0.9015	0.9020	0.9025
8.0	0.9031	0.9036	0.9042	0.9047	0.9053	0.9058	0.9063	0.9069	0.9074	0.9079
8.1	0.9085	0.9090	0.9096	0.9101	0.9106	0.9112	0.9117	0.9122	0.9128	0.9133
8.2	0.9138	0.9143	0.9149	0.9154	0.9159	0.9165	0.9170	0.9175	0.9180	0.9186
8.3	0.9191	0.9196	0.9201	0.9206	0.9212	0.9217	0.9222	0.9227	0.9232	0.9238
8.4	0.9243	0.9248	0.9253	0.9258	0.9263	0.9269	0.9274	0.9279	0.9284	0.9289
8.5	0.9294	0.9299	0.9304	0.9309	0.9315	0.9320	0.9325	0.9330	0.9335	0.9340
8.6	0.9345	0.9350	0.9355	0.9360	0.9365	0.9370	0.9375	0.9380	0.9385	0.9390
8.7	0.9395	0.9400	0.9405	0.9410	0.9415	0.9420	0.9425	0.9430	0.9435	0.9440
8.8	0.9445	0.9450	0.9455	0.9460	0.9465	0.9469	0.9474	0.9479	0.9484	0.9489
8.9	0.9494	0.9499	0.9504	0.9509	0.9513	0.9518	0.9523	0.9528	0.9533	0.9538
9.0	0.9542	0.9547	0.9552	0.9557	0.9562	0.9566	0.9571	0.9576	0.9581	0.9586
9.1	0.9590	0.9595	0.9600	0.9605	0.9609	0.9614	0.9619	0.9624	0.9628	0.9633
9.2	0.9638	0.9643	0.9647	0.9652	0.9657	0.9661	0.9666	0.9671	0.9675	0.9680
9.3	0.9685	0.9689	0.9694	0.9699	0.9703	0.9708	0.9713	0.9717	0.9722	0.9727
9.4	0.9731	0.9736	0.9741	0.9745	0.9750	0.9754	0.9759	0.9763	0.9768	0.9773
9.5	0.9777	0.9782	0.9786	0.9791	0.9795	0.9800	0.9805	0.9809	0.9814	0.9818
9.6	0.9823	0.9827	0.9832	0.9836	0.9841	0.9845	0.9850	0.9854	0.9859	0.9863
9.7	0.9868	0.9872	0.9877	0.9881	0.9886	0.9890	0.9894	0.9899	0.9903	0.9908
9.8	0.9912	0.9917	0.9921	0.9926	0.9930	0.9934	0.9939	0.9943	0.9948	0.9952
9.9	0.9956	0.9961	0.9965	0.9969	0.9974	0.9978	0.9983	0.9987	0.9991	0.9996

TABLE C3 Exponential functions

x	e^x	e^{-x}	x	e^x	e^{-x}
0.00	1.0000	1.0000	3.0	20.086	0.0498
0.05	1.0513	0.9512	3.1	22.198	0.0450
0.10	1.1052	0.9048	3.2	24.533	0.0408
0.15	1.1618	0.8607	3.3	27.113	0.0369
0.20	1.2214	0.8187	3.4	29.964	0.0334
0.25	1.2840	0.7788	3.5	33.115	0.0302
0.30	1.3499	0.7408	3.6	36.598	0.0273
0.35	1.4191	0.7047	3.7	40.447	0.0247
0.40	1.4918	0.6703	3.8	44.701	0.0224
0.45	1.5683	0.6376	3.9	49.402	0.0202
0.50	1.6487	0.6065	4.0	54.598	0.0183
0.55	1.7333	0.5769	4.1	60.340	0.0166
0.60	1.8221	0.5488	4.2	66.686	0.0150
0.65	1.9155	0.5220	4.3	73.700	0.0136
0.70	2.0138	0.4966	4.4	81.451	0.0123
0.75	2.1170	0.4724	4.5	90.017	0.0111
0.80	2.2255	0.4493	4.6	99.484	0.0101
0.85	2.3396	0.4274	4.7	109.95	0.0091
0.90	2.4596	0.4066	4.8	121.51	0.0082
0.95	2.5857	0.3867	4.9	134.29	0.0074
1.0	2.7183	0.3679	5.0	148.41	0.0067
1.1	3.0042	0.3329	5.1	164.02	0.0061
1.2	3.3201	0.3012	5.2	181.27	0.0055
1.3	3.6693	0.2725	5.3	200.34	0.0050
1.4	4.0552	0.2466	5.4	221.41	0.0045
1.5	4.4817	0.2231	5.5	244.69	0.0041
1.6	4.9530	0.2019	5.6	270.43	0.0037
1.7	5.4739	0.1827	5.7	298.87	0.0033
1.8	6.0496	0.1653	5.8	330.30	0.0030
1.9	6.6859	0.1496	5.9	365.04	0.0027
2.0	7.3891	0.1353	6.0	403.43	0.0025
2.1	8.1662	0.1225	6.5	665.14	0.0015
2.2	9.0250	0.1108	7.0	1096.6	0.0009
2.3	9.9742	0.1003	7.5	1808.0	0.0006
2.4	11.023	0.0907	8.0	2981.0	0.0003
2.5	12.182	0.0821	8.5	4914.8	0.0002
2.6	13.464	0.0743	9.0	8103.1	0.0001
2.7	14.880	0.0672	9.5	13,360	0.00007
2.8	16.445	0.0608	10.0	22,026	0.00004
2.9	18.174	0.0550			

APPENDIX D

Trigonometric Tables

In this appendix we present tables of trigonometric functions and give examples illustrating their use. Table D1 is used mainly for angles in degrees, although it can be used for angles in radians given to four decimal places. When radians are given to two decimal places, it is easier to use Table D2. Keep in mind that the entries in the body of the tables are approximations.

The presentation is divided into two parts: The first part discusses angles in right triangles, while the second part discusses general angles.

Part 1 Angles in Right Triangles

When an angle is in degrees, we use Table D1; an excerpt from Table D1 is given in Table 1 on the next page.

EXAMPLE 1 Find (a) $\sin 38° 20'$ (b) $\csc 38° 20'$ (c) $\cot 38° 40'$

Solution

(a) We look in the column on the left labeled "degrees" at the top to find the angle $38° 20'$, then look across that row to the column with "$\sin x$" at the top. The entry in that row and column is 0.6202. Thus $\sin 38° 20' \approx 0.6202$.

(b) Proceeding as in (a), we find $\csc 38° 20' \approx 1.612$. Since $\csc \theta = \dfrac{1}{\sin \theta}$, you can check that $1.612 \approx \dfrac{1}{0.6202}$.

(c) Proceeding as in (a), $\cot 38° 40' \approx 1.250$. ■

TABLE 1

x	x degrees	sin x	cos x	tan x	cot x	sec x	csc x		
⋮	⋮	⋮	⋮	⋮	⋮	⋮	⋮		
0.6603	50	0.6134	0.7898	0.7766	1.288	1.266	1.630	10	0.9105
0.6632	38° 00′	0.6157	0.7880	0.7813	1.280	1.269	1.624	52° 00′	0.9076
0.6661	10	0.6180	0.7862	0.7860	1.272	1.272	1.618	50	0.9047
0.6690	20	0.6202	0.7844	0.7907	1.265	1.275	1.612	40	0.9018
0.6720	30	0.6225	0.7826	0.7954	1.257	1.278	1.606	30	0.8988
0.6749	40	0.6248	0.7808	0.8002	1.250	1.281	1.601	20	0.8959
0.6778	50	0.6271	0.7790	0.8050	1.242	1.284	1.595	10	0.8930
0.6807	39° 00′	0.6293	0.7771	0.8098	1.235	1.287	1.589	51° 00′	0.8901
		⋮	⋮	⋮	⋮	⋮	⋮	⋮	⋮
		cos x	sin x	cot x	tan x	csc x	sec x	x degrees	

When $0° \leq \theta \leq 45°$, the captions at the top of the columns apply.

When $45° \leq \theta \leq 90°$, it is not necessary to have additional tables, because of the identities

$$\cos\theta = \sin(90° - \theta) \quad \text{or} \quad \sin\theta = \cos(90° - \theta)$$
$$\csc\theta = \sec(90° - \theta) \quad \text{or} \quad \sec\theta = \csc(90° - \theta)$$
$$\cot\theta = \tan(90° - \theta) \quad \text{or} \quad \tan\theta = \cot(90° - \theta)$$

For instance, $\cos 51° 40' = \sin(90° - 51° 40') = \sin 38° 20'$, a value already in Table D1 as we have seen in Example 1(a). However, it is not necessary to recall the identity and go through the computation; just use the captions at the *bottom* of the columns.

EXAMPLE 2 Find (a) $\cos 51° 40'$ (b) $\tan 51° 20'$

Solution (a) We look in the column on the *right* labeled "degrees" at the *bottom* until we find $51° 40'$, then look across the row to that column with "cos x" at the *bottom*. The entry in that row and column is 0.6202. Thus $\cos(51° 40') \approx 0.6202$. You can see the results of Example 1(a) and Example 2(a) illustrate the identity $\cos\theta = \sin(90° - \theta)$.

(b) Proceeding as in (a), $\tan 51° 20' \approx 1.250$. You can see the results of Example 1(c) and Example 2(b) illustrate the identity $\tan\theta = \cot(90° - \theta)$. ■

If the angle is in radians to two decimal places and $0 \leq \theta \leq \frac{\pi}{2}$, we use Table D2.

EXAMPLE 3 Find (a) sin 0.82 (b) sec 1.36

Solution (a) Since the angle is given in radians to two decimal places, we use Table D2.

We look in the column on the left labeled "x" at the top to find the angle 0.82, then look across that row to the column with "$\sin x$" at the top. The entry in that row and column is 0.7311, so $\sin 0.82 \approx 0.7311$.

(b) Proceeding as in (a), we find $\sec 1.36 \approx 4.779$. ■

INTERPOLATION

Linear interpolation can be used to find the values of trigonometric functions for angles between the entries in either Table D1 or Table D2. The fundamental idea behind linear interpolation is that small changes in an angle lead to corresponding (approximately) proportional changes in the values of a function. The next example illustrates the process.

EXAMPLE 4 Use linear interpolation to find (a) $\cos 78° \, 36'$ (b) $\tan 0.364$

Solution (a) Since $78° \, 36'$ is between $78° \, 30'$ and $78° \, 40'$, we look up the values of cosine at these two numbers and arrange our work in the following table:

		x	$\cos x$		
10	6	$78° \, 30'$	0.1994	d	-0.0029
		$78° \, 36'$	?		
		$78° \, 40'$	0.1965		

Warning: Be careful of negative signs. Notice on the left we take the bottom number minus the top number, so on the right we must also take the bottom number minus the top number. Here, we get a negative number since cosine is decreasing on the interval $0° \leq x \leq 90°$.

From the assumption that corresponding differences are (approximately) proportional, we obtain

$$\frac{6}{10} \approx \frac{d}{-0.0029} \quad \text{or} \quad d \approx -0.0029(0.6) = -0.00174$$

We round d off to four decimal places and obtain

$$\cos 78° \, 36' \approx \cos 78° \, 30' + d \approx 0.1994 - 0.0017 = 0.1977$$

(b) To find $\tan 0.364$ we shall use Table D2 and interpolate. (Note that we could use Table D1, but we would have to interpolate there, too.) As in part (a), we set up our work in the following table:

		x	$\tan x$		
0.010	0.004	0.360	0.3764	d	0.0115
		0.364	?		
		0.370	0.3879		

From this we obtain

$$\frac{0.004}{0.010} \approx \frac{d}{0.0115} \qquad \text{or} \qquad d \approx 0.0115(0.4) = 0.00460$$

We round d off to four decimal places and obtain

$$\tan 0.364 \approx \tan 0.360 + d \approx 0.3764 + 0.0046 = 0.3810 \qquad \blacksquare$$

INVERSE TRIGONOMETRIC FUNCTIONS

Relationships between trigonometric functions and their inverse functions are stated as equivalences such as

$$y = \sin x \qquad \text{if and only if} \qquad \sin^{-1} y = x$$

for $-1 \leq y \leq 1$ and $-90° \leq x \leq 90°$ or $-\frac{\pi}{2} \leq x \leq \frac{\pi}{2}$. For angles between $0°$ and $90°$ $\left(\text{or between } 0 \text{ and } \frac{\pi}{2} \text{ radians}\right)$, given an x our tables give us a y. Consequently, to evaluate an inverse trigonometric function, if we are given a y we can go backward in the tables to obtain x. We use Table D1 if we want x in degrees. In the next example, we use Table D2 for radians, though we could use Table D1.

EXAMPLE 5 Find $\tan^{-1} 2.427$ in both (a) radians and (b) degrees.

Solution (a) To find $x = \tan^{-1} 2.427$ in radians, we look in the columns labled "tan x" in Table D2 to find an x such that $\tan x \approx 2.427$. We find 2.427 in the row with 1.18 on the left. We conclude $\tan 1.18 \approx 2.427$ so $\tan^{-1} 2.427 \approx 1.18$.

(b) To find $x = \tan^{-1} 2.427$ in degrees, we look for 2.427 in the column labeled "tan x" at the *bottom* of Table D1 and for the corresponding x at the *right* side of the table. Unfortunately, 2.427 is between two numbers in Table D1, so we must interpolate as in Example 4. We set up the following table:

		x	$\tan x$		
10	d	68° 30′	2.414	0.013	0.020
		?	2.427		
		68° 40′	2.434		

This time we know the right side of the table completely and are filling in the left side. From this table we obtain

$$\frac{d}{10} \approx \frac{0.013}{0.020} \qquad \text{or} \qquad d \approx 10\frac{0.013}{0.020} = 6.5$$

We round d off and conclude

$$\tan^{-1} 2.427 \approx \tan^{-1} 2.414 + d \approx 68° \, 30' + 7' = 68° \, 37' \quad ■$$

Part 2 General Angles

When the angle θ is not between 0° and 90° (or between 0 and $\frac{\pi}{2}$ radians), it is necessary to use the reference angle ϕ, of θ. As described in the text, to evaluate a trigonometric function at θ, we can evaluate that function at ϕ and attach the appropriate sign.

EXAMPLE 6 Find (a) $\sin 261° \, 50'$ (b) $\tan 5.12$

Solution (a) Since $261° \, 50'$ is in Q_3, its reference angle is $\phi = 261° \, 50' - 180° = 81° \, 50'$. By Table D1, $\sin 81° \, 50' \approx 0.9899$. Since sine is negative in Q_3, we see

$$\sin 261° \, 50' = -\sin 81° \, 50' \approx -0.9899$$

(b) To find $\tan 5.12$, we first recognize that 5.12 is not in the interval $\left[0, \frac{\pi}{2}\right]$, since $\frac{\pi}{2} \approx \frac{3.14}{2} = 1.57$. Indeed, 5.12 is in the interval $\left[\frac{3\pi}{2}, 2\pi\right]$, so it is in Q_4. The reference angle of 5.12 is

$$\phi = 2\pi - 5.12 \approx 6.28 - 5.12 = 1.16$$

From Table D2 we obtain $\tan \phi \approx \tan 1.16 \approx 2.296$. Since tangent is negative in Q_4, we conclude

$$\tan 5.12 = -\tan \phi \approx -2.296$$

Note that we computed ϕ to two decimal places, since Table D2 lists angles to two decimal places. The limit of Table D2 is three decimal places, and we could compute

$$\phi = 2\pi - 5.12 \approx 6.283 - 5.12 = 1.163$$

Now to find $\tan \phi$ we must use interpolation, obtaining the result

$$\tan \phi \approx 2.315 \qquad \text{so} \qquad \tan 5.12 \approx -2.315$$

Of course, if 5.12 was obtained as an approximation, such interpolation makes little sense. ■

INVERSES REVISITED

To find the value of inverse trigonometric functions at positive numbers, we can pretend we are looking for an angle of a right triangle, so we can do it exactly the way that was described earlier. To find the value at negative numbers requires a little more care. To see why, compare the following two statements for $\sin^{-1} y$ and $\cos^{-1} y$.

Suppose $-1 \le y \le 1$. Then

$$x = \sin^{-1} y \quad \text{if and only if} \quad y = \sin x \text{ and } -\frac{\pi}{2} \le x \le \frac{\pi}{2}$$

$$x = \cos^{-1} y \quad \text{if and only if} \quad y = \cos x \text{ and } 0 \le x \le \pi$$

Notice the two different intervals restricting x. Both intervals contain $\left[0, \frac{\pi}{2}\right]$ which corresponds to the interval $0 \le y \le 1$. However, for $-1 \le y < 0$, to see how to compute x, we consider the two cases separately.

For $x = \sin^{-1} y$, $-1 \le y < 0$, we have $-\frac{\pi}{2} \le x < 0$. Let $r = \sin^{-1} |y|$. Since $\sin(-u) = -\sin u$ and $|y| = -y$, we have

(1) $$\sin(-x) = -\sin x = -y = |y| = \sin r$$

But

(2) $$0 < -x \le \frac{\pi}{2} \quad \left(\text{since } -\frac{\pi}{2} \le x < 0\right) \quad \text{and} \quad 0 < r \le \frac{\pi}{2}$$

Together (1) and (2) imply $-x = r$. Hence r is the reference angle of x and to find x we simply compute r in the usual way and take its negative.

EXAMPLE 7 Find $\sin^{-1}(-0.7243)$ in both radians and degrees.

Solution We first find $r = \sin^{-1}(+0.7243)$. In radians, we use Table D2 and see $\sin 0.81 \approx 0.7243$. Thus

$$r \approx 0.81 \quad \text{and hence} \quad x = -r \approx -0.81$$

In degrees, we use Table D1 and must interpolate to obtain $\sin 46° 25' \approx 0.7243$. Thus

$$r \approx 46° 25' \quad \text{and hence} \quad x = -r \approx -46° 25'$$ ■

For $x = \cos^{-1} y$, $-1 \le y < 0$, we have $\frac{\pi}{2} < x \le \pi$. Let $r = \cos^{-1} |y|$. Since $\cos(\pi - u) = -\cos u$ and $|y| = -y$, we have

(3) $$\cos(\pi - x) = -\cos x = -y = |y| = \cos r$$

But

(4) $$0 \le \pi - x < \frac{\pi}{2} \quad \left(\text{since } \frac{\pi}{2} < x \le \pi\right) \quad \text{and} \quad 0 \le r < \frac{\pi}{2}$$

Together (3) and (4) imply $\pi - x = r$. Hence r is the reference angle for x, and to find x we find r in the usual way and compute $x = \pi - r$.

EXAMPLE 8 Find $\cos^{-1}(-0.9872)$ in both radians and degrees.

Solution We first find $r = \cos^{-1}(+0.9872)$. In radians, we use Table D2 and see $\cos 0.16 \approx 0.9872$. Thus

$$r \approx 0.16 \qquad \text{and hence} \qquad x \approx \pi - 0.16 \approx 2.98$$

In degrees, we use Table D1 and see $\cos 9°\,10' \approx 0.9872$. Thus

$$r \approx 9°\,10' \qquad \text{and hence} \qquad x \approx 180° - 9°\,10' = 170°\,50'$$ ■

The remaining four trigonometric functions are handled in one of the above two ways. If $y < 0$, then either $-\frac{\pi}{2} < x < 0$ (for $\tan^{-1} y$ and $\csc^{-1} y$) or $\frac{\pi}{2} < x < \pi$ (for $\cot^{-1} y$ and $\sec^{-1} y$). We use $|y|$ to obtain r in the usual way. Then r is the reference angle and either $x = -r$ (for $\tan^{-1} y$ and $\csc^{-1} y$) or $x = \pi - r$ (for $\cot^{-1} y$ and $\sec^{-1} y$).

TABLE D1 Values of the Trigonometric Functions

x	x degrees	sin x	cos x	tan x	cot x	sec x	csc x		
0.0000	**0° 00′**	0.0000	1.0000	0.0000	—	1.000	—	**90° 00′**	1.5708
0.0029	10	0.0029	1.0000	0.0029	343.8	1.000	343.8	50	1.5679
0.0058	20	0.0058	1.0000	0.0058	171.9	1.000	171.9	40	1.5650
0.0087	30	0.0087	1.0000	0.0087	114.6	1.000	114.6	30	1.5621
0.0116	40	0.0116	0.9999	0.0116	85.94	1.000	85.95	20	1.5592
0.0145	50	0.0145	0.9999	0.0145	68.75	1.000	68.76	10	1.5563
0.0175	**1° 00′**	0.0175	0.9998	0.0175	57.29	1.000	57.30	**89° 00′**	1.5533
0.0204	10	0.0204	0.9998	0.0204	49.10	1.000	49.11	50	1.5504
0.0233	20	0.0233	0.9997	0.0233	42.96	1.000	42.98	40	1.5475
0.0262	30	0.0262	0.9997	0.0262	38.19	1.000	38.20	30	1.5446
0.0291	40	0.0291	0.9996	0.0291	34.37	1.000	34.38	20	1.5417
0.0320	50	0.0320	0.9995	0.0320	31.24	1.001	31.26	10	1.5388
0.0349	**2° 00′**	0.0349	0.9994	0.0349	28.64	1.001	28.65	**88° 00′**	1.5359
0.0378	10	0.0378	0.9993	0.0378	26.43	1.001	26.45	50	1.5330
0.0407	20	0.0407	0.9992	0.0407	24.54	1.001	24.56	40	1.5301
0.0436	30	0.0436	0.9990	0.0437	22.90	1.001	22.93	30	1.5272
0.0465	40	0.0465	0.9989	0.0466	21.47	1.001	21.49	20	1.5243
0.0495	50	0.0494	0.9988	0.0495	20.21	1.001	20.23	10	1.5213
0.0524	**3° 00′**	0.0523	0.9986	0.0524	19.08	1.001	19.11	**87° 00′**	1.5184
0.0553	10	0.0552	0.9985	0.0553	18.07	1.002	18.10	50	1.5155
0.0582	20	0.0581	0.9983	0.0582	17.17	1.002	17.20	40	1.5126
0.0611	30	0.0610	0.9981	0.0612	16.35	1.002	16.38	30	1.5097
0.0640	40	0.0640	0.9980	0.0641	15.60	1.002	15.64	20	1.5068
0.0669	50	0.0669	0.9978	0.0670	14.92	1.002	14.96	10	1.5039
0.0698	**4° 00′**	0.0698	0.9976	0.0699	14.30	1.002	14.34	**86° 00′**	1.5010
0.0727	10	0.0727	0.9974	0.0729	13.73	1.003	13.76	50	1.4981
0.0756	20	0.0756	0.9971	0.0758	13.20	1.003	13.23	40	1.4952
0.0785	30	0.0785	0.9969	0.0787	12.71	1.003	12.75	30	1.4923
0.0814	40	0.0814	0.9967	0.0816	12.25	1.003	12.29	20	1.4893
0.0844	50	0.0843	0.9964	0.0846	11.83	1.004	11.87	10	1.4864
0.0873	**5° 00′**	0.0872	0.9962	0.0875	11.43	1.004	11.47	**85° 00′**	1.4835
0.0902	10	0.0901	0.9959	0.0904	11.06	1.004	11.10	50	1.4806
0.0931	20	0.0929	0.9957	0.0934	10.71	1.004	10.76	40	1.4777
0.0960	30	0.0958	0.9954	0.0963	10.39	1.005	10.43	30	1.4748
0.0989	40	0.0987	0.9951	0.0992	10.08	1.005	10.13	20	1.4719
0.1018	50	0.1016	0.9948	0.1022	9.788	1.005	9.839	10	1.4690
0.1047	**6° 00′**	0.1045	0.9945	0.1051	9.514	1.006	9.567	**84° 00′**	1.4661
0.1076	10	0.1074	0.9942	0.1080	9.255	1.006	9.309	50	1.4632
0.1105	20	0.1103	0.9939	0.1110	9.010	1.006	9.065	40	1.4603
0.1134	30	0.1132	0.9936	0.1139	8.777	1.006	8.834	30	1.4573
0.1164	40	0.1161	0.9932	0.1169	8.556	1.007	8.614	20	1.4544
0.1193	50	0.1190	0.9929	0.1198	8.345	1.007	8.405	10	1.4515
0.1222	**7° 00′**	0.1219	0.9925	0.1228	8.144	1.008	8.206	**83° 00′**	1.4486
		cos x	sin x	cot x	tan x	csc x	sec x	x degrees	x

TABLE D1 Values of the Trigonometric Functions (*continued*)

x	x degrees	sin x	cos x	tan x	cot x	sec x	csc x		
0.1222	**7° 00′**	0.1219	0.9925	0.1228	8.144	1.008	8.206	**83° 00′**	1.4486
0.1251	10	0.1248	0.9922	0.1257	7.953	1.008	8.016	50	1.4457
0.1280	20	0.1276	0.9918	0.1287	7.770	1.008	7.834	40	1.4428
0.1309	30	0.1305	0.9914	0.1317	7.596	1.009	7.661	30	1.4399
0.1338	40	0.1334	0.9911	0.1346	7.429	1.009	7.496	20	1.4370
0.1367	50	0.1363	0.9907	0.1376	7.269	1.009	7.337	10	1.4341
0.1396	**8° 00′**	0.1392	0.9903	0.1405	7.115	1.010	7.185	**82° 00′**	1.4312
0.1425	10	0.1421	0.9899	0.1435	6.968	1.010	7.040	50	1.4283
0.1454	20	0.1449	0.9894	0.1465	6.827	1.011	6.900	40	1.4254
0.1484	30	0.1478	0.9890	0.1495	6.691	1.011	6.765	30	1.4224
0.1513	40	0.1507	0.9886	0.1524	6.561	1.012	6.636	20	1.4195
0.1542	50	0.1536	0.9881	0.1554	6.435	1.012	6.512	10	1.4166
0.1571	**9° 00′**	0.1564	0.9877	0.1584	6.314	1.012	6.392	**81° 00′**	1.4137
0.1600	10	0.1593	0.9872	0.1614	6.197	1.013	6.277	50	1.4108
0.1629	20	0.1622	0.9868	0.1644	6.084	1.013	6.166	40	1.4079
0.1658	30	0.1650	0.9863	0.1673	5.976	1.014	6.059	30	1.4050
0.1687	40	0.1679	0.9858	0.1703	5.871	1.014	5.955	20	1.4021
0.1716	50	0.1708	0.9853	0.1733	5.769	1.015	5.855	10	1.3992
0.1745	**10° 00′**	0.1736	0.9848	0.1763	5.671	1.015	5.759	**80° 00′**	1.3963
0.1774	10	0.1765	0.9843	0.1793	5.576	1.016	5.665	50	1.3934
0.1804	20	0.1794	0.9838	0.1823	5.485	1.016	5.575	40	1.3904
0.1833	30	0.1822	0.9833	0.1853	5.396	1.017	5.487	30	1.3875
0.1862	40	0.1851	0.9827	0.1883	5.309	1.018	5.403	20	1.3846
0.1891	50	0.1880	0.9822	0.1914	5.226	1.018	5.320	10	1.3817
0.1920	**11° 00′**	0.1908	0.9816	0.1944	5.145	1.019	5.241	**79° 00′**	1.3788
0.1949	10	0.1937	0.9811	0.1974	5.066	1.019	5.164	50	1.3759
0.1978	20	0.1965	0.9805	0.2004	4.989	1.020	5.089	40	1.3730
0.2007	30	0.1994	0.9799	0.2035	4.915	1.020	5.016	30	1.3701
0.2036	40	0.2022	0.9793	0.2065	4.843	1.021	4.945	20	1.3672
0.2065	50	0.2051	0.9787	0.2095	4.773	1.022	4.876	10	1.3643
0.2094	**12° 00′**	0.2079	0.9781	0.2126	4.705	1.022	4.810	**78° 00′**	1.3614
0.2123	10	0.2108	0.9775	0.2156	4.638	1.023	4.745	50	1.3584
0.2153	20	0.2136	0.9769	0.2186	4.574	1.024	4.682	40	1.3555
0.2182	30	0.2164	0.9763	0.2217	4.511	1.024	4.620	30	1.3526
0.2211	40	0.2193	0.9757	0.2247	4.449	1.025	4.560	20	1.3497
0.2240	50	0.2221	0.9750	0.2278	4.390	1.026	4.502	10	1.3468
0.2269	**13° 00′**	0.2250	0.9744	0.2309	4.331	1.026	4.445	**77° 00′**	1.3439
0.2298	10	0.2278	0.9737	0.2339	4.275	1.027	4.390	50	1.3410
0.2327	20	0.2306	0.9730	0.2370	4.219	1.028	4.336	40	1.3381
0.2356	30	0.2334	0.9724	0.2401	4.165	1.028	4.284	30	1.3352
0.2385	40	0.2363	0.9717	0.2432	4.113	1.029	4.232	20	1.3323
0.2414	50	0.2391	0.9710	0.2462	4.061	1.030	4.182	10	1.3294
0.2443	**14° 00′**	0.2419	0.9703	0.2493	4.011	1.031	4.134	**76° 00′**	1.3265
		cos x	sin x	cot x	tan x	csc x	sec x	x degrees	x

TABLE D1 Values of the Trigonometric Functions (*continued*)

x	x degrees	$\sin x$	$\cos x$	$\tan x$	$\cot x$	$\sec x$	$\csc x$		
0.2443	**14° 00′**	0.2419	0.9703	0.2493	4.011	1.031	4.134	**76° 00′**	1.3265
0.2473	10	0.2447	0.9696	0.2524	3.962	1.031	4.086	50	1.3235
0.2502	20	0.2476	0.9689	0.2555	3.914	1.032	4.039	40	1.3206
0.2531	30	0.2504	0.9681	0.2586	3.867	1.033	3.994	30	1.3177
0.2560	40	0.2532	0.9674	0.2617	3.821	1.034	3.950	20	1.3148
0.2589	50	0.2560	0.9667	0.2648	3.776	1.034	3.906	10	1.3119
0.2618	**15° 00′**	0.2588	0.9659	0.2679	3.732	1.035	3.864	**75° 00′**	1.3090
0.2647	10	0.2616	0.9652	0.2711	3.689	1.036	3.822	50	1.3061
0.2676	20	0.2644	0.9644	0.2742	3.647	1.037	3.782	40	1.3032
0.2705	30	0.2672	0.9636	0.2773	3.606	1.038	3.742	30	1.3003
0.2734	40	0.2700	0.9628	0.2805	3.566	1.039	3.703	20	1.2974
0.2763	50	0.2728	0.9621	0.2836	3.526	1.039	3.665	10	1.2945
0.2793	**16° 00′**	0.2756	0.9613	0.2867	3.487	1.040	3.628	**74° 00′**	1.2915
0.2822	10	0.2784	0.9605	0.2899	3.450	1.041	3.592	50	1.2886
0.2851	20	0.2812	0.9596	0.2931	3.412	1.042	3.556	40	1.2857
0.2880	30	0.2840	0.9588	0.2962	3.376	1.043	3.521	30	1.2828
0.2909	40	0.2868	0.9580	0.2994	3.340	1.044	3.487	20	1.2799
0.2938	50	0.2896	0.9572	0.3026	3.305	1.045	3.453	10	1.2770
0.2967	**17° 00′**	0.2924	0.9563	0.3057	3.271	1.046	3.420	**73° 00′**	1.2741
0.2996	10	0.2952	0.9555	0.3089	3.237	1.047	3.388	50	1.2712
0.3025	20	0.2979	0.9546	0.3121	3.204	1.048	3.356	40	1.2683
0.3054	30	0.3007	0.9537	0.3153	3.172	1.049	3.326	30	1.2654
0.3083	40	0.3035	0.9528	0.3185	3.140	1.049	3.295	20	1.2625
0.3113	50	0.3062	0.9520	0.3217	3.108	1.050	3.265	10	1.2595
0.3142	**18° 00′**	0.3090	0.9511	0.3249	3.078	1.051	3.236	**72° 00′**	1.2566
0.3171	10	0.3118	0.9502	0.3281	3.047	1.052	3.207	50	1.2537
0.3200	20	0.3145	0.9492	0.3314	3.018	1.053	3.179	40	1.2508
0.3229	30	0.3173	0.9483	0.3346	2.989	1.054	3.152	30	1.2479
0.3258	40	0.3201	0.9474	0.3378	2.960	1.056	3.124	20	1.2450
0.3287	50	0.3228	0.9465	0.3411	2.932	1.057	3.098	10	1.2421
0.3316	**19° 00′**	0.3256	0.9455	0.3443	2.904	1.058	3.072	**71° 00′**	1.2392
0.3345	10	0.3283	0.9446	0.3476	2.877	1.059	3.046	50	1.2363
0.3374	20	0.3311	0.9436	0.3508	2.850	1.060	3.021	40	1.2334
0.3403	30	0.3338	0.9426	0.3541	2.824	1.061	2.996	30	1.2305
0.3432	40	0.3365	0.9417	0.3574	2.798	1.062	2.971	20	1.2275
0.3462	50	0.3393	0.9407	0.3607	2.773	1.063	2.947	10	1.2246
0.3491	**20° 00′**	0.3420	0.9397	0.3640	2.747	1.064	2.924	**70° 00′**	1.2217
0.3520	10	0.3448	0.9387	0.3673	2.723	1.065	2.901	50	1.2188
0.3549	20	0.3475	0.9377	0.3706	2.699	1.066	2.878	40	1.2159
0.3578	30	0.3502	0.9367	0.3739	2.675	1.068	2.855	30	1.2130
0.3607	40	0.3529	0.9356	0.3772	2.651	1.069	2.833	20	1.2101
0.3636	50	0.3557	0.9346	0.3805	2.628	1.070	2.812	10	1.2072
0.3665	**21° 00′**	0.3584	0.9336	0.3839	2.605	1.071	2.790	**69° 00′**	1.2043
		$\cos x$	$\sin x$	$\cot x$	$\tan x$	$\csc x$	$\sec x$	x degrees	x

x	x degrees	sin x	cos x	tan x	cot x	sec x	csc x		
0.3665	**21° 00′**	0.3584	0.9336	0.3839	2.605	1.071	2.790	**69° 00′**	1.2043
0.3694	10	0.3611	0.9325	0.3872	2.583	1.072	2.769	50	1.2014
0.3723	20	0.3638	0.9315	0.3906	2.560	1.074	2.749	40	1.1985
0.3752	30	0.3665	0.9304	0.3939	2.539	1.075	2.729	30	1.1956
0.3782	40	0.3692	0.9293	0.3973	2.517	1.076	2.709	20	1.1926
0.3811	50	0.3719	0.9283	0.4006	2.496	1.077	2.689	10	1.1897
0.3840	**22° 00′**	0.3746	0.9272	0.4040	2.475	1.079	2.669	**68° 00′**	1.1868
0.3869	10	0.3773	0.9261	0.4074	2.455	1.080	2.650	50	1.1839
0.3898	20	0.3800	0.9250	0.4108	2.434	1.081	2.632	40	1.1810
0.3927	30	0.3827	0.9239	0.4142	2.414	1.082	2.613	30	1.1781
0.3956	40	0.3854	0.9228	0.4176	2.394	1.084	2.595	20	1.1752
0.3985	50	0.3881	0.9216	0.4210	2.375	1.085	2.577	10	1.1723
0.4014	**23° 00′**	0.3907	0.9205	0.4245	2.356	1.086	2.559	**67° 00′**	1.1694
0.4043	10	0.3934	0.9194	0.4279	2.337	1.088	2.542	50	1.1665
0.4072	20	0.3961	0.9182	0.4314	2.318	1.089	2.525	40	1.1636
0.4102	30	0.3987	0.9171	0.4348	2.300	1.090	2.508	30	1.1606
0.4131	40	0.4014	0.9159	0.4383	2.282	1.092	2.491	20	1.1577
0.4160	50	0.4041	0.9147	0.4417	2.264	1.093	2.475	10	1.1548
0.4189	**24° 00′**	0.4067	0.9135	0.4452	2.246	1.095	2.459	**66° 00′**	1.1519
0.4218	10	0.4094	0.9124	0.4487	2.229	1.096	2.443	50	1.1490
0.4247	20	0.4120	0.9112	0.4522	2.211	1.097	2.427	40	1.1461
0.4276	30	0.4147	0.9100	0.4557	2.194	1.099	2.411	30	1.1432
0.4305	40	0.4173	0.9088	0.4592	2.177	1.100	2.396	20	1.1403
0.4334	50	0.4200	0.9075	0.4628	2.161	1.102	2.381	10	1.1374
0.4363	**25° 00′**	0.4226	0.9063	0.4663	2.145	1.103	2.366	**65° 00′**	1.1345
0.4392	10	0.4253	0.9051	0.4699	2.128	1.105	2.352	50	1.1316
0.4422	20	0.4279	0.9038	0.4734	2.112	1.106	2.337	40	1.1286
0.4451	30	0.4305	0.9026	0.4770	2.097	1.108	2.323	30	1.1257
0.4480	40	0.4331	0.9013	0.4806	2.081	1.109	2.309	20	1.1228
0.4509	50	0.4358	0.9001	0.4841	2.066	1.111	2.295	10	1.1199
0.4538	**26° 00′**	0.4384	0.8988	0.4877	2.050	1.113	2.281	**64° 00′**	1.1170
0.4567	10	0.4410	0.8975	0.4913	2.035	1.114	2.268	50	1.1141
0.4596	20	0.4436	0.8962	0.4950	2.020	1.116	2.254	40	1.1112
0.4625	30	0.4462	0.8949	0.4986	2.006	1.117	2.241	30	1.1083
0.4654	40	0.4488	0.8936	0.5022	1.991	1.119	2.228	20	1.1054
0.4683	50	0.4514	0.8923	0.5059	1.977	1.121	2.215	10	1.1025
0.4712	**27° 00′**	0.4540	0.8910	0.5095	1.963	1.122	2.203	**63° 00′**	1.0996
0.4741	10	0.4566	0.8897	0.5132	1.949	1.124	2.190	50	1.0966
0.4771	20	0.4592	0.8884	0.5169	1.935	1.126	2.178	40	1.0937
0.4800	30	0.4617	0.8870	0.5206	1.921	1.127	2.166	30	1.0908
0.4829	40	0.4643	0.8857	0.5243	1.907	1.129	2.154	20	1.0879
0.4858	50	0.4669	0.8843	0.5280	1.894	1.131	2.142	10	1.0850
0.4887	**28° 00′**	0.4695	0.8829	0.5317	1.881	1.133	2.130	**62° 00′**	1.0821
		cos x	sin x	cot x	tan x	csc x	sec x	x degrees	x

TABLE D1 Values of the Trigonometric Functions (*continued*)

x	x degrees	sin x	cos x	tan x	cot x	sec x	csc x		
0.4887	**28° 00′**	0.4695	0.8829	0.5317	1.881	1.133	2.130	**62° 00′**	1.0821
0.4916	10	0.4720	0.8816	0.5354	1.868	1.134	2.118	50	1.0792
0.4945	20	0.4746	0.8802	0.5392	1.855	1.136	2.107	40	1.0763
0.4974	30	0.4772	0.8788	0.5430	1.842	1.138	2.096	30	1.0734
0.5003	40	0.4797	0.8774	0.5467	1.829	1.140	2.085	20	1.0705
0.5032	50	0.4823	0.8760	0.5505	1.816	1.142	2.074	10	1.0676
0.5061	**29° 00′**	0.4848	0.8746	0.5543	1.804	1.143	2.063	**61° 00′**	1.0647
0.5091	10	0.4874	0.8732	0.5581	1.792	1.145	2.052	50	1.0617
0.5120	20	0.4899	0.8718	0.5619	1.780	1.147	2.041	40	1.0588
0.5149	30	0.4924	0.8704	0.5658	1.767	1.149	2.031	30	1.0559
0.5178	40	0.4950	0.8689	0.5696	1.756	1.151	2.020	20	1.0530
0.5207	50	0.4975	0.8675	0.5735	1.744	1.153	2.010	10	1.0501
0.5236	**30° 00′**	0.5000	0.8660	0.5774	1.732	1.155	2.000	**60° 00′**	1.0472
0.5265	10	0.5025	0.8646	0.5812	1.720	1.157	1.990	50	1.0443
0.5294	20	0.5050	0.8631	0.5851	1.709	1.159	1.980	40	1.0414
0.5323	30	0.5075	0.8616	0.5890	1.698	1.161	1.970	30	1.0385
0.5352	40	0.5100	0.8601	0.5930	1.686	1.163	1.961	20	1.0356
0.5381	50	0.5125	0.8587	0.5969	1.675	1.165	1.951	10	1.0327
0.5411	**31° 00′**	0.5150	0.8572	0.6009	1.664	1.167	1.942	**59° 00′**	1.0297
0.5440	10	0.5175	0.8557	0.6048	1.653	1.169	1.932	50	1.0268
0.5469	20	0.5200	0.8542	0.6088	1.643	1.171	1.923	40	1.0239
0.5498	30	0.5225	0.8526	0.6128	1.632	1.173	1.914	30	1.0210
0.5527	40	0.5250	0.8511	0.6168	1.621	1.175	1.905	20	1.0181
0.5556	50	0.5275	0.8496	0.6208	1.611	1.177	1.896	10	1.0152
0.5585	**32° 00′**	0.5299	0.8480	0.6249	1.600	1.179	1.887	**58° 00′**	1.0123
0.5614	10	0.5324	0.8465	0.6289	1.590	1.181	1.878	50	1.0094
0.5643	20	0.5348	0.8450	0.6330	1.580	1.184	1.870	40	1.0065
0.5672	30	0.5373	0.8434	0.6371	1.570	1.186	1.861	30	1.0036
0.5701	40	0.5398	0.8418	0.6412	1.560	1.188	1.853	20	1.0007
0.5730	50	0.5422	0.8403	0.6453	1.550	1.190	1.844	10	0.9977
0.5760	**33° 00′**	0.5446	0.8387	0.6494	1.540	1.192	1.836	**57° 00**	0.9948
0.5789	10	0.5471	0.8371	0.6536	1.530	1.195	1.828	50	0.9919
0.5818	20	0.5495	0.8355	0.6577	1.520	1.197	1.820	40	0.9890
0.5847	30	0.5519	0.8339	0.6619	1.511	1.199	1.812	30	0.9861
0.5876	40	0.5544	0.8323	0.6661	1.501	1.202	1.804	20	0.9832
0.5905	50	0.5568	0.8307	0.6703	1.492	1.204	1.796	10	0.9803
0.5934	**34° 00′**	0.5592	0.8290	0.6745	1.483	1.206	1.788	**56° 00′**	0.9774
0.5963	10	0.5616	0.8274	0.6787	1.473	1.209	1.781	50	0.9745
0.5992	20	0.5640	0.8258	0.6830	1.464	1.211	1.773	40	0.9716
0.6021	30	0.5664	0.8241	0.6873	1.455	1.213	1.766	30	0.9687
0.6050	40	0.5688	0.8225	0.6916	1.446	1.216	1.758	20	0.9657
0.6080	50	0.5712	0.8208	0.6959	1.437	1.218	1.751	10	0.9628
0.6109	**35° 00′**	0.5736	0.8192	0.7002	1.428	1.221	1.743	**55° 00′**	0.9599
		cos x	sin x	cot x	tan x	csc x	sec x	x degrees	x

TABLE D1 Values of the Trigonometric Functions (*continued*)

x	x degrees	sin x	cos x	tan x	cot x	sec x	csc x		
0.6109	**35° 00′**	0.5736	0.8192	0.7002	1.428	1.221	1.743	**55° 00′**	0.9599
0.6138	10	0.5760	0.8175	0.7046	1.419	1.223	1.736	50	0.9570
0.6167	20	0.5783	0.8158	0.7089	1.411	1.226	1.729	40	0.9541
0.6196	30	0.5807	0.8141	0.7133	1.402	1.228	1.722	30	0.9512
0.6225	40	0.5831	0.8124	0.7177	1.393	1.231	1.715	20	0.9483
0.6254	50	0.5854	0.8107	0.7221	1.385	1.233	1.708	10	0.9454
0.6283	**36° 00′**	0.5878	0.8090	0.7265	1.376	1.236	1.701	**54° 00′**	0.9425
0.6312	10	0.5901	0.8073	0.7310	1.368	1.239	1.695	50	0.9396
0.6341	20	0.5925	0.8056	0.7355	1.360	1.241	1.688	40	0.9367
0.6370	30	0.5948	0.8039	0.7400	1.351	1.244	1.681	30	0.9338
0.6400	40	0.5972	0.8021	0.7445	1.343	1.247	1.675	20	0.9308
0.6429	50	0.5995	0.8004	0.7490	1.335	1.249	1.668	10	0.9279
0.6458	**37° 00′**	0.6018	0.7986	0.7536	1.327	1.252	1.662	**53° 00′**	0.9250
0.6487	10	0.6041	0.7969	0.7581	1.319	1.255	1.655	50	0.9221
0.6516	20	0.6065	0.7951	0.7627	1.311	1.258	1.649	40	0.9192
0.6545	30	0.6088	0.7934	0.7673	1.303	1.260	1.643	30	0.9163
0.6574	40	0.6111	0.7916	0.7720	1.295	1.263	1.636	20	0.9134
0.6603	50	0.6134	0.7898	0.7766	1.288	1.266	1.630	10	0.9105
0.6632	**38° 00′**	0.6157	0.7880	0.7813	1.280	1.269	1.624	**52° 00′**	0.9076
0.6661	10	0.6180	0.7862	0.7860	1.272	1.272	1.618	50	0.9047
0.6690	20	0.6202	0.7844	0.7907	1.265	1.275	1.612	40	0.9018
0.6720	30	0.6225	0.7826	0.7954	1.257	1.278	1.606	30	0.8988
0.6749	40	0.6248	0.7808	0.8002	1.250	1.281	1.601	20	0.8959
0.6778	50	0.6271	0.7790	0.8050	1.242	1.284	1.595	10	0.8930
0.6807	**39° 00′**	0.6293	0.7771	0.8098	1.235	1.287	1.589	**51° 00′**	0.8901
0.6836	10	0.6316	0.7753	0.8146	1.228	1.290	1.583	50	0.8872
0.6865	20	0.6338	0.7735	0.8195	1.220	1.293	1.578	40	0.8843
0.6894	30	0.6361	0.7716	0.8243	1.213	1.296	1.572	30	0.8814
0.6923	40	0.6383	0.7698	0.8292	1.206	1.299	1.567	20	0.8785
0.6952	50	0.6406	0.7679	0.8342	1.199	1.302	1.561	10	0.8756
0.6981	**40° 00′**	0.6428	0.7660	0.8391	1.192	1.305	1.556	**50° 00′**	0.8727
0.7010	10	0.6450	0.7642	0.8441	1.185	1.309	1.550	50	0.8698
0.7039	20	0.6472	0.7623	0.8491	1.178	1.312	1.545	40	0.8668
0.7069	30	0.6494	0.7604	0.8541	1.171	1.315	1.540	30	0.8639
0.7098	40	0.6517	0.7585	0.8591	1.164	1.318	1.535	20	0.8610
0.7127	50	0.6539	0.7566	0.8642	1.157	1.322	1.529	10	0.8581
0.7156	**41° 00′**	0.6561	0.7547	0.8693	1.150	1.325	1.524	**49° 00′**	0.8552
0.7185	10	0.6583	0.7528	0.8744	1.144	1.328	1.519	50	0.8523
0.7214	20	0.6604	0.7509	0.8796	1.137	1.332	1.514	40	0.8494
0.7243	30	0.6626	0.7490	0.8847	1.130	1.335	1.509	30	0.8465
0.7272	40	0.6648	0.7470	0.8899	1.124	1.339	1.504	20	0.8436
0.7301	50	0.6670	0.7451	0.8952	1.117	1.342	1.499	10	0.8407
0.7330	**42° 00′**	0.6691	0.7431	0.9004	1.111	1.346	1.494	**48° 00′**	0.8378
		cos x	sin x	cot x	tan x	csc x	sec x	x degrees	x

TABLE D1 Values of the Trigonometric Functions (*continued*)

x	x degrees	sin x	cos x	tan x	cot x	sec x	csc x		
0.7330	**42° 00′**	0.6691	0.7431	0.9004	1.111	1.346	1.494	**48° 00′**	0.8378
0.7359	10	0.6713	0.7412	0.9057	1.104	1.349	1.490	50	0.8348
0.7389	20	0.6734	0.7392	0.9110	1.098	1.353	1.485	40	0.8319
0.7418	30	0.6756	0.7373	0.9163	1.091	1.356	1.480	30	0.8290
0.7447	40	0.6777	0.7353	0.9217	1.085	1.360	1.476	20	0.8261
0.7476	50	0.6799	0.7333	0.9271	1.079	1.364	1.471	10	0.8232
0.7505	**43° 00′**	0.6820	0.7314	0.9325	1.072	1.367	1.466	**47° 00′**	0.8203
0.7534	10	0.6841	0.7294	0.9380	1.066	1.371	1.462	50	0.8174
0.7563	20	0.6862	0.7274	0.9435	1.060	1.375	1.457	40	0.8145
0.7592	30	0.6884	0.7254	0.9490	1.054	1.379	1.453	30	0.8116
0.7621	40	0.6905	0.7234	0.9545	1.048	1.382	1.448	20	0.8087
0.7650	50	0.6926	0.7214	0.9601	1.042	1.386	1.444	10	0.8058
0.7679	**44° 00′**	0.6947	0.7193	0.9657	1.036	1.390	1.440	**46° 00′**	0.8029
0.7709	10	0.6967	0.7173	0.9713	1.030	1.394	1.435	50	0.7999
0.7738	20	0.6988	0.7153	0.9770	1.024	1.398	1.431	40	0.7970
0.7767	30	0.7009	0.7133	0.9827	1.018	1.402	1.427	30	0.7941
0.7796	40	0.7030	0.7112	0.9884	1.012	1.406	1.423	20	0.7912
0.7825	50	0.7050	0.7092	0.9942	1.006	1.410	1.418	10	0.7883
0.7854	**45° 00′**	0.7071	0.7071	1.0000	1.0000	1.414	1.414	**45° 00′**	0.7854
		cos x	sin x	cot x	tan x	csc x	sec x	x degrees	x

TABLE D2 Trigonometric Functions of Radians and Real Numbers

x	sin x	cos x	tan x	cot x	sec x	csc x
0.00	0.0000	1.0000	0.0000	—	1.000	—
0.01	0.0100	1.0000	0.0100	99.997	1.000	100.00
0.02	0.0200	0.9998	0.0200	49.993	1.000	50.00
0.03	0.0300	0.9996	0.0300	33.323	1.000	33.34
0.04	0.0400	0.9992	0.0400	24.987	1.001	25.01
0.05	0.0500	0.9988	0.0500	19.983	1.001	20.10
0.06	0.0600	0.9982	0.0601	16.647	1.002	16.68
0.07	0.0699	0.9976	0.0701	14.262	1.002	14.30
0.08	0.0799	0.9968	0.0802	12.473	1.003	12.51
0.09	0.0899	0.9960	0.0902	11.081	1.004	11.13
0.10	0.0998	0.9950	0.1003	9.967	1.005	10.02
0.11	0.1098	0.9940	0.1104	9.054	1.006	9.109
0.12	0.1197	0.9928	0.1206	8.293	1.007	8.353
0.13	0.1296	0.9916	0.1307	7.649	1.009	7.714
0.14	0.1395	0.9902	0.1409	7.096	1.010	7.166
0.15	0.1494	0.9888	0.1511	6.617	1.011	6.692
0.16	0.1593	0.9872	0.1614	6.197	1.013	6.277
0.17	0.1692	0.9856	0.1717	5.826	1.015	5.911
0.18	0.1790	0.9838	0.1820	5.495	1.016	5.586
0.19	0.1889	0.9820	0.1923	5.200	1.018	5.295
0.20	0.1987	0.9801	0.2027	4.933	1.020	5.033
0.21	0.2085	0.9780	0.2131	4.692	1.022	4.797
0.22	0.2182	0.9759	0.2236	4.472	1.025	4.582
0.23	0.2280	0.9737	0.2341	4.271	1.027	4.386
0.24	0.2377	0.9713	0.2447	4.086	1.030	4.207
0.25	0.2474	0.9689	0.2553	3.916	1.032	4.042
0.26	0.2571	0.9664	0.2660	3.759	1.035	3.890
0.27	0.2667	0.9638	0.2768	3.613	1.038	3.749
0.28	0.2764	0.9611	0.2876	3.478	1.041	3.619
0.29	0.2860	0.9582	0.2984	3.351	1.044	3.497
0.30	0.2955	0.9553	0.3093	3.233	1.047	3.384
0.31	0.3051	0.9523	0.3203	3.122	1.050	3.278
0.32	0.3146	0.9492	0.3314	3.018	1.053	3.179
0.33	0.3240	0.9460	0.3425	2.920	1.057	3.086
0.34	0.3335	0.9428	0.3537	2.827	1.061	2.999
0.35	0.3429	0.9394	0.3650	2.740	1.065	2.916
0.36	0.3523	0.9359	0.3764	2.657	1.068	2.839
0.37	0.3616	0.9323	0.3879	2.578	1.073	2.765
0.38	0.3709	0.9287	0.3994	2.504	1.077	2.696
0.39	0.3802	0.9249	0.4111	2.433	1.081	2.630

TABLE D2 Trigonometric Functions of Radians and Real Numbers *(continued)*

x	$\sin x$	$\cos x$	$\tan x$	$\cot x$	$\sec x$	$\csc x$
0.40	0.3894	0.9211	0.4228	2.365	1.086	2.568
0.41	0.3986	0.9171	0.4346	2.301	1.090	2.509
0.42	0.4078	0.9131	0.4466	2.239	1.095	2.452
0.43	0.4169	0.9090	0.4586	2.180	1.100	2.399
0.44	0.4259	0.9048	0.4708	2.124	1.105	2.348
0.45	0.4350	0.9004	0.4831	2.070	1.111	2.299
0.46	0.4439	0.8961	0.4954	2.018	1.116	2.253
0.47	0.4529	0.8916	0.5080	1.969	1.122	2.208
0.48	0.4618	0.8870	0.5206	1.921	1.127	2.166
0.49	0.4706	0.8823	0.5334	1.875	1.133	2.125
0.50	0.4794	0.8776	0.5463	1.830	1.139	2.086
0.51	0.4882	0.8727	0.5594	1.788	1.146	2.048
0.52	0.4969	0.8678	0.5726	1.747	1.152	2.013
0.53	0.5055	0.8628	0.5859	1.707	1.159	1.978
0.54	0.5141	0.8577	0.5994	1.668	1.166	1.945
0.55	0.5227	0.8525	0.6131	1.631	1.173	1.913
0.56	0.5312	0.8473	0.6269	1.595	1.180	1.883
0.57	0.5396	0.8419	0.6410	1.560	1.188	1.853
0.58	0.5480	0.8365	0.6552	1.526	1.196	1.825
0.59	0.5564	0.8309	0.6696	1.494	1.203	1.797
0.60	0.5646	0.8253	0.6841	1.462	1.212	1.771
0.61	0.5729	0.8196	0.6989	1.431	1.220	1.746
0.62	0.5810	0.8139	0.7139	1.401	1.229	1.721
0.63	0.5891	0.8080	0.7291	1.372	1.238	1.697
0.64	0.5972	0.8021	0.7445	1.343	1.247	1.674
0.65	0.6052	0.7961	0.7602	1.315	1.256	1.652
0.66	0.6131	0.7900	0.7761	1.288	1.266	1.631
0.67	0.6210	0.7838	0.7923	1.262	1.276	1.610
0.68	0.6288	0.7776	0.8087	1.237	1.286	1.590
0.69	0.6365	0.7712	0.8253	1.212	1.297	1.571
0.70	0.6442	0.7648	0.8423	1.187	1.307	1.552
0.71	0.6518	0.7584	0.8595	1.163	1.319	1.534
0.72	0.6594	0.7518	0.8771	1.140	1.330	1.517
0.73	0.6669	0.7452	0.8949	1.117	1.342	1.500
0.74	0.6743	0.7385	0.9131	1.095	1.354	1.483
0.75	0.6816	0.7317	0.9316	1.073	1.367	1.467
0.76	0.6889	0.7248	0.9505	1.052	1.380	1.452
0.77	0.6961	0.7179	0.9697	1.031	1.393	1.437
0.78	0.7033	0.7109	0.9893	1.011	1.407	1.422
0.79	0.7104	0.7038	1.009	0.9908	1.421	1.408

x	$\sin x$	$\cos x$	$\tan x$	$\cot x$	$\sec x$	$\csc x$
0.80	0.7174	0.6967	1.030	0.9712	1.435	1.394
0.81	0.7243	0.6895	1.050	0.9520	1.450	1.381
0.82	0.7311	0.6822	1.072	0.9331	1.466	1.368
0.83	0.7379	0.6749	1.093	0.9146	1.482	1.355
0.84	0.7446	0.6675	1.116	0.8964	1.498	1.343
0.85	0.7513	0.6600	1.138	0.8785	1.515	1.331
0.86	0.7578	0.6524	1.162	0.8609	1.533	1.320
0.87	0.7643	0.6448	1.185	0.8437	1.551	1.308
0.88	0.7707	0.6372	1.210	0.8267	1.569	1.297
0.89	0.7771	0.6294	1.235	0.8100	1.589	1.287
0.90	0.7833	0.6216	1.260	0.7936	1.609	1.277
0.91	0.7895	0.6137	1.286	0.7774	1.629	1.267
0.92	0.7956	0.6058	1.313	0.7615	1.651	1.257
0.93	0.8016	0.5978	1.341	0.7458	1.673	1.247
0.94	0.8076	0.5898	1.369	0.7303	1.696	1.238
0.95	0.8134	0.5817	1.398	0.7151	1.719	1.229
0.96	0.8192	0.5735	1.428	0.7001	1.744	1.221
0.97	0.8249	0.5653	1.459	0.6853	1.769	1.212
0.98	0.8305	0.5570	1.491	0.6707	1.795	1.204
0.99	0.8360	0.5487	1.524	0.6563	1.823	1.196
1.00	0.8415	0.5403	1.557	0.6421	1.851	1.188
1.01	0.8468	0.5319	1.592	0.6281	1.880	1.181
1.02	0.8521	0.5234	1.628	0.6142	1.911	1.174
1.03	0.8573	0.5148	1.665	0.6005	1.942	1.166
1.04	0.8624	0.5062	1.704	0.5870	1.975	1.160
1.05	0.8674	0.4976	1.743	0.5736	2.010	1.153
1.06	0.8724	0.4889	1.784	0.5604	2.046	1.146
1.07	0.8772	0.4801	1.827	0.5473	2.083	1.140
1.08	0.8820	0.4713	1.871	0.5344	2.122	1.134
1.09	0.8866	0.4625	1.917	0.5216	2.162	1.128
1.10	0.8912	0.4536	1.965	0.5090	2.205	1.122
1.11	0.8957	0.4447	2.014	0.4964	2.249	1.116
1.12	0.9001	0.4357	2.066	0.4840	2.295	1.111
1.13	0.9044	0.4267	2.120	0.4718	2.344	1.106
1.14	0.9086	0.4176	2.176	0.4596	2.395	1.101
1.15	0.9128	0.4085	2.234	0.4475	2.448	1.096
1.16	0.9168	0.3993	2.296	0.4356	2.504	1.091
1.17	0.9208	0.3902	2.360	0.4237	2.563	1.086
1.18	0.9246	0.3809	2.427	0.4120	2.625	1.082
1.19	0.9284	0.3717	2.498	0.4003	2.691	1.077

TABLE D2 Trigonometric Functions of Radians and Real Numbers *(continued)*

x	$\sin x$	$\cos x$	$\tan x$	$\cot x$	$\sec x$	$\csc x$
1.20	0.9320	0.3624	2.572	0.3888	2.760	1.073
1.21	0.9356	0.3530	2.650	0.3773	2.833	1.069
1.22	0.9391	0.3436	2.733	0.3659	2.910	1.065
1.23	0.9425	0.3342	2.820	0.3546	2.992	1.061
1.24	0.9458	0.3248	2.912	0.3434	3.079	1.057
1.25	0.9490	0.3153	3.010	0.3323	3.171	1.054
1.26	0.9521	0.3058	3.113	0.3212	3.270	1.050
1.27	0.9551	0.2963	3.224	0.3102	3.375	1.047
1.28	0.9580	0.2867	3.341	0.2993	3.488	1.044
1.29	0.9608	0.2771	3.467	0.2884	3.609	1.041
1.30	0.9636	0.2675	3.602	0.2776	3.738	1.038
1.31	0.9662	0.2579	3.747	0.2669	3.878	1.035
1.32	0.9687	0.2482	3.903	0.2562	4.029	1.032
1.33	0.9711	0.2385	4.072	0.2456	4.193	1.030
1.34	0.9735	0.2288	4.256	0.2350	4.372	1.027
1.35	0.9757	0.2190	4.455	0.2245	4.566	1.025
1.36	0.9779	0.2092	4.673	0.2140	4.779	1.023
1.37	0.9799	0.1994	4.913	0.2035	5.014	1.021
1.38	0.9819	0.1896	5.177	0.1931	5.273	1.018
1.39	0.9837	0.1798	5.471	0.1828	5.561	1.017
1.40	0.9854	0.1700	5.798	0.1725	5.883	1.015
1.41	0.9871	0.1601	6.165	0.1622	6.246	1.013
1.42	0.9887	0.1502	6.581	0.1519	6.657	1.011
1.43	0.9901	0.1403	7.055	0.1417	7.126	1.010
1.44	0.9915	0.1304	7.602	0.1315	7.667	1.009
1.45	0.9927	0.1205	8.238	0.1214	8.299	1.007
1.46	0.9939	0.1106	8.989	0.1113	9.044	1.006
1.47	0.9949	0.1006	9.887	0.1011	9.938	1.005
1.48	0.9959	0.0907	10.983	0.0910	11.029	1.004
1.49	0.9967	0.0807	12.350	0.0810	12.390	1.003
1.50	0.9975	0.0707	14.101	0.0709	14.137	1.003
1.51	0.9982	0.0608	16.428	0.0609	16.458	1.002
1.52	0.9987	0.0508	19.670	0.0508	19.695	1.001
1.53	0.9992	0.0408	24.498	0.0408	24.519	1.001
1.54	0.9995	0.0308	32.461	0.0308	32.476	1.000
1.55	0.9998	0.0208	48.078	0.0208	48.089	1.000
1.56	0.9999	0.0108	92.620	0.0108	92.626	1.000
1.57	1.0000	0.0008	1255.8	0.0008	1255.8	1.000

APPENDIX E

Background Material

Background Material

This appendix is a review of some of the more important background material needed for the study of trigonometry.

E.1 The Real Numbers, The Plane, and Graphs

A **real number** is a number that can be expressed as a (possibly infinite) decimal, for example,

$$3\tfrac{1}{2} = 3.5, \qquad -\tfrac{7}{8} = -0.875, \qquad \tfrac{2}{3} = 0.6666\ldots, \qquad \sqrt{2} = 1.414\ldots$$

The collection of all real numbers is denoted by $\mathbb{R}$.

There are several kinds of numbers that are very important. The **positive integers** or **natural numbers** are the numbers 1, 2, 3, 4, ... They may be obtained by beginning at 1 and successively adding 1. Their negatives −1, −2, −3, −4, . . . are called the **negative integers**. The **integers** are the positive integers and the negative integers together with the number zero. Calculators can handle integers easily, within the limits of the machine. (That is, if the integers are not too large or take too many digits to write.)

A number is a **rational** number if it can be expressed as the ratio of two integers. For example,

$$3\frac{1}{2} = \frac{7}{2}, \qquad -5 = \frac{-5}{1}, \qquad 23.41 = \frac{2341}{100}, \qquad 0.666\ldots = \frac{2}{3}$$

are all rational numbers. An **irrational number** is a real number that is not rational, for example, $\sqrt{2}$ and π.

In order to help visualize real numbers and their relative positions, we often form what is called a **real number line** (or, for short, a **number line** or a **real line**). This is a line where each point is associated with a unique real number (called the **coordinate** of the point), and vice versa. Recall that the number line is usually formed by taking a horizontal line, choosing any point on it, and labeling that point 0 (that point is called the **origin**). Then pick a point to the right of the origin and label it 1. The distance between these points is 1 unit. The points 2, 3, 4, $\sqrt{2}$, 1.5, π, etc., units to the right of the origin are labeled 2, 3, 4, $\sqrt{2}$, 1.5, π, etc., respectively. The points 1, 2, 3, $\frac{15}{8}$, etc., units to the left of the origin are labeled -1, -2, -3, $-\frac{15}{8}$, etc., respectively. See Fig. 1.

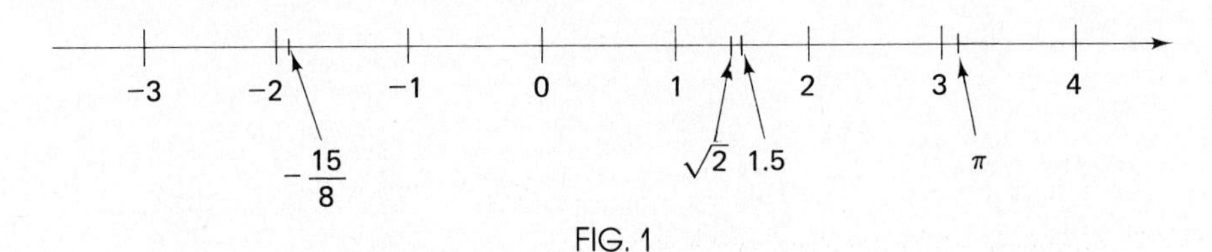

FIG. 1

We tend to identify points with their coordinates and say, for example, "3 is to the right of 2" as an abbreviation for "the point whose coordinate is 3 is to the right of the point whose coordinate is 2."

On the real line, the distance between two points is given by the absolute value of the difference of their coordinates. See Fig. 2.

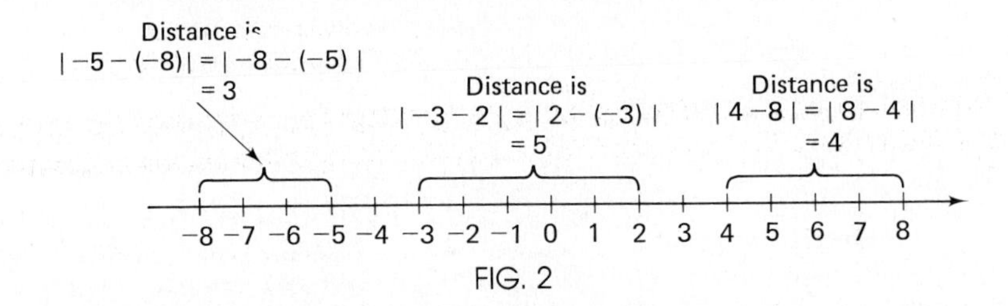

FIG. 2

We will frequently be dealing with intervals on the real line, and there is standard notation to denote the various types. There are four types of finite intervals.

(1) Notation*

$$(a, b) = \{x \mid a < x < b\}, \qquad [a, b] = \{x \mid a \leq x \leq b\}$$
$$(a, b] = \{x \mid a < x \leq b\}, \qquad [a, b) = \{x \mid a \leq x < b\}$$

*The notation $\{x \mid x$ satisfies a certain condition$\}$ means "the set of all x such that x satisfies a certain condition."

If $a < b$, they are sketched as follows:

(a, b): 

$[a, b]$: 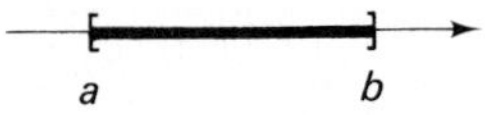

$(a, b]$: 

$[a, b)$: 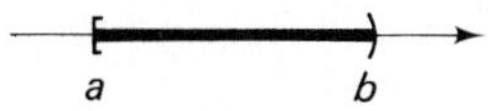

The interval (a, b) is called the **open interval** from a to b; $[a, b]$ is called the **closed interval** from a to b. The other two intervals are called **half-open** (or **half-closed**) intervals.

Note that all four intervals are the *empty set,* or *null set,* denoted by $\emptyset$, if $a > b$.

There are also infinite intervals.

(2) Notation

$$(-\infty, a) = \{x \mid x < a\}, \qquad (a, \infty) = \{x \mid x > a\}$$
$$(-\infty, a] = \{x \mid x \leq a\}, \qquad [a, \infty) = \{x \mid x \geq a\}$$

The symbols ∞ and $-\infty$ are read "infinity" and "minus infinity." It is important to remember these symbols are *notational devices* and do *not* represent any real numbers. These intervals are sketched as follows:

$(-\infty, a)$: 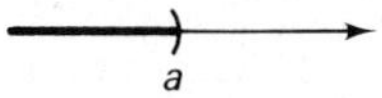

(a, ∞): 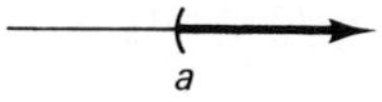

$(-\infty, a]$: 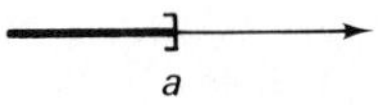

$[a, \infty)$: 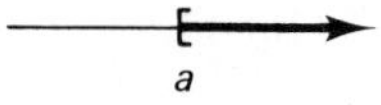

THE PLANE

Just as we find it convenient to put coordinates on the number line, we often find it useful to coordinatize the plane. A **rectangular** or **Cartesian coordinate system** is introduced into the plane using two perpendicular lines, usually one horizontal and the other vertical. Coordinates are associated with each line so that the origin O on each line is their point of intersection, and usually increasing numbers go to the right on the horizontal line and up on the vertical line. The two lines are called **coordinate axes** and the common point O is called the **origin** of the plane. Usually the horizontal axis is called the ***x*-axis** and the vertical axis the ***y*-axis**, and they are labeled x and y, respectively. The coordinate axes divide the plane into four regions, called the **first, second, third,** and **fourth quadrants,** and may be labeled Q_1, Q_2, Q_3, Q_4, respectively, as shown in Fig. 3.

Let P be a point in the plane. Suppose the vertical and horizontal lines through P intersect the x-axis at a and the y-axis at b, respectively. Then associate P with the ordered pair (a, b). In this way, each point P in the plane is associated with a unique ordered pair of real numbers (a, b) and conversely. The number a is called the ***x*-coordinate**, or **abscissa**, of P, and b is called the

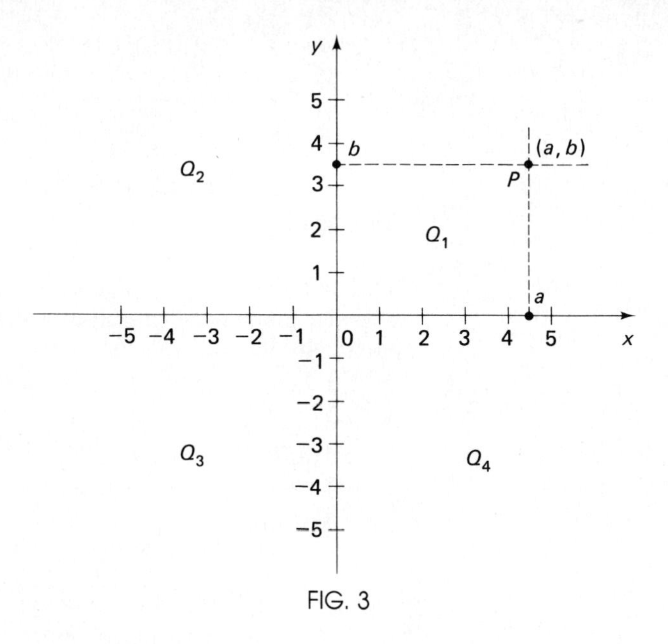

FIG. 3

y-coordinate, or **ordinate**, of P. Together, a and b are the **coordinates** of P. We sometimes identify points and their coordinates and write "$P = (a, b)$" when we really mean "P is the point whose coordinates are a and b." Fig. 4 shows the coordinates of several points.

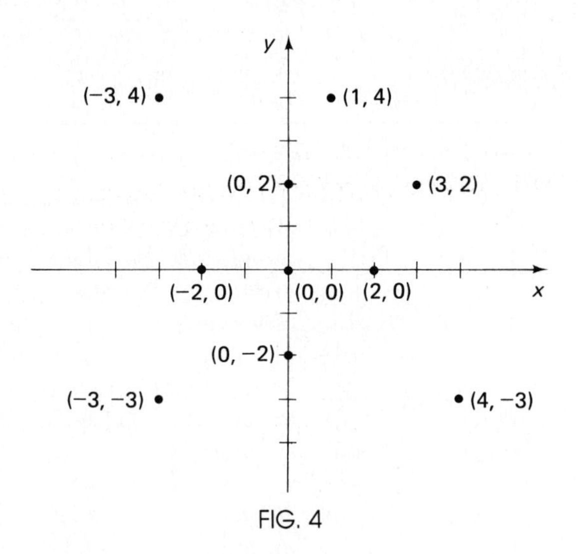

FIG. 4

We now derive the formula for finding the distance between two points in the plane, $P = (x_1, y_1)$ and $Q = (x_2, y_2)$. We shall use the following notation:

The distance between the points P and Q is denoted by $d(P, Q)$.

We now turn to the general case.

Let R be the point (x_2, y_1), and consider the right triangle PQR. See Fig. 5 for two possibilities.

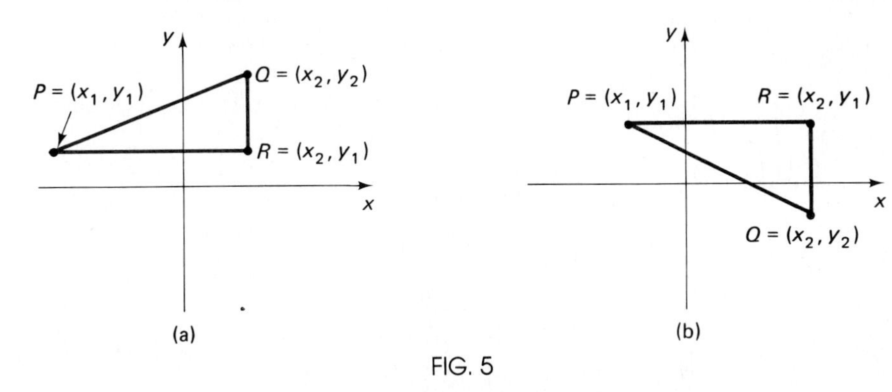

FIG. 5

By the Pythagorean theorem,

$$d(P,Q)^2 = d(P,R)^2 + d(R,Q)^2$$

However, $d(P,R) = |x_2 - x_1|$ and $d(R,Q) = |y_2 - y_1|$. Since $|u|^2 = u^2$ for any real number u,

$$d(P,Q)^2 = (x_2 - x_1)^2 + (y_2 - y_1)^2$$

or

(3) Distance Formula If $P = (x_1, y_1)$ and $Q = (x_2, y_2)$, then

$$d(P,Q) = \sqrt{(x_2 - x_1)^2 + (y_2 - y_1)^2}$$

GRAPHS

Subsets of the plane, i.e., sets of ordered pairs of real numbers, arise in many different contexts, and it becomes convenient to have a name for such sets.

(4) Definition A **relation** is a set of ordered pairs of real numbers.

Usually relations are given in terms of conditions that the ordered pairs must satisfy. For example,

$$\{(x, y) \mid y \leq x\}, \qquad \{(x, y) \mid y = x^2 + 1\},$$
$$\{(x, y) \mid |x| < 1 \text{ or } |y| < 2\}$$

In most cases, the set notation is omitted, and we just write "the relation $y \leq x$" or "the relation $y = x^2 + 1$," etc. However, sometimes we just have to list the members of the set, for example $\{(1,2),(3,5),(-2,6),(1,-4)\}$.

To understand a particular relation, it is helpful to sketch its graph.

> **(5) Definition** The **graph of a relation** is the set of points in the plane whose coordinates are in the relation.

EXAMPLE 1 Sketch the graph of the following relations:

(a) $y \leq x$ (b) $y = x^3 + 1$ (c) $x = y^2$

Solution See Fig. 6.

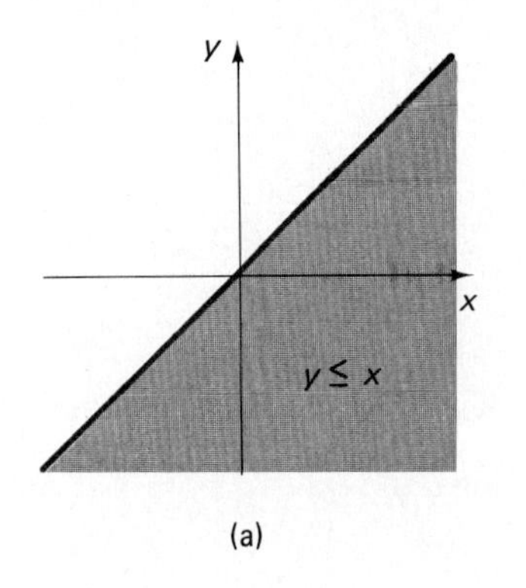

(a)

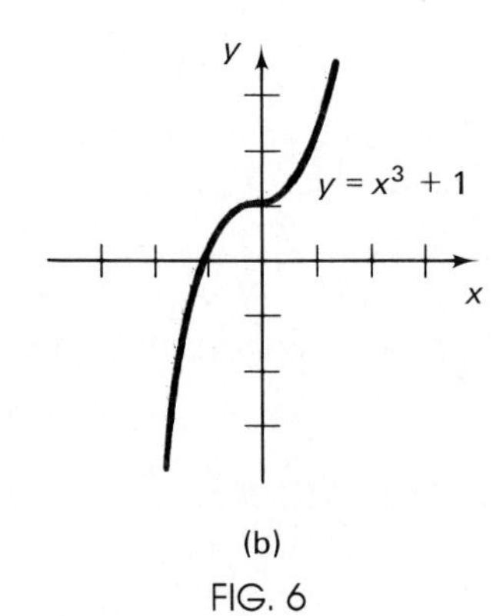

(b)

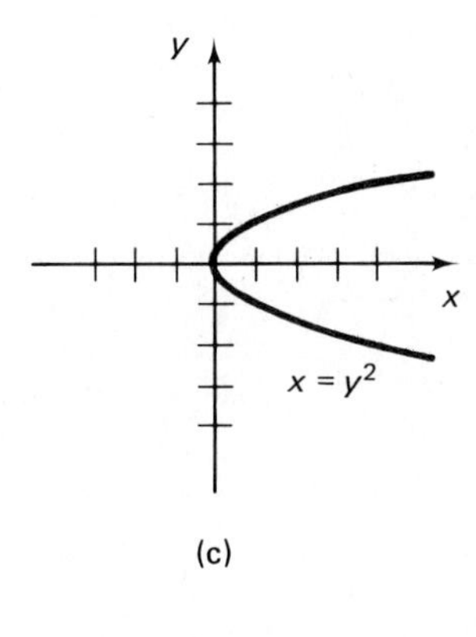

(c)

FIG. 6

■

EXERCISES

In Exercises 1–4, express the inequality in interval notation, and sketch the graph of the interval.

1. **a.** $-2 < x < 3$ **b.** $-2 < x \leq 3$ **c.** $-2 \leq x < 3$ **d.** $-2 \leq x \leq 3$

2. **a.** $x < 5$ **b.** $x \leq 5$ **c.** $x > 5$ **d.** $x \geq 5$

3. **a.** $x < -3$ and $-7 < x$ **b.** $x < -3$ or $-7 < x$ **c.** $x < -7$ and $-3 < x$ **d.** $x < -7$ or $-3 < x$

4. **a.** $x \leq 5$ and $-2 < x$ **b.** $x \leq 5$ or $-2 < x$ **c.** $x < -2$ and $5 \leq x$ **d.** $x < -2$ or $5 \leq x$

In Exercises 5 and 6, plot the points on a rectangular coordinate system.

5. $A = (5, -3)$, $B = (-5, -3)$, $C = (3, 2)$, $D = (-3, 2)$, $E = (-2, 0)$, $F = (0, -2)$

6. $A = (4, 5)$, $B = (-4, 5)$, $C = (3, -2)$, $D = (-3, -2)$, $E = (3, 0)$, $F = (0, 3)$

In Exercises 7–14, plot the points and then find the distance between the given points.

7. $(5, 2), (-7, -3)$

8. $(-4, 3), (2, 2)$

9. $(7, -2), (7, 5)$

10. $(-3, 7), (0, -5)$

11. $(a + b, a - b), (b - a, a + b)$

12. $(c^{1/2}, d^{1/2}), (-c^{1/2}, -d^{1/2})$

13. $(s^{3/2}, \frac{3}{2}), (s^{-3/2}, -\frac{1}{2})$

14. $(t, |t|), (-|t|, t)$

15. The points (1, 2) and (1, 6) are vertices of a square. Find the other two vertices. (*Note:* There are three possible pairs.)
16. Answer Exercise 15 if the two vertices are (n, p) and (n, q) $q > p$.
17. Find all points on the y-axis that are a distance of 5 from the point (3, 5).
18. Find all points on the x-axis that are a distance of 13 from the point (3, 5).
19. Find all points on the line $x = -2$ that are a distance of 13 from the point (3, 5).
20. Find all points on the x-axis equidistant from the points $(1, -2)$ and $(3, 1)$.

In Exercises 21–52, sketch the graphs of the relation.

21. $y = -x$
22. $y < -x$
23. $y \geq -x$
24. $y = 3$
25. $y = -4$
26. $x = -7$
27. $x = 1$
28. $xy > 0$
29. $xy \leq 0$
30. $(x - 3)(y + 4) = 0$
31. $(x + 2)(y - 3) = 0$
32. $|x - 2| < 3$
33. $|x + 1| \geq 4$
34. $|y + 2| > 1$
35. $|y - 7| < 2$
36. $|x - 3| < 1$ and $|y + 1| < 2$
37. $|x + 4| \geq 4$ and $|y - 1| \leq 2$
38. $|x - 2| \leq 3$ or $|y + 4| < 1$
39. $|x - 7| \geq 1$ or $|y - 2| \geq 2$
40. $y = x^2 - 1$
41. $x = y^2 + 1$
42. $y = |x + 1|$
43. $y \leq |x| + 1$
44. $x = |y| - 1$
45. $y = -x^3$
46. $y \geq x^3 + 1$
47. $y = -\dfrac{1}{x}$
48. $y = \dfrac{1}{x^2}$
49. $y = x + |x|$
50. $y = |x| - x$
51. $|x| + |y| = 0$
52. $|y| = |x|$

E.2 Functions

The concept of a function is one of the most central concepts in modern mathematics. It is also used frequently in everyday life. For instance, the statements "each student in the course will be assigned a grade at the end of the course" and "each item in the store has a price marked on it" describe functions. If we analyze these statements, we shall find the essential ingredients of a function.

For the first statement, there is a set of students, a set of possible grades, and a rule which assigns to each member of the first set (to each student) a unique member of the second set (a grade). For the second statement, there is a set of items for sale, a set of possible prices, and a rule that assigns to each member of the first set (to each item) a unique member of the second set (a price).

(6) Definition A **function** f from a set D to a set R is a rule that assigns to each element of D a unique element of R. The set D is called the **domain** of the function, while the set R is called the **range** of the function.

Besides f, we also use the letters F, g, G, etc., to denote functions.

If a is an element of D, then the unique element in R which f assigns to a is called the **value of f at a** or the **image of a under f** and is denoted by $f(a)$. We will understand that the range R is the set of all values of the function.* We sometimes represent functions pictorially as in the accompanying figure.

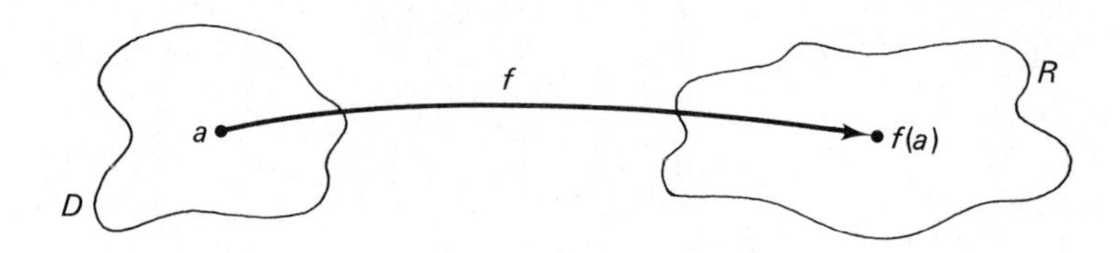

We often think of x as representing an arbitrary element of D and y representing the corresponding value of f at x. We then call x the **independent variable,** because a value for x can be selected arbitrarily from D, and call y the **dependent variable,** because the value of y depends on the value of x selected. We also write

$$y = f(x)$$

which is read "y is a function of x" or "y is f of x."

The **rule** of a function gives the value of the function at each element of the domain. Often the rule is a formula, but it can be other things, such as a list of ordered pairs, a table, or a set of instructions.

You can think of some of the buttons on your calculator as providing rules for different functions. One that you already know about is [√]. If you press [2] [√], you obtain [1.4142135]; your calculator is giving you the (approximate) value of the square root function at the number 2. In this way, the calculator illustrates what has been one traditional description of a function: A function is like a black box into which you can put any number from the domain and out of which comes the corresponding value in the range. See Fig. 7.

If the rule of a function f is a formula giving y in terms of x, say $y = x^2 + 2x$, then the statements "the function $y = x^2 + 2x$," "$y = f(x)$ where $f(x) = x^2 + 2x$," or just "$f(x) = x^2 + 2x$" all indicate the same thing. Given any of these, to find the value of f at a number, you substitute that number for x wherever x occurs in the formula (and then simplify).

* Some texts allow the range to be larger than the set of all values of the function.

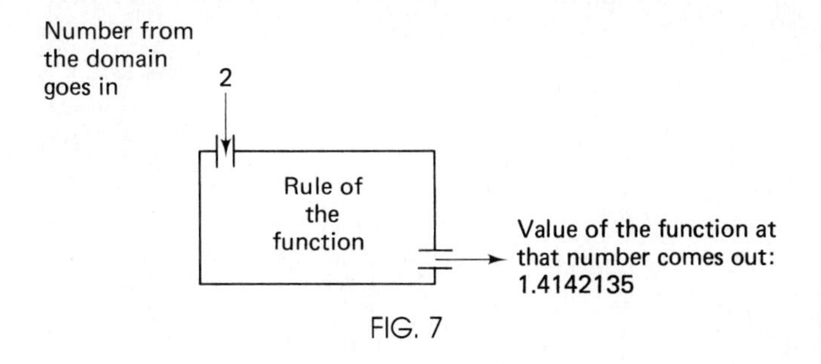

FIG. 7

EXAMPLE 1 If f is a function with domain the set of all real numbers and with rule $f(x) = x^2 + 2x$, find the following values of f:

(a) $f(3)$; (b) $f(-11.21)$; (c) $f\left(\frac{1}{a}\right)$, $a \neq 0$; (d) $f(-1 + h)$

(e) $f(f(3))$

Solution (a) $f(3) = 3^2 + 2(3) = 9 + 6 = 15$

(b) $f(-11.21) = (-11.21)^2 + 2(-11.21) = 125.6641 - 22.42$
$= 103.2441$

(c) $f\left(\frac{1}{a}\right) = \left(\frac{1}{a}\right)^2 + 2\frac{1}{a} = \frac{1}{a^2} + \frac{2}{a} = \frac{1 + 2a}{a^2}$

(d) $f(-1 + h) = (-1 + h)^2 + 2(-1 + h) = 1 - 2h + h^2 - 2 + 2h$
$= h^2 - 1$

(e) By part (a), $f(3) = 15$, so $f(f(3)) = f(15) = 15^2 + 2(15) = 255$ ■

There is a standard notation indicating a function together with its domain and range.

> **(7) Notation** We write "$f: D \rightarrow R$" for "f is a function with domain D and range R" or, equivalently, "f is a function from D to R."

For example, a statement like

"Let $f: \mathbb{R} \rightarrow [1, \infty)$ by $f(x) = x^2 + 1$"

gives the domain ($\mathbb{R}$), range ($[1, \infty)$), and rule ($f(x) = x^2 + 1$). However, since you can compute the range from the domain and the rule, the range is often not given. In fact, most of the time the domain is also not given. All that is written is something like

"Let $f(x) = x^2 + 1$" or "Let $y = x^2 + 1$"

When this is written, the domain is understood to be the largest possible set of real numbers for which the rule applies (i.e., for which the rule gives a real number).

The graph of a function leads us to a new way of thinking about functions. If f is a function, its graph is the set of all points in the plane whose coordinates satisfy the equation $y = f(x)$. Thus the function f determines a relation, namely, the set of all ordered pairs (x, y) where $y = f(x)$. From the ordered pairs, you can easily see the domain and range. The domain D is the set of first elements from the ordered pairs, and the range R is the set of second elements. The special property that functions have, namely that the rule assigns to each element of D a *unique* element of R, means that no two different ordered pairs in the relation have the same first element. With this in mind, we give an alternate definition of a function.

(8) (Alternate) Definition A **function** is a set of ordered pairs such that no two different pairs have the same first element.

Since every function is a relation, it is natural to ask how to recognize whether or not a particular relation is a function. It is easy to do this from its graph. If there are two or more points of the graph of a relation on the same vertical line, then the coordinates of these points have the same first coordinate but different second coordinates. These ordered pairs are in the relation, so the relation is not a function. If there is no vertical line which contains two points of the graph of the relation, the relation is a function. This is the **vertical line test.**

EXAMPLE 2 From the graphs in Fig. 8, determine which of the relations are functions.

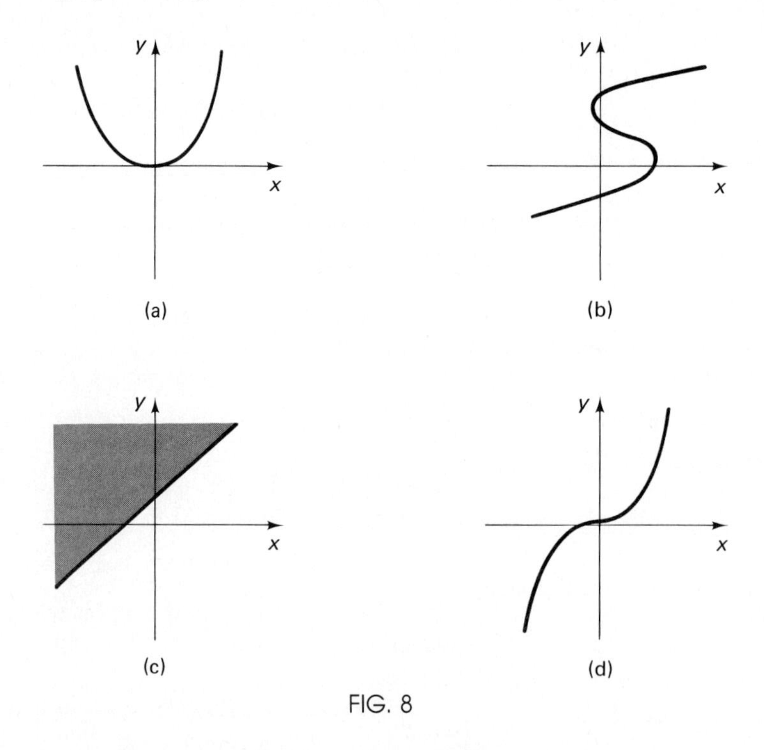

FIG. 8

Solution We see in (a) and (d) that there is no vertical line containing two points of these graphs. So (a) and (d) "pass" the vertical line test and hence are functions. In (b) and (c), however, there are vertical lines that contain two (or more) points of the graph. Hence (b) and (c) "fail" the vertical line test and are not functions. ■

When functions are used in applied situations, often the relationships are described in words, and the formula or rule for the function has to be found.

EXAMPLE 3 Find the rule for the function described, and give the domain: A piece of wire is 60 inches long. Part of the wire is cut off and bent into a rectangle whose length is twice its width. Find the area of the rectangle as a function of the width.

Solution Let w be the width of the rectangle. Then the length is $2w$, and the area $A = l \cdot w = 2w \cdot w$ or $A = 2w^2$. Since $w + 2w + w + 2w \leq 60$, $6w \leq 60$ or $w \leq 10$. Thus the domain is $0 \leq w \leq 10$, and the rule is $A = 2w^2$.

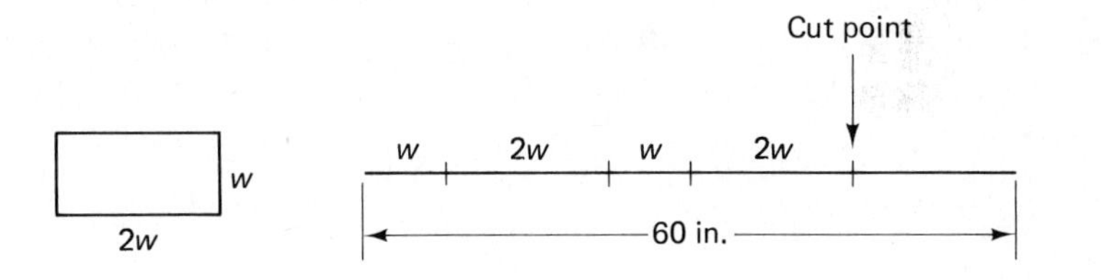

■

EXERCISES

1. If $f(x) = 3x^2 - 2x + 1$, find the values $f(-3)$, $f(-1)$, $f(0)$, and $f(2)$.

2. If $g(x) = x^3 + 1$, find the values $g(-2)$, $g(-1)$, $g(0)$, and $g(1)$.

3. If $h(x) = \sqrt{x + 1}$, find the values $h(8)$, $h(3)$, $h(0)$, and $h(-1)$.

4. If $k(x) = \dfrac{x}{x + 1}$, find $k(-2)$, $k(0)$ $k(3)$, and $k(9)$.

5. If $F(x) = 7x + 3$, find
   a. $F(0)$
   b. $F(-3)$
   c. $F(1/a)$
   d. $F(a + 3)$
   e. $\dfrac{F(2 + h) - F(2)}{h}$

6. If $G(x) = 3x^2 - 2$, find
   a. $G(-1)$
   b. $G(2)$
   c. $G(2a)$
   d. $G(a - 1)$
   e. $\dfrac{G(-1 + h) - G(-1)}{h}$

7. If $H(x) = 3|x| + 2x$, find
   a. $H(-2)$
   b. $H(2)$
   c. $H(\frac{1}{2})$
   d. $H(a^2)$
   e. $H(H(-2))$

8. If $K(x) = \sqrt{2x + 3}$, find
   a. $K(3)$
   b. $K(11)$
   c. $K(-1)$
   d. $K\left(\dfrac{1}{a^2}\right)$
   e. $K(K(11))$

9. If the graph of f is the sketch shown, estimate the given values.

a. $f(-2)$ **b.** $f(-1)$
c. $f(0)$ **d.** $f(1)$
e. $f(2)$

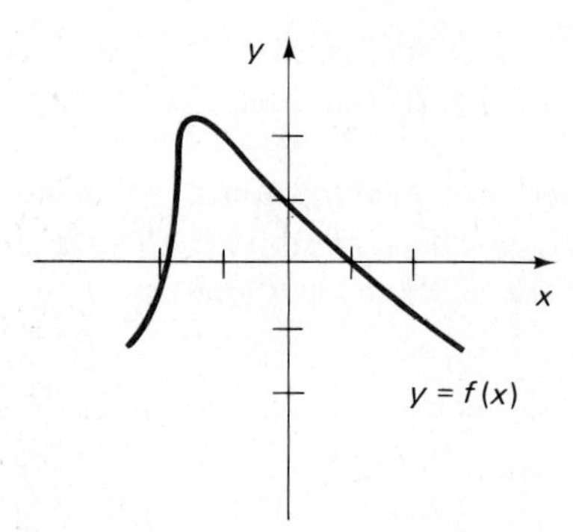

10. If the graph of g is the sketch shown, estimate the given values.

a. $g(-2)$ **b.** $g(-1)$
c. $g(0)$ **d.** $g(1)$
e. $g(2)$

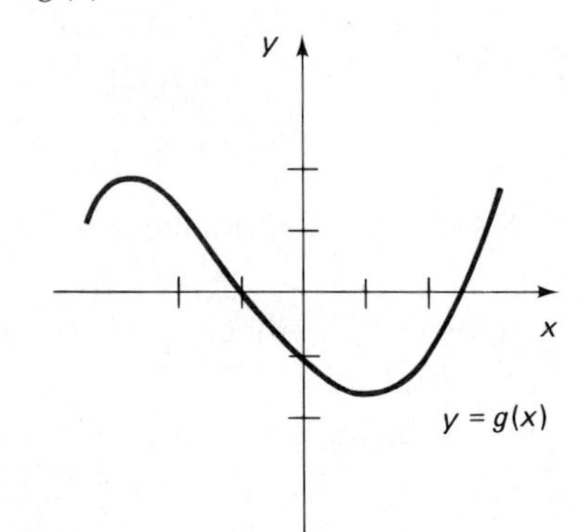

11. The graph of $y = \sqrt{x}$ is shown. Use the graph to estimate the given values.

a. $\sqrt{2}$ **b.** $\sqrt{4.5}$
c. $\sqrt{0.8}$ **d.** $\sqrt{\pi}$
e. $\sqrt{\sqrt{2}}$

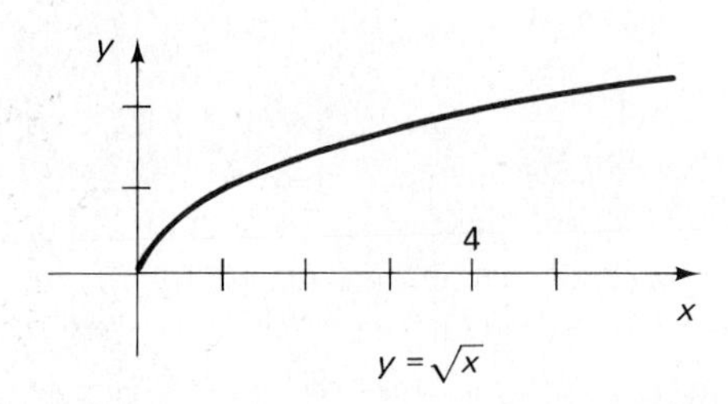

12. Use the graph of $y = \sqrt[3]{x}$ as shown to estimate the given values.

a. $\sqrt[3]{2}$ **b.** $\sqrt[3]{6}$
c. $\sqrt[3]{0.5}$ **d.** $\sqrt[3]{-4}$
e. $\sqrt[3]{-5}$

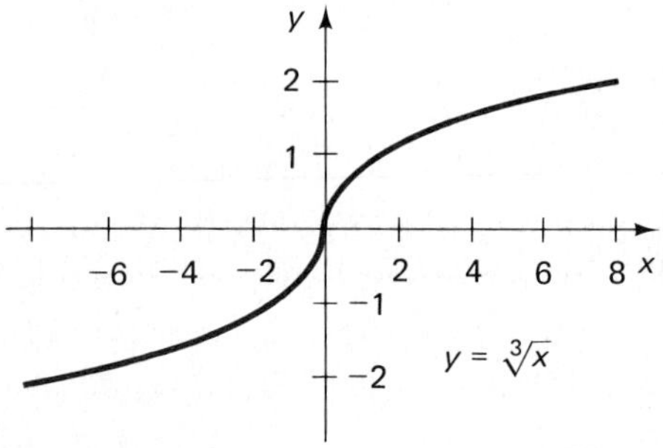

y, 2, 1, −6, −4, −2, 2, 4, 6, 8, x, −1, −2, $y = \sqrt[3]{x}$

13. Which of the following are graphs of functions?

a.

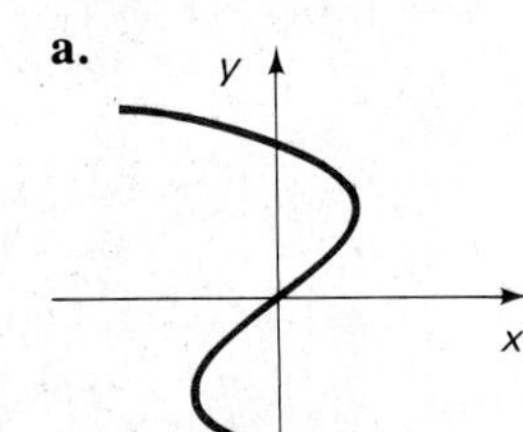

b.

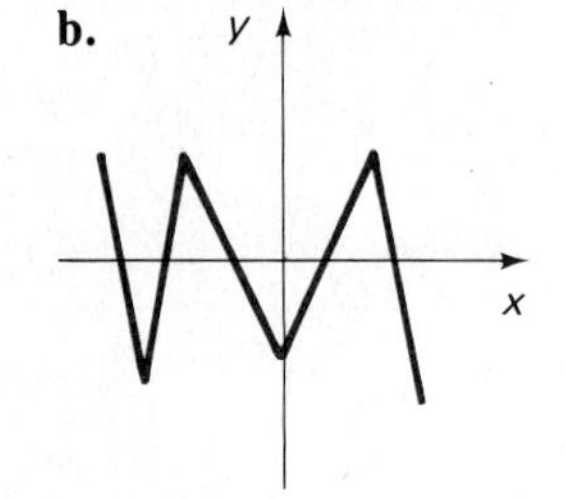

c.

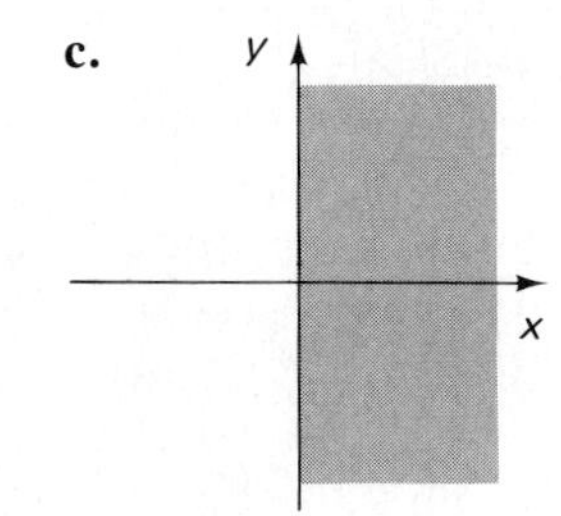

d.

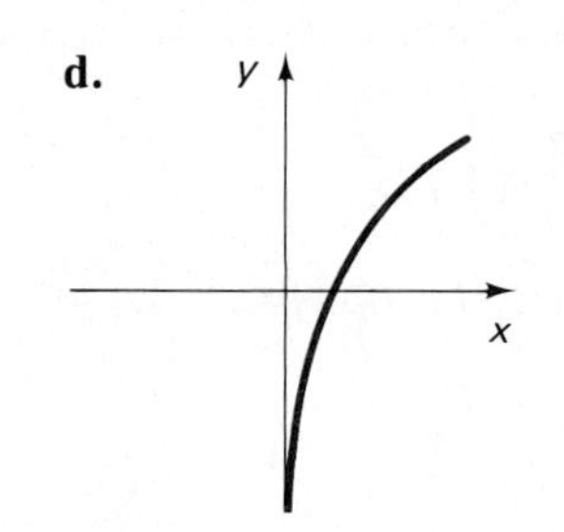

14. Which of the following are graphs of functions?

a.

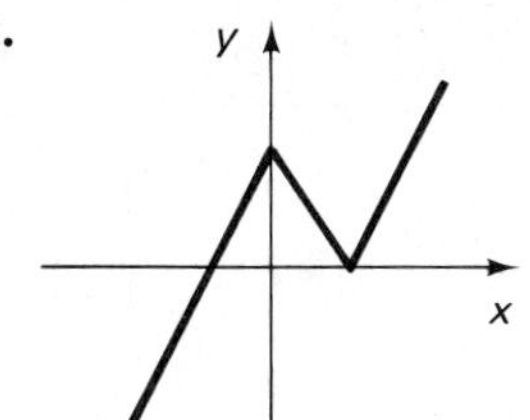

b.

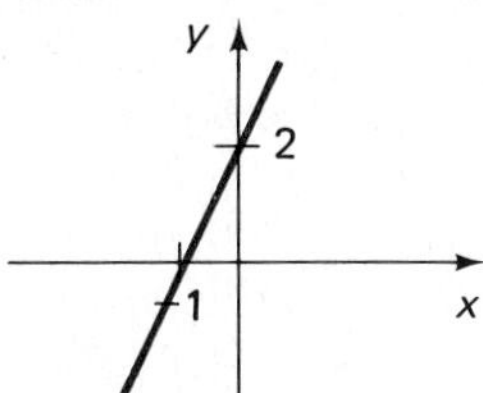

c.

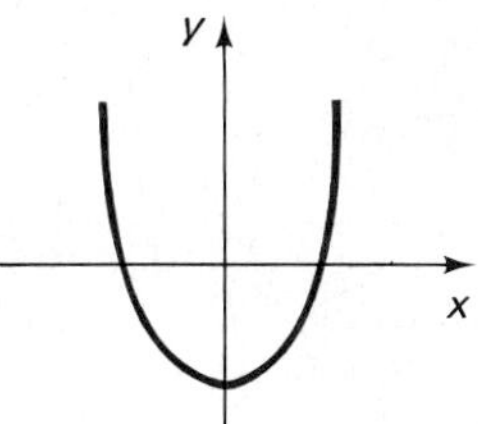

d.

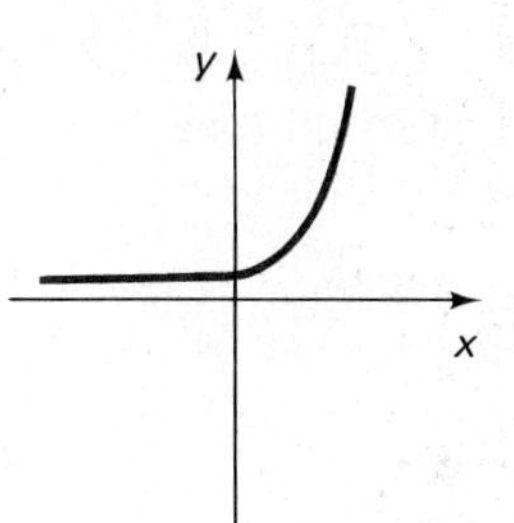

15. Reflect each of the graphs in Exercise 13 through the x-axis, and determine which of these reflected graphs is a function.

16. Reflect each of the graphs in Exercise 14 through the y-axis, and determine which of these reflected graphs is a function.

17. Find a formula that expresses the radius of a circle as a function of its circumference.

18. Find a formula that expresses the area of a circle as a function of its circumference.

19. Find a formula that expresses the area of a cube as a function of its volume.

20. Find a formula that expresses the volume of a sphere as a function of its surface area.

21. An open box is to be made from a rectangular piece of tin 12 inches long and 10 inches wide by cutting pieces x inches square from each corner and bending up the sides.

a. Find a formula that expresses the volume V of the box as a function of x.

b. Find the domain of the function.

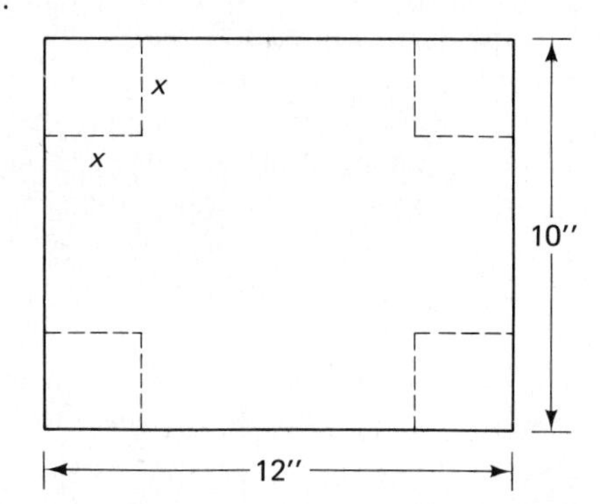

22. A balloon is released from the ground 200 meters from an observer and rises steadily at the rate of 3 meters per second.

a. Find a formula which expresses the distance d between the observer and the balloon as a function of the time t after the balloon was released.

b. What is the value of d twenty seconds after the balloon was released?

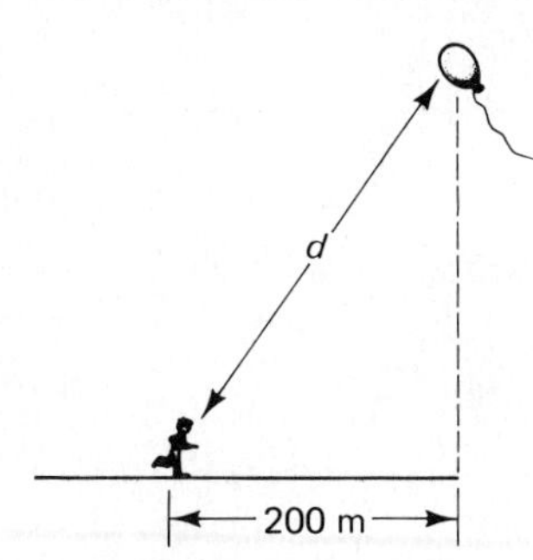

23. A man is in a rowboat 1 mile from a straight shore. He starts out rowing toward a point x miles up the shore. Suppose he rows at a steady 3 miles per hour.

a. Find a formula that expresses the time T that it takes him to reach the shore as a function of x.

b. How long does it take him if x is 2 miles?

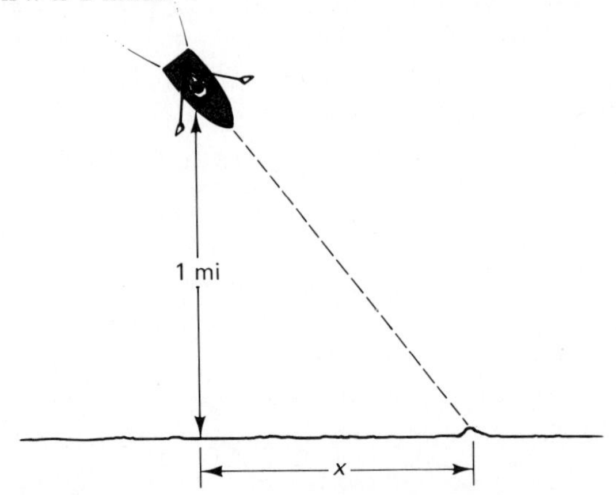

24. A rectangular region of 600 square meters is to be fenced. On three of the sides the fencing used will cost \$2 per meter, and on the fourth side the fencing will cost \$5 per meter.

a. If x represents the length of the fourth side, find a formula that expresses the total cost C as a function of x.

b. Find the total cost when $x = 30$, 60, or 100.

25. Two ships leave a port on the same morning, one at 8 A.M. heading south at 15 miles per hour and the other at 10 A.M. heading west at 20 miles per hour. Find formulas expressing the distance d between them as a function of the time t, in hours, after 8A.M. Include the domain of each formula.

26. A manufacturer sells a certain item at a price of \$30 each if fewer than 20 items are ordered. If 20 or more items are ordered (up to 300), the price per article is \$31 less 5 cents times the total number ordered. Find formulas that express the total cost as a function of the number n of items ordered. Include the domain of each formula.

27. A travel agency offers a tour. It charges \$20 per person if fewer than 25 people go. If 25 people or more, up to a maximum of 110, take the tour, they charge each person \$22.40 less 10 cents times the number of people who go. Find formulas that express the total charge C as a function of the number n who go. Include the domain of each formula.

28. A piece of wire, 90 centimeters long, is cut once, and each piece is bent into a square. Find a formula giving the total area A of the two squares as a function of the distance x that the cut is made from one end.

E.3 Composite and Inverse Functions

In this section we first discuss an important method of combining two given functions to form a third function. Suppose that X, Y, and Z are sets of real numbers, and suppose that f is a function from X to Y and that g is a function from Y to Z. Then we can naturally form a function from X to Z as follows (see Fig. 9):

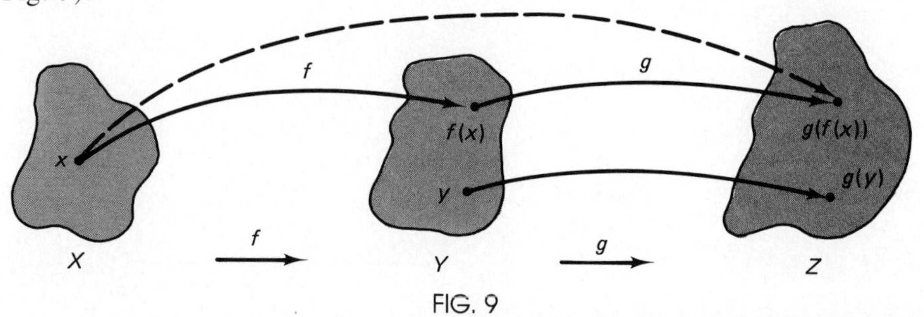

FIG. 9

For every number x in X, f associates a number $f(x)$ in Y.

For every number y in Y, g associates a number $g(y)$ in Z; in particular $f(x)$ is in Y, so g associates with $f(x)$ the number $g(f(x))$ in Z. By associating with each x in X the number $g(f(x))$ in Z, we obtain what is called the **composite function** of f and g. This function is denoted by $g \circ f$.

> **(9) Definition** If f is a function from X to Y and g is a function from Y to Z, then the **composite function** $g \circ f$ is the function from X to Z defined by
>
> $$(g \circ f)(x) = g(f(x)) \qquad \text{for every } x \text{ in } X$$

EXAMPLE 1 If $f(x) = x^2$ and $g(x) = x + 1$, find $(g \circ f)(x)$ and $(f \circ g)(x)$.

Solution We apply the definitions and then simplify:

$$\begin{aligned}
(g \circ f)(x) &= g(f(x)) && \text{(def. of } g \circ f) \\
&= g(x^2) && \text{(def. of } f) \\
&= x^2 + 1 && \text{(def. of } g) \\
(f \circ g)(x) &= f(g(x)) && \text{(def. of } f \circ g) \\
&= f(x + 1) && \text{(def. of } g) \\
&= (x + 1)^2 && \text{(def. of } f)
\end{aligned}$$

Note that $(g \circ f)(x)$ and $(f \circ g)(x)$ are different. This is often the case. ■

A calculator illustrates composites nicely. The composite of functions on a calculator "means" using one button (or buttons) and then another, and of course the order is important.

EXAMPLE 2 Suppose $f(x) = x + 3$ and $g(x) = \sqrt{x}$. Then to use a calculator to compute $(g \circ f)(7) = g(f(7))$, press

		$f(7)$	$g(f(7))$
Alg.:	[7]	[+] [3] [=]	[√‾]
RPN:	[7]	[ENT] [3] [+]	[√‾]

getting $\sqrt{7 + 3} \approx 3.16228$. To use the calculator to compute $(f \circ g)(7) = f(g(7))$, press

		$g(7)$	$f(g(7))$
Alg.:	[7]	[√‾]	[+] [3] [=]
RPN:	[7]	[√‾]	[ENT] [3] [+]

getting $\sqrt{7} + 3 \approx 5.64575$. ■

We now turn to the very important topic of inverse functions. We begin with an example. Let $f(x) = x^3$ and $g(x) = \sqrt[3]{x}$. Since

$$g(f(x)) = \sqrt[3]{x^3} = x \qquad \text{and} \qquad f(g(x)) = (\sqrt[3]{x})^3 = x$$

we have

$$(g \circ f)(x) = x, \quad \text{all } x \qquad \text{and} \qquad (f \circ g)(x) = x, \quad \text{all } x$$

Thus applying the composite in either direction gets us back to where we started. When this happens, we say g is the **inverse function** of f, because g does the inverse of (i.e., reverses) what f does.

(10) Definition Suppose f and g are functions such that

$$(g \circ f)(x) = g(f(x)) = x \qquad \text{for all } x \text{ in the domain of } f, \qquad \text{and}$$
$$(f \circ g)(x) = f(g(x)) = x \qquad \text{for all } x \text{ in the domain of } g$$

Then we say g is the **inverse function** of f, and we write $g = f^{-1}$. (f^{-1} is read "f inverse.")

Thus suppose $f: U \to V$ and $f(u) = v$. If g is f^{-1}, then $g: V \to U$ and $g(v) = u$. See Fig. 10.

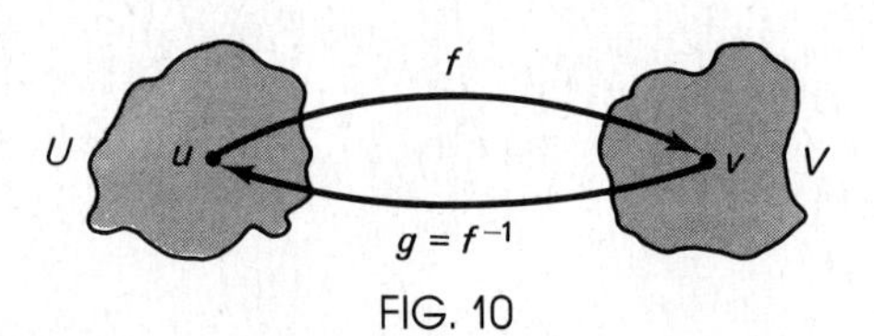

FIG. 10

Note the following:

(11) The domain of f is the range of f^{-1} and vice versa.

It will be useful to rewrite the equations in (10) using f^{-1}.

(12) If f has an inverse function f^{-1}, then

$$f^{-1}(f(x)) = x \qquad \text{for all } x \text{ in the domain of } f, \qquad \text{and}$$
$$f(f^{-1}(x)) = x \qquad \text{for all } x \text{ in the domain of } f^{-1}$$

Sometimes an inverse function can be found easily using (12).

EXAMPLE 3 By the preceding discussion, if $f(x) = x^3$, then $f^{-1}(x) = x^{1/3}$, since

$$f^{-1}(f(x)) = f^{-1}(x^3) = (x^3)^{1/3} = x$$

and

$$f(f^{-1}(x)) = f(x^{1/3}) = (x^{1/3})^3 = x \qquad \blacksquare$$

EXAMPLE 4 If $g(x) = x + 2$, then $g^{-1}(x) = x - 2$ since

$$g^{-1}(g(x)) = g^{-1}(x + 2) = (x + 2) - 2 = x$$

and

$$g(g^{-1}(x)) = g(x - 2) = (x - 2) + 2 = x \quad \blacksquare$$

Warning: $f^{-1}(x)$ is not $\dfrac{1}{f(x)}$, as can be seen in Examples 3 and 4. If we want $\dfrac{1}{f(x)}$, we write $[f(x)]^{-1}$.

It is very interesting and instructive to see how your calculator handles certain inverse functions. Your calculator has special keys:

$$\boxed{\text{sin}}, \quad \boxed{\text{cos}}, \quad \boxed{\text{tan}}, \quad \boxed{\text{log}}, \quad \boxed{\text{ln}}$$

We shall use them here *only* to illustrate (12). (Of course, labeling varies on different calculators; for example $\boxed{\text{ln}}$ might be $\boxed{\text{ln } x}$, $\boxed{\text{LN}}$, etc.) These five keys represent five functions which also have inverse functions (when the domain is appropriately restricted). To compute the inverse functions, your calculator has a key labeled

$$\boxed{\text{INV}} \quad \text{or} \quad \boxed{\text{ARC}} \quad \text{or} \quad \boxed{\text{F}}$$

(and there are a few other variations). If $g(x) = \sin x$, to compute $g^{-1}(0.5)$ press

$$\boxed{.5}\ \boxed{\text{INV}}\ \boxed{\text{sin}}$$

The other inverse functions work similarly (with a few exceptions for $\boxed{\text{log}}$ and $\boxed{\text{ln}}$). We can now use a calculator to illustrate (12). To avoid difficulties that we do not need to consider for this illustration, we shall always start with numbers between 0 and 1.

EXAMPLE 5 If $h(x) = \cos x$, use your calculator to compute (a) $h^{-1}(h(0.2))$ and (b) $h(h^{-1}(0.3))$.

Solution Of course, we should end up with what we started with. For (a), press

$$\boxed{.2}\ \underbrace{\boxed{\text{cos}}}_{h(0.2)}\ \underbrace{\boxed{\text{INV}}\ \boxed{\text{cos}}}_{h^{-1}(h(0.2))}$$

getting 0.2. For (b), press

$$\boxed{.3}\ \underbrace{\boxed{\text{INV}}\ \boxed{\text{cos}}}_{h^{-1}(0.3)}\ \underbrace{\boxed{\text{cos}}}_{h(h^{-1}(0.3)}$$

getting 0.3. $\blacksquare$

We now have three important questions to answer concerning inverse functions. Suppose we start with a function f.

1. When does f have an inverse function f^{-1}?
2. How do you find the rule for f^{-1} from the rule for f?
3. How do you find the graph for f^{-1} from the graph for f?

To give comprehensive and integrated answers to these three questions, we turn to the ordered pair definition of a function (8). First, consider again the function $f(x) = x^3$ and its inverse $f^{-1}(x) = \sqrt[3]{x}$. Since $f(2) = 2^3 = 8$, (2, 8) is an ordered pair in f. However, because $2^3 = 8$, it follows that $\sqrt[3]{8} = 2$, so that $f^{-1}(8) = 2$ and (8, 2) is in f^{-1}. This illustrates the general situation.

(13) f^{-1} is the set of all ordered pairs (b, a) such that (a, b) is an ordered pair in f.

This immediately tells us about the graphs. By (13), the graph of the inverse is the set of all points in the plane with coordinates (b, a), where (a, b) is in the original function. When $a \neq b$, the relationship between (a, b) and (b, a) is that they are *symmetric with respect to the line* $y = x$. See Fig. 11. When

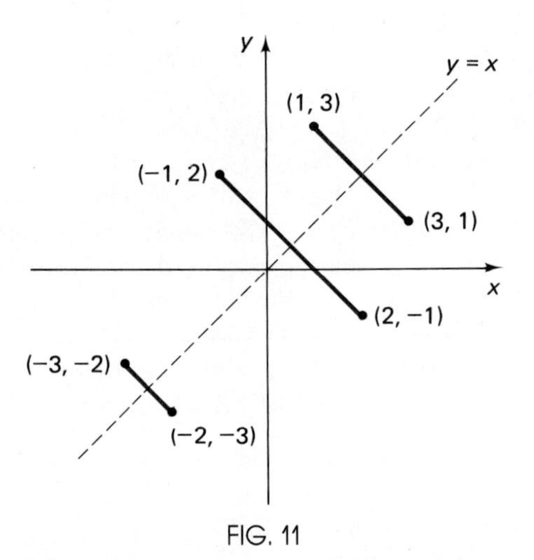

FIG. 11

we start with a graph and replace each point of that graph with the point symmetric to it with respect to the line $y = x$, we say we have *reflected the graph through the line* $y = x$. Hence:

(14) The graph of an inverse function is just the graph of the original function reflected through the line $y = x$.

When we reflect the graph of a function through the line $y = x$, the resulting graph may or may not pass the vertical line test. Hence it may or may not be the graph of a function. A function f has an inverse function f^{-1} if reflecting the graph of f through the line $y = x$ results in a graph of a function (and this will then be the graph of f^{-1}). There is a geometric test to determine this. The test corresponds to doing the vertical line test after reflecting the graph through the line $y = x$.

(15) The Horizontal Line Test

1. If there are two or more points of the graph of a given function on the same horizontal line, the function does not have an inverse function.
2. If there is no horizontal line which contains two or more points of the graph of the function, it has an inverse function.

EXAMPLE 6 If we examine Fig. 12, we see that the function graphed in Fig. 12(a) "fails" the horizontal line test; i.e., there is a horizontal line (e.g., l or l') which intersects the graph in two or more places. So the reflected graph is not the graph of a function, as can be seen in Fig. 12(b). On the other hand, the function graphed in Fig. 12(c) "passes" the horizontal line test, so reflecting its graph through the line $y = x$ yields the graph of a function, as can be seen in Fig. 12(d).

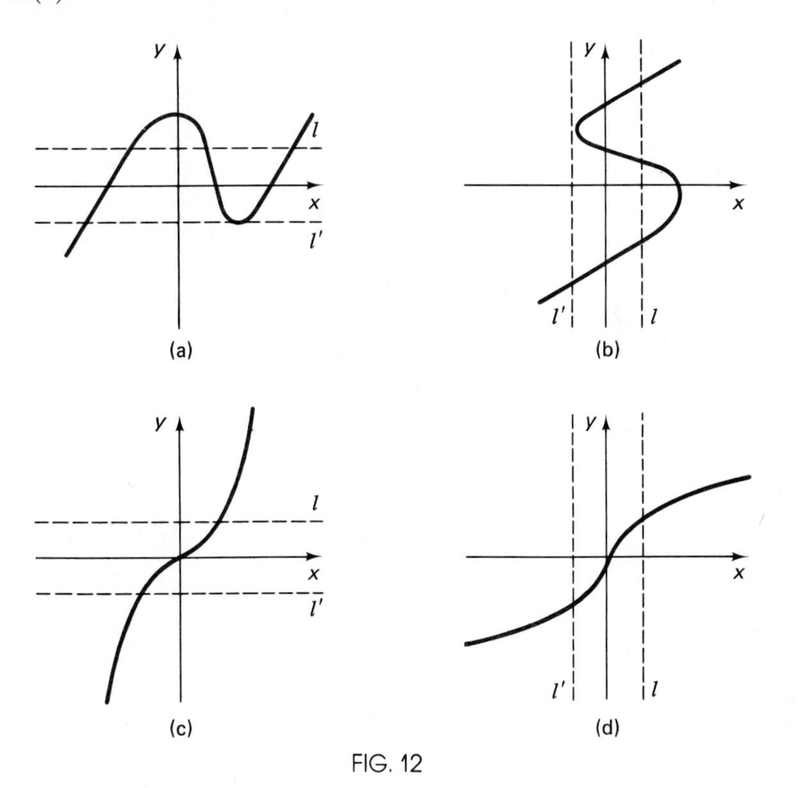

FIG. 12 ■

There are two important and common types of functions that have inverse functions. A function f is called **strictly increasing** if whenever a and b are in the domain of f and $a < b$, then $f(a) < f(b)$; it is **strictly decreasing** if $f(a) > f(b)$ whenever $a < b$ and a and b are in the domain of f. See Fig. 13. Clearly, such functions pass the horizontal line test.

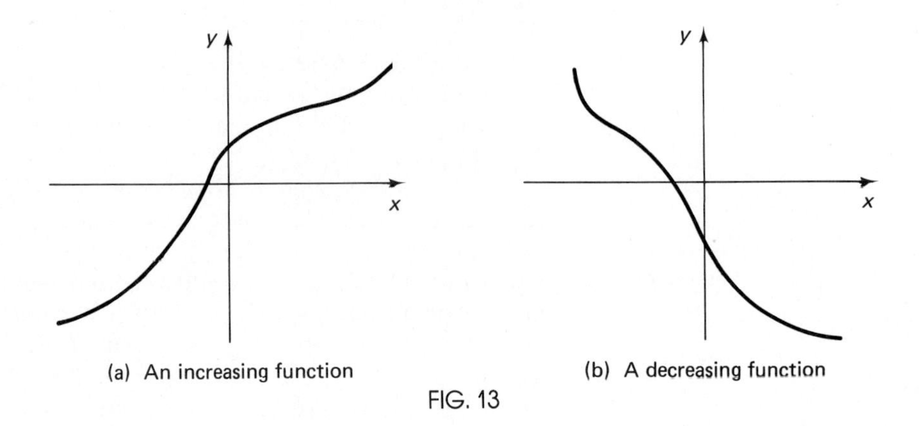

FIG. 13

You should note that there are many functions that have inverse functions but which are neither strictly increasing nor strictly decreasing. See Example 8.

We are now ready to see how to obtain the rule for the inverse from the original function. Suppose that we start with a function f with rule $y = f(x)$ and that f has an inverse function f^{-1}. Then f is the set of all ordered pairs whose coordinates satisfy the equation $y = f(x)$. The ordered pairs of f^{-1} are obtained by interchanging all the x- and y-coordinates of f. Thus if we start with the equation $y = f(x)$ and interchange x and y, the new equation obtained must determine f^{-1}. Hence if we solve this new equation for y, we get an equation that both determines f^{-1} and gives a formula for y in terms of x. Such a formula is exactly what we mean by a rule for f^{-1}, so the last equation is of the form $y = f^{-1}(x)$.

EXAMPLE 7 If $f(x) = -2x + 4$, find $f^{-1}(x)$. Graph $y = f(x)$ and $y = f^{-1}(x)$.

Solution Start with $y = f(x)$:

$$y = -2x + 4$$

Switch x and y (this equation determines the inverse function):

$$x = -2y + 4$$

Solve for y:

$$2y = -x + 4$$

$$y = -\tfrac{1}{2}x + 2$$

The last equation is of the form $y = f^{-1}(x)$, so we can just read off $f^{-1}(x)$:

$$f^{-1}(x) = -\tfrac{1}{2}x + 2$$

The graphs are in Fig. 14.

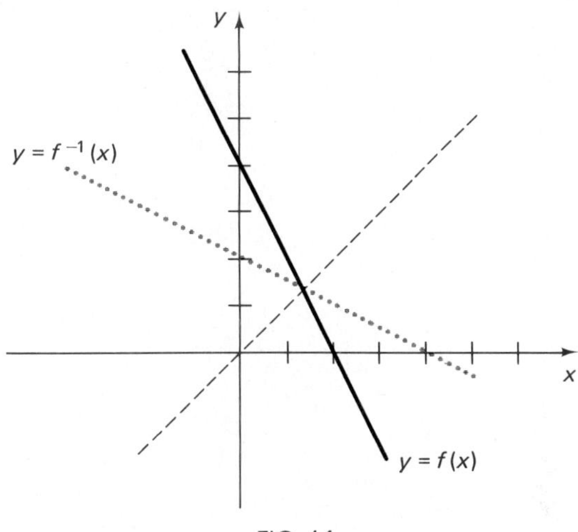

FIG. 14

■

EXAMPLE 8 If $h(x) = \dfrac{1}{x - 1}$, find $h^{-1}(x)$. Graph.

Solution Start with $y = h(x)$:

$$y = \frac{1}{x - 1}$$

Switch x and y:

$$x = \frac{1}{y - 1}$$

Solve for y:

$$y - 1 = \frac{1}{x}, \qquad y = \frac{1}{x} + 1$$

Read off $h^{-1}(x)$:

$$h^{-1}(x) = \frac{1}{x} + 1$$

The graphs are in Fig. 15. Note that you can see from the graphs that

$$h: \quad \{\text{nos.} \neq 0\} \to \{\text{nos.} \neq 1\}$$

and

$$h^{-1}: \quad \{\text{nos.} \neq 1\} \to \{\text{nos.} \neq 0\}$$

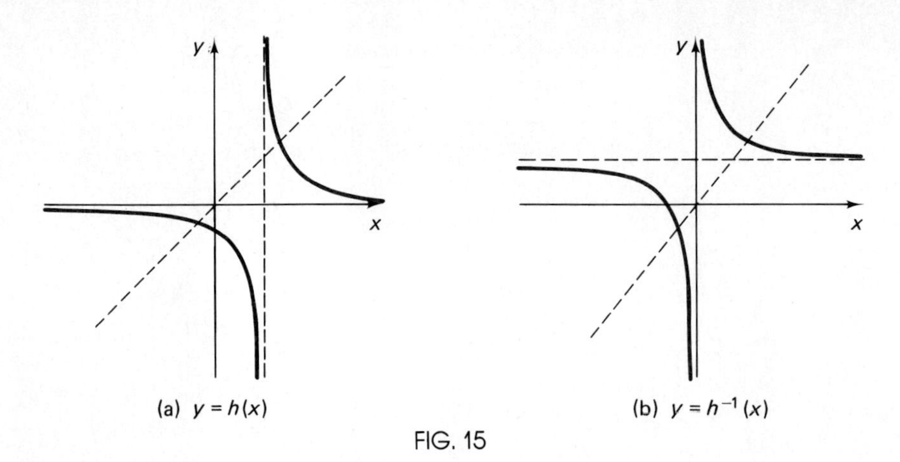

(a) $y = h(x)$ (b) $y = h^{-1}(x)$

FIG. 15

This illustrates (11): The domain of a function is the range of its inverse and vice versa. ■

Sometimes a given function does not have an inverse, but you can obtain a function with the same rule that does have an inverse by restricting the domain.

EXAMPLE 9 Determine which of the functions have inverse functions:
(a) $f(x) = x^2$, all x; (b) $g(x) = x^2$, $x \geq 0$; (c) $h(x) = x^2$, $x \leq 0$.

Solution These functions are graphed in Fig. 16. It is easy to see that f fails the horizontal line test but that g and h pass it. Thus g and h have inverse functions. They are $g^{-1}(x) = +\sqrt{x}$ and $h^{-1}(x) = -\sqrt{x}$. Note that you can read off the domains and ranges from the graphs, again illustrating that (11) holds:

$$g\colon\ [0, \infty) \to [0, \infty) \qquad \text{and} \qquad g^{-1}\colon\ [0, \infty) \to [0, \infty)$$
$$h\colon\ (-\infty, 0] \to [0, \infty) \qquad \text{and} \qquad h^{-1}\colon\ [0, \infty) \to (-\infty, 0]$$

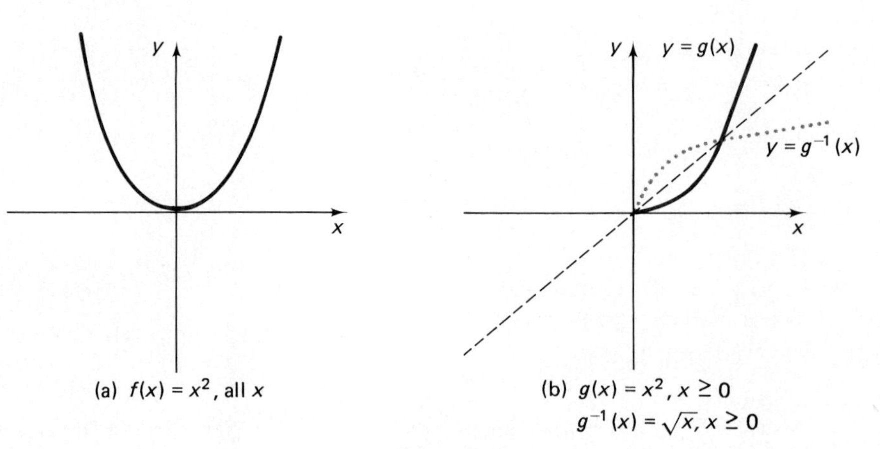

(a) $f(x) = x^2$, all x (b) $g(x) = x^2$, $x \geq 0$
$g^{-1}(x) = \sqrt{x}$, $x \geq 0$

FIG. 16

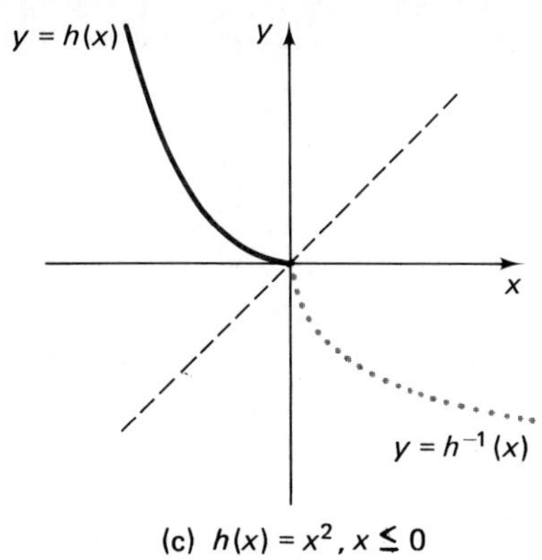

(c) $h(x) = x^2, x \le 0$
$h^{-1}(x) = -\sqrt{x}, x \le 0$

FIG. 16

■

EXERCISES

In Exercises 1–12, find $(f \circ g)(x)$ and $(g \circ f)(x)$.

1. $f(x) = 2x - 3,\ g(x) = 4x + 1$
2. $f(x) = -x + 1,\ g(x) = 2x - 1$
3. $f(x) = 3x^2 + 1,\ g(x) = 2x$
4. $f(x) = 1 - x^2,\ g(x) = 4x - 1$
5. $f(x) = x^2 - 1,\ g(x) = \sqrt{x + 2},\ x \ge -2$
6. $f(x) = x^4 + 4,\ g(x) = \sqrt{x - 4},\ x \ge 4$
7. $f(x) = x^3 - 1,\ g(x) = 2\sqrt[3]{x}$
8. $f(x) = 8x^3,\ g(x) = \sqrt[3]{x} + 1$
9. $f(x) = \dfrac{1}{x - 3},\ x \ne 3,\ g(x) = \dfrac{1}{x},\ x \ne 0$
10. $f(x) = \dfrac{x}{x - 4},\ x \ne 4,\ g(x) = x + 4$
11. $f(x) = \dfrac{1}{x^2},\ x \ne 0,\ g(x) = x^2$
12. $f(x) = x - 4,\ g(x) = x + 5$

In Exercises 13–20, for the given function F, find functions f and g such that $F(x) = f(g(x))$. For example, if $F(x) = \sqrt{x + 1}$, then $f(x) = \sqrt{x}$ and $g(x) = x + 1$. Be careful of the order, for if $F(x) = \sqrt{x} + 1$, then $f(x) = x + 1$ and $g(x) = \sqrt{x}$.

13. $F(x) = (x + 1)^2$
14. $F(x) = 5x - 3$
15. $F(x) = x^2 + 1$
16. $F(x) = 5(x - 3)$
17. $F(x) = \sqrt[3]{\dfrac{1}{x}}$
18. $F(x) = 4^{x+3}$
19. $F(x) = \dfrac{1}{\sqrt[3]{x}}$
20. $F(x) = 4^x + 3$

In Exercises 21–28, assume that $f(x) = x + 1$, $g(x) = x^2$, $h(x) = -x$, and $k(x) = \sqrt{x}$, and use your calculator, where necessary, to compute the answers.

21. $(f \circ g)(0.5)$ and $(g \circ f)(0.5)$
22. $(g \circ h)(0.4)$ and $(h \circ g)(0.4)$
23. $(f \circ h)(0.8)$ and $(h \circ f)(0.8)$
24. $(f \circ k)(0.1)$ and $(k \circ f)(0.1)$
25. $(g \circ k)(0.7)$ and $(k \circ g)(0.7)$
26. $(g \circ h)(0.9)$ and $(h \circ g)(0.9)$
27. $(f \circ h \circ g)(0.4)$ and $(h \circ f \circ g)(0.4)$
28. $(k \circ g \circ f)(0.3)$ and $(g \circ k \circ f)(0.3)$

In Exercises 29–38, use the horizontal line test to determine if the function whose graph is given has an inverse function.

29.

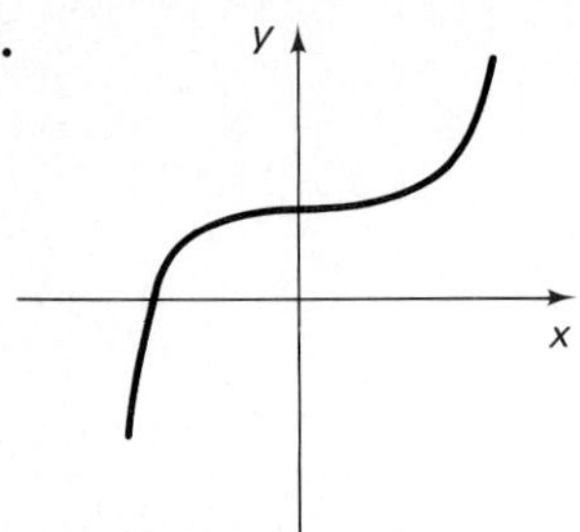

30.

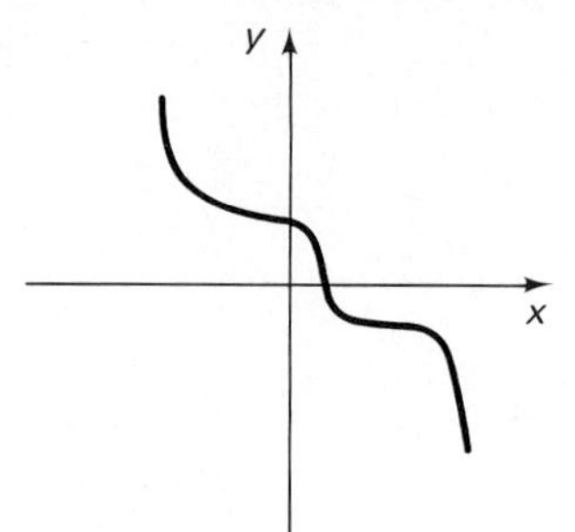

31.

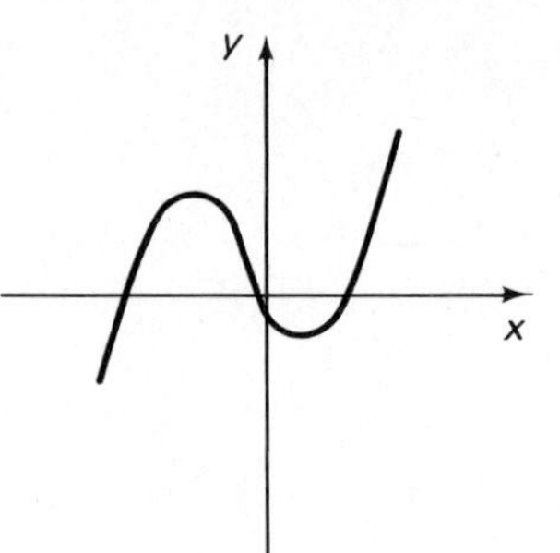

32.

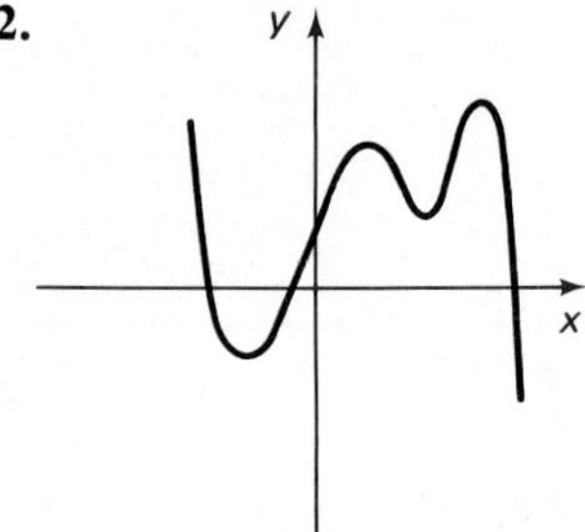

33.

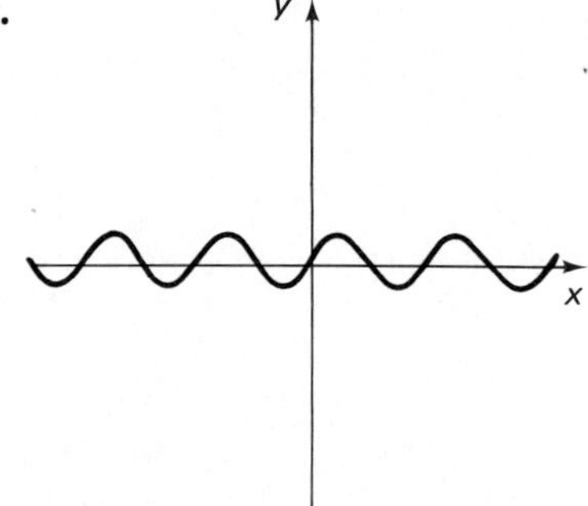

34.

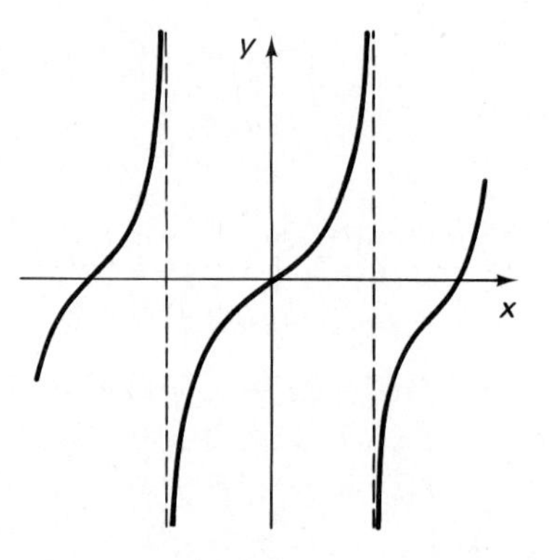

35.

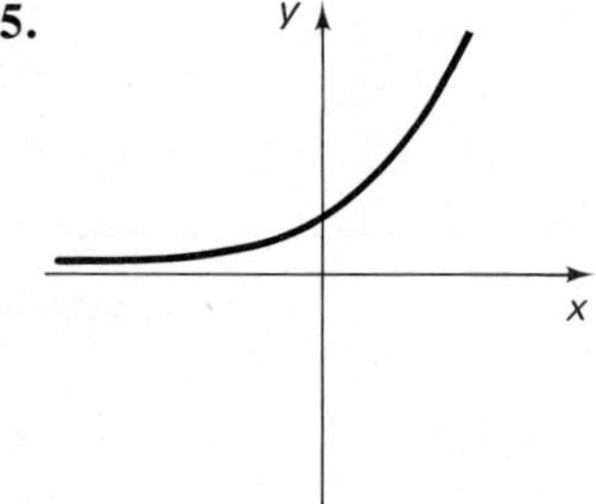

36.

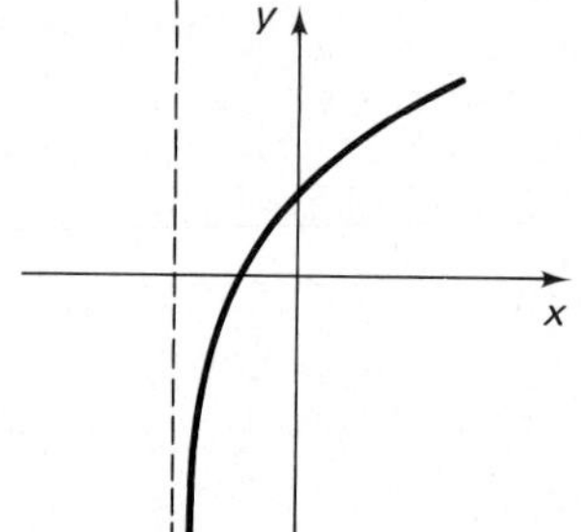

37.

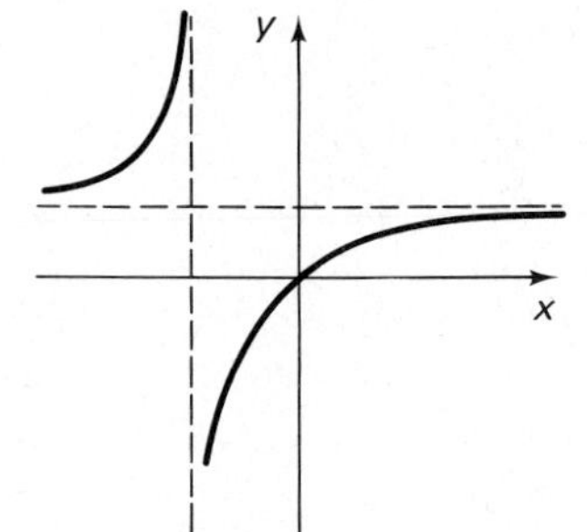

38.

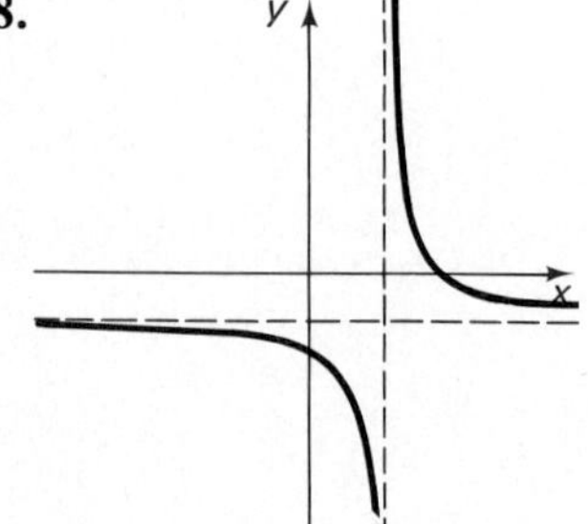

In Exercises 39–52, find the rule for the inverse and graph the function and its inverse.

39. $f(x) = x - 1$

40. $g(x) = x + 2$

41 $h(x) = -\frac{1}{2}x$

42. $k(x) = -3x$

43. $F(x) = 2x - 3$

44. $G(x) = \frac{1}{4}x + \frac{1}{2}$

45. $H(x) = -3x + \frac{1}{3}$

46. $K(x) = -2x - 4$

47. $f(x) = \dfrac{1}{x - 2}, x \neq 2$

48. $g(x) = \dfrac{1}{x + 3}, x \neq -3$

49. $h(x) = (x + 1)^2, x \geq -1$

50. $k(x) = \sqrt{x + 2}, x \geq -2$

51. $F(x) = (x - 2)^2 - 1, x \leq 2$

52. $G(x) = \sqrt{2 - x} + 1, x \leq 2$

In Exercises 53–60, for the given function f, find the rule for f^{-1}, and show that $f(f^{-1}(x)) = x$ and that $f^{-1}(f(x)) = x$ whenever this is defined.

53. $f(x) = x + 1$

54. $f(x) = 5x$

55. $f(x) = 2x + 3$

56. $f(x) = \frac{1}{3}x - \frac{5}{6}$

57. $f(x) = \dfrac{x}{x - 1}, x \neq 1$

58. $f(x) = \dfrac{x + 1}{x}, x \neq 0$

59. $f(x) = (x - 1)^2, x \geq 1$

60. $f(x) = (2x + 3)^2, x \leq -\frac{3}{2}$

In Exercises 61–66, use your calculator, if necessary, to show that $f^{-1}(f(a)) = a$ and that $f(f^{-1}(a)) = a$ for the given function f and number a.

61. $f(x) = x^2, a = 0.11$

62. $f(x) = \sqrt{x}, a = \pi$

63. $f(x) = \sin x, a = 0.92$

64. $f(x) = \cos x, a = 0.37$

65. $f(x) = \tan x, a = 0.18$

66. $f(x) = \ln x, a = 0.56$

In Exercises 67–72, from the given function and its graph, determine its domain and range. Then find the graph, domain, and range for its inverse.

67. $f(x) = x^2 - 1, x \geq 0$

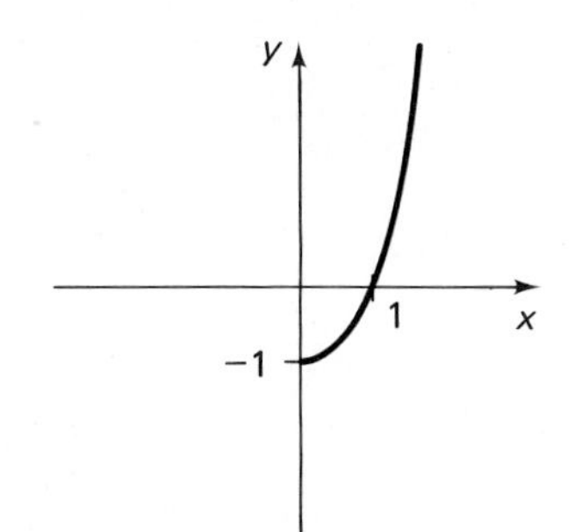

68. $g(x) = 1 - x^2, x \leq 0$

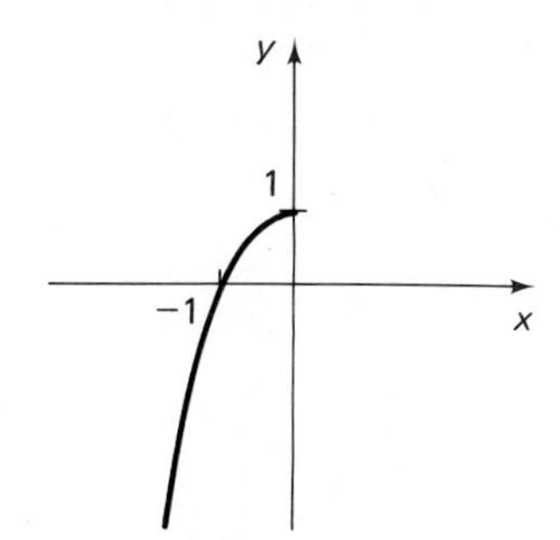

69. $h(x) = \frac{1}{x} - 2,\ x > 0$

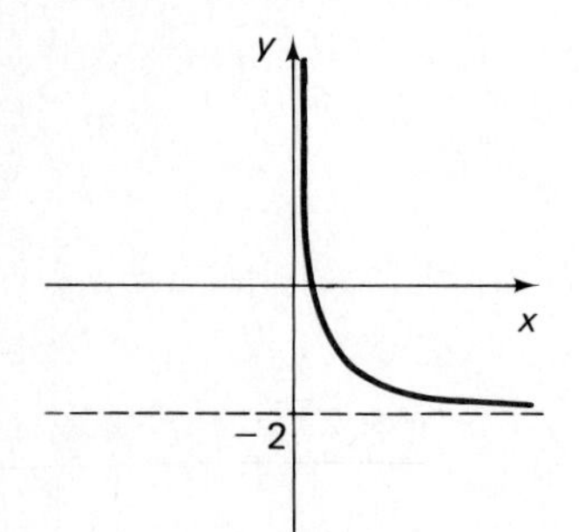

70. $k(x) = \frac{1}{x + 1},\ x < -1$

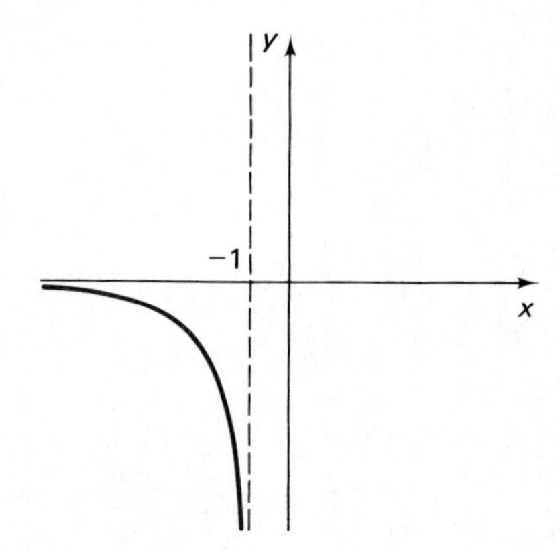

71. $F(x) = \frac{6 - x}{3 + x},\ x \neq -3$

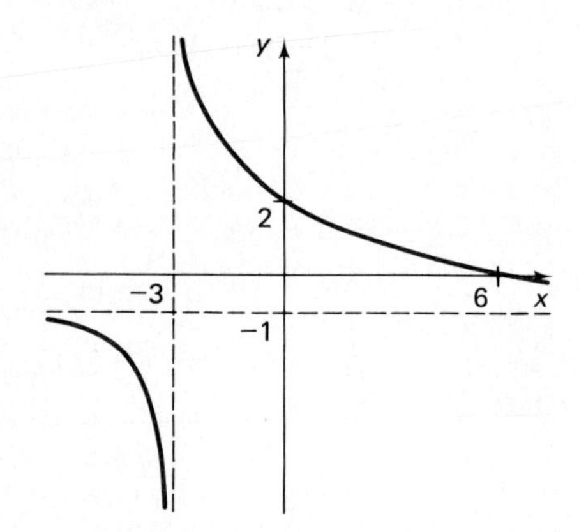

72. $G(x) = \frac{x + 2}{2x - 1},\ x \neq \frac{1}{2}$

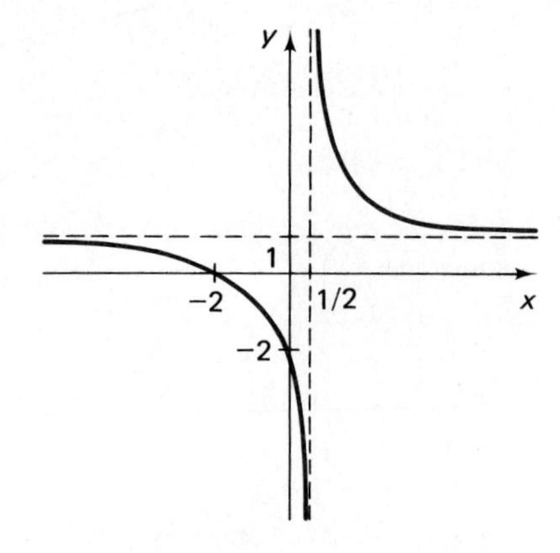

73. Let $f(x) = ax + b$ and $g(x) = cx + d$, $a \neq 0$, $c \neq 0$.
a. Assume $a = b = 1$. If $(f \circ g)(x) = (g \circ f)(x)$, all x, what can you say about c and d?
b. Assume $a = 1$, $b = 2$. If $(f \circ g)(x) = (g \circ f)(x)$, all x, what can you say about c and d?
c. If $a = c$ and $f(g(x)) = g(f(x))$, all x, what can you say about b and d?

APPENDIX F

Proof of the Formulas in Section 5.6

We first need the following fact:

> **(1) Theorem** If n is a positive integer and $s \neq 1$, then
> $$1 + s + s^2 + \cdots + s^{n-1} = \frac{s^n - 1}{s - 1}$$

Proof. By multiplying out, we see that

$$(s - 1)(s^{n-1} + s^{n-2} + \cdots + s^2 + s + 1) = s^n - 1$$

If $s \neq 1$, then $s - 1 \neq 0$ so we can divide both sides by $s - 1$. This yields

$$s^{n-1} + s^{n-2} + \cdots + s^2 + s + 1 = \frac{s^n - 1}{s - 1}$$

which is what we wanted to prove. ■

Derivation of Formula 1: $A = P\left(1 + \frac{1}{r}\right)[(1 + r)^n - 1]$, where A is the amount of money you have after n time periods of investing P dollars at the beginning of each period. The rate of interest per period is r and all interest is compounded each period.

Recall that if you invest P dollars at the interest rate r per period for k periods, compounded each period, it is worth $P(1 + r)^k$ at the end of those k periods. Now, you invest P dollars at the beginning of the first period. It will be worth $P(1 + r)^n$ after n periods. At the beginning of the second period you again invest P. But it is only invested for $n - 1$ periods, so it will be worth $P(1 + r)^{n-1}$ at the end. At the beginning of the third period, you again invest P; it is invested for $n - 2$ periods, so it will be worth $P(1 + r)^{n-2}$ at the end. You continue until the beginning of the last period when you invest P, and at the end it is worth $P(1 + r)$. Thus, all together you will have

$$A = P(1 + r)^n + P(1 + r)^{n-1} + P(1 + r)^{n-2} + \cdots + P(1 + r)$$

We wish to use (1), so we factor out a $P(1 + r)$,

$$A = P(1 + r)[(1 + r)^{n-1} + (1 + r)^{n-2} + \cdots + (1 + r) + 1]$$

By (1) (letting $s = 1 + r$) we have

$$A = P(1 + r)\frac{(1 + r)^n - 1}{(1 + r) - 1} = P\left(\frac{1 + r}{r}\right)[(1 + r)^n - 1]$$

$$= P\left(\frac{1}{r} + 1\right)[(1 + r)^n - 1] \quad \blacksquare$$

Derivation of Formula 2: $A = P\dfrac{1 - (1 + r)^{-n}}{r}$, where A is the amount of money (borrowed) at the beginning, P is the amount of each payment at the end of each period, and r is the interest rate per period compounded each period.

It often happens in mathematics that the direct approach to solving a problem is difficult if not impossible. Occasionally, if you change your viewpoint, an indirect approach can yield a relatively simple solution. This is one of the ways in which mathematics can be beautiful.

In this case, if we were to try a direct derivation of Formula 2, we would proceed as follows:

We have borrowed A. At the end of the first time period, rA is the interest due. Of our first payment, then, rA is for interest, and the remainder, $P - rA$, is used to reduce the amount of our debt to $A - (P - rA)$. At the end of the second time period, the interest due on our current debt is $r[A - (P - rA)]$. Thus after the second payment, the debt is reduced to $A - (P - rA) - \{P - r[A - (P - rA)]\}$. If we were to continue this, trying to find a general pattern to the debt owed after n time periods (and hence when it is zero), most of us would find it very difficult. Thus we change our viewpoint. We say that instead of paying the interest due and reducing the principal with our payment, we pretend that we put each payment into a savings account that pays r rate of interest per period, compounded each period. Then at the end of the n periods, we pay back the whole amount A plus the interest it has earned for the whole time. This does not cost us any more, because the additional interest we have to pay is exactly paid back by the savings account; the lender gets exactly the same in the long run if we pretend he is the savings bank! Mathematically, we can now argue similarly to the proof of Formula 1. The first payment earns interest for $(n - 1)$ periods; the second payment, for $(n - 2)$ periods; etc., up to the last payment which, in this case, is made at the end of the nth period so it does not earn any interest at all. Meanwhile the original amount borrowed, A, earns interest for all n periods. Hence, the equation is simply

$$P(1 + r)^{n-1} + P(1 + r)^{n-2} + \cdots + P(1 + r) + P = A(1 + r)^n,$$

$$P[(1 + r)^{n-1} + (1 + r)^{n-2} + \cdots + (1 + r) + 1] = A(1 + r)^n$$

$$P\,\frac{(1 + r)^n - 1}{1 + r - 1} = A(1 + r)^n, \text{ by (1)}$$

So

$$A = P\,\frac{[(1 + r)^n - 1]/(1 + r)^n}{r} = P\,\frac{1 - (1 + r)^{-n}}{r} \qquad \blacksquare$$

Answers to Odd-Numbered Exercises

CHAPTER 1

Section 1.1

1. 480°, $8\pi/3$ **3.** 540°, 3π **5.** −120°, $-2\pi/3$ **7.** −320°, $-16\pi/9$ **9.** 14.42° **11.** −24.25° **13.** 231.533° **15.** 84.1019° **17.** −132.694° **19.** 1227.08° **21.** 9°27′36″ **23.** −75°36′48.6″ **25.** 431°58′48″ **27.** $3\pi/4 \approx 2.35619$ **29.** $5\pi/12 \approx 1.30900$ **31.** $2\pi/3 \approx 2.09440$ **33.** $\pi \approx 3.14159$ **35.** −1.49051 **37.** 5.43214 **39.** −0.806342 **41.** 15.0348 **43.** 210° **45.** 225° **47.** 1620° **49.** 450° **51.** 171.887° **53.** 45.8366° **55.** 469.825° **57.** 19.9962° **59.** 143°14′22.0″ **61.** 39°4′32.6″ **63.** −91°40′23.7″ **65.** −5°12′50.1″ **67.** 100°, 1.75 **69.** 47.5 m **71.** **a.** 120°, $2\pi/3$ **b.** 39.7935 in. **73.** **a.** 6°, 0.104720 **b.** 0.0785398 in. **75.** 2950 ft **77.** 3500 mi **79.** 391.111 rad

For the remainder of the book, when a question asks for all trigonometric functions of θ, the answers are given in the order $\sin\theta$, $\cos\theta$, $\tan\theta$, $\csc\theta$, $\sec\theta$, $\cot\theta$.

Section 1.2

1. 5, 6, 10, $\frac{4}{5}$, $\frac{4}{5}$ **3.** 39, 5, 13, $\frac{5}{12}$, $\frac{5}{12}$ **5.** $\sqrt{13}$, $24/\sqrt{13}$, $36/\sqrt{13}$, 2/3, 2/3 **7.** $(-\frac{3}{5}, \frac{4}{5})$ **9.** (4, −8) **11.** $\frac{4}{5}, \frac{3}{5}, \frac{4}{3}, \frac{5}{4}, \frac{5}{3}, \frac{3}{4}$ **13.** $\frac{7}{25}, \frac{24}{25}, \frac{7}{24}, \frac{25}{7}, \frac{25}{24}, \frac{24}{7}$ **15.** $1/\sqrt{2}$, $1/\sqrt{2}$, 1, $\sqrt{2}$, $\sqrt{2}$, 1 **17.** $\sqrt{11}/6$, 5/6, $\sqrt{11}/5$, $6/\sqrt{11}$, 6/5, $5/\sqrt{11}$ **19.** $\frac{12}{13}, \frac{5}{13}, \frac{12}{5}, \frac{13}{12}, \frac{13}{5}, \frac{5}{12}$ **21.** $1/\sqrt{2}$, $1/\sqrt{2}$, 1, $\sqrt{2}$, $\sqrt{2}$, 1 **23.** $\sqrt{3}/2$, 1/2, $\sqrt{3}$, $2/\sqrt{3}$, 2, $1/\sqrt{3}$ **25.** 2/3, $\sqrt{5}/3$, $2/\sqrt{5}$, 3/2, $3/\sqrt{5}$, $\sqrt{5}/2$ **27.** $1 + \cot^2\theta = \csc^2\theta$, $\tan^2\theta + 1 = \sec^2\theta$ **29.** 5.90 ft **31.** 56 ft **33.** 6.4 ft

Section 1.3

		sin θ	cos θ	tan θ	csc θ	sec θ	cot θ
1. **3.**	30°, $\pi/6$	$1/2$	$\sqrt{3}/2$	$1/\sqrt{3}$	2	$2/\sqrt{3}$	$\sqrt{3}$
	60°, $\pi/3$	$\sqrt{3}/2$	$1/2$	$\sqrt{3}$	$2/\sqrt{3}$	2	$1/\sqrt{3}$

5. $\sqrt{3}$ **7.** 1.53987 **9.** 0.826590 **11.** 1.11740 **13.** $1/\sqrt{2}$ **15.** $\sqrt{3}/2$ **17.** 0.841471 **19.** 0.647859 **21.** 0.726543 **23.** 1.12207 **25.** 2.36522 **27.** 1.23607 **29.** 0.977076 **31.** 2.04950 **33.** $4/\sqrt{2}$ **35.** 13.4116 **37.** 0.250220 **39.** −1.76948 **41.** 36 ft **43.** 235,075 sq yd **45.** $1/\sqrt{2}$ **47.** 25.1725 m

Section 1.4

1. 30°, $\pi/6$ **3.** 60°, $\pi/3$ **5.** 60°, $\pi/3$ **7.** 45°, $\pi/4$ **9.** 45°, $\pi/4$ **11.** 51.7575°, 0.903339 **13.** 84.2608°, 1.47063 **15.** 53.1301°, 0.927296 **17.** 36.8699°, 0.643501 **19.** 23.5782°, 0.411517 **21.** 11.5370°, 0.201358 **23.** 11.3099°, 0.197396 **25.** 38.6598°, 0.674741 **27.** 71.5651°, 1.24905 **29.** 89.4271°, 1.56080 **31.** 0.6, 36.8699° **33.** 4.75, 78.1113° **35.** 0.189, 79.1056° **37.** 26° **39.** 38° **41.** 0.381 **43.** 72°

Section 1.5

1. 3.32257 ft, 12.8374 ft **3.** 12.2257 m, 26.5174 m **5.** 1.56846 in., 3.24036 in. **7.** 14.2808 ft, 12.9826 ft **9.** 6.13562 cm, 49.0850 cm **11.** 80.4562 in., 257.872 in. **13.** 137.589 ft, 166.071 ft **15.** 24.0965°, 65.9035° **17.** 36.2931°, 53.7069° **19.** 23.8830°, 66.1170° **21.** 11.5 ft **23.** 13.33 mi, 400.0 mph **25.** 598.8 m **27.** 47 ft **29.** 53.35° **31.** 10.7907, 12.4600 **33.** 1136 ft **35.** 140 ft **37.** 3.6 mi **39.** 9221 yd **41.** 63.5 ft **43.** 188 mph

Review

1. 62.305° **3.** 4817.87° **5.** −187.595° **7.** $\pi/3$ **9.** 1.46608 **11.** −8.22609 **13.** 120° **15.** 131.780° **17.** −2.69290° **19.** $1/\sqrt{5}$, $2/\sqrt{5}$, $1/2$, $\sqrt{5}$, $\sqrt{5}/2$, 2 **21.** $1/4$, $\sqrt{15}/4$, $1/\sqrt{15}$, 4, $4/\sqrt{15}$, $\sqrt{15}$ **23.** 0.0174524 **25.** 1.13897 **27.** 1.12433 **29.** 0.251183 **31.** 0.508968 **33.** −5.42284 **35.** 693 ft **37.** 30°, $\pi/6$ **39.** 28.6854°, 0.500655 **41.** 45°, $\pi/4$ **43.** 60°, $\pi/3$ **45.** 3.98, 4.88 **47.** 1.20 **49.** 31.2 ft **51.** 326 ft

CHAPTER 2

Section 2.1

1. $\frac{3}{5}, -\frac{4}{5}, -\frac{3}{4}, \frac{5}{3}, -\frac{5}{4}, -\frac{4}{3}$ **3.** $3/\sqrt{13}$, $2/\sqrt{13}$, $3/2$, $\sqrt{13}/3$, $\sqrt{13}/2$, $2/3$ **5.** $-3/\sqrt{10}$, $-1/\sqrt{10}$, 3, $-\sqrt{10}/3$, $-\sqrt{10}$, $1/3$ **7.** $-2/\sqrt{13}$, $3/\sqrt{13}$, $-2/3$, $-\sqrt{13}/2$, $\sqrt{13}/3$, $-3/2$ **9.** $\pm 3\sqrt{5}/7$, $-2/7$, $\mp 3\sqrt{5}/2$, $\pm 7/3\sqrt{5}$, $-7/2$, $\mp 2/3\sqrt{5}$ **11.** $\pm\sqrt{3}/2$, 0.5, $\pm\sqrt{3}$, $\pm 2/\sqrt{3}$, 2, $\pm 1/\sqrt{3}$ **13.** $\pm 3/\sqrt{10}$, $\pm 1/\sqrt{10}$, 3, $\pm\sqrt{10}/3$, $\pm\sqrt{10}$, $1/3$ **15.** $\pm 4/\sqrt{17}$, $\mp 1/\sqrt{17}$, -4, $\pm\sqrt{17}/4$, $\mp\sqrt{17}$, $-1/4$ **17.** Q_3, −, −, + **19.** Q_4, −, +, − **21.** Q_2, +, −, − **23.** Q_4, −, +, − **25.** $\pm\frac{24}{25}, -\frac{7}{25}, \mp\frac{24}{7}, \pm\frac{25}{24}, -\frac{25}{7}, \mp\frac{7}{24}$ **27.** $\pm\frac{5}{13}, \pm\frac{12}{13}, \frac{5}{12}, \pm\frac{13}{5}, \pm\frac{13}{12}, \frac{12}{5}$ **29.** $\pm\frac{12}{13}, \frac{5}{13}, \pm\frac{12}{5}, \pm\frac{13}{12}, \frac{13}{5}, \pm\frac{5}{12}$ **31.** $-\sqrt{5}/3$, $-2/3$, $\sqrt{5}/2$, $-3/\sqrt{5}$, $-3/2$, $2/\sqrt{5}$ **33.** $-2/\sqrt{5}$, $-1/\sqrt{5}$, 2, $-\sqrt{5}/2$, $-\sqrt{5}$, $1/2$ **35.** $-\frac{3}{5}, \frac{4}{5}, -\frac{3}{4}, -\frac{5}{3}, \frac{5}{4}, -\frac{4}{3}$ **37.** $-1/4$, $\sqrt{15}/4$, $-1/\sqrt{15}$, -4, $4/\sqrt{15}$, $-\sqrt{15}$ **39.** $-2/5$, $-\sqrt{21}/5$, $2/\sqrt{21}$, $-5/2$, $-5/\sqrt{21}$, $\sqrt{21}/2$ **41.** $\sqrt{3}/2$, $1/2$, $\sqrt{3}$, $2/\sqrt{3}$, 2, $1/\sqrt{3}$ **43.** $\frac{3}{5}, -\frac{4}{5}, -\frac{3}{4}, \frac{5}{3}, -\frac{5}{4}, -\frac{4}{3}$ **45.** $-\frac{72}{7}$, 10.7143 **47.** $\frac{27}{4}$, 27.8310

Section 2.2

1. (−1, 0) **3.** (0, 1) **5.** (−1, 0) **7.** (0, −1) **9.** (0, 1) **11.** (1, 0) **13.** (0, −1) **15.** $(\sqrt{3}/2, 1/2)$ **17.** $(-\sqrt{2}/2, \sqrt{2}/2)$ **19.** $(-\sqrt{3}/2, -1/2)$ **21.** $(\sqrt{2}/2, -\sqrt{2}/2)$ **23.** $(-\sqrt{3}/2, -1/2)$

25. 1, 0, —, 1, —, 0 **27.** −1, 0, —, −1, —, 0 **29.** 1, 0, —, 1, —, 0 **31.** $1/\sqrt{2}, 1/\sqrt{2}, 1, \sqrt{2}, \sqrt{2}, 1$
33. $\frac{1}{2}, \frac{\sqrt{3}}{2}, 1/\sqrt{3}, 2, 2/\sqrt{3}, \sqrt{3}$ **35.** $1/\sqrt{2}, -1/\sqrt{2}, -1, \sqrt{2}, -\sqrt{2}, -1$
37. $-1/\sqrt{2}, 1/\sqrt{2}, -1, -\sqrt{2}, \sqrt{2}, -1$ **39.** $-1/2, -\sqrt{3}/2, 1/\sqrt{3}, -2, -2/\sqrt{3}, \sqrt{3}$

Section 2.3

1. $5/13, \pm 12/13, \pm 5/12, 13/5, \pm 13/12, \pm 12/5$ **3.** $\pm 4/5, \mp 3/5, -4/3, \pm 5/4, \mp 5/3, -3/4$
5. $-24/25, \pm 7/25, \mp 24/7, -25/24, \pm 25/7, \mp 7/24$ **7.** $1/3, -2\sqrt{2}/3, -1/2\sqrt{2}, 3, -3/2\sqrt{2}, -2\sqrt{2}$
9. $\sqrt{3}/2, -1/2, -\sqrt{3}, 2/\sqrt{3}, -2, -1/\sqrt{3}$ **11.** $-1/\sqrt{2}, 1/\sqrt{2}, -1, -\sqrt{2}, \sqrt{2}, -1$
13. $2/3, -\sqrt{5}/3, -2/\sqrt{5}, 3/2, -3/\sqrt{5}, -\sqrt{5}/2$ **15.** $-\sqrt{3}/2, 1/2, -\sqrt{3}, -2/\sqrt{3}, 2, -1/\sqrt{3}$
17. $-2/\sqrt{5}, -1/\sqrt{5}, 2, -\sqrt{5}/2, -\sqrt{5}, 1/2$ **19.** $-1/2, -\sqrt{3}/2, 1/\sqrt{3}, -2, -2/\sqrt{3}, \sqrt{3}$ **21.** 0 **23.** 0
25. −1 **27.** 1 **29.** $\sqrt{2}$ **31.** −1 **33.** $-\sqrt{3}$ **35.** $\sqrt{3}/2$ **37.** $\sqrt{2}$ **39.** $-\frac{1}{2}$

Section 2.4

1. Q_3, 57° **3.** Q_2, 77° **5.** Q_1, 4° **7.** Q_1, 28°45′ **9.** Q_2, $2\pi/9$ **11.** Q_3, $\pi/5$ **13.** Q_2, 1.14159
15. Q_2, 0.707963 **17.** $2\pi/9$ **19.** $2\pi/5$ **21.** $\pi - 2 \approx 1.14159$ **23.** $2\pi - 6 \approx 0.283185$
25.

$\pi/6$	$\pi/4$	$\pi/3$	$2\pi/3$	$3\pi/4$	$5\pi/6$	$7\pi/6$	$5\pi/4$	$4\pi/3$	$5\pi/3$	$7\pi/4$	$11\pi/6$
$1/2$	$1/\sqrt{2}$	$\sqrt{3}/2$	$\sqrt{3}/2$	$1/\sqrt{2}$	$1/2$	$-1/2$	$-1/\sqrt{2}$	$-\sqrt{3}/2$	$-\sqrt{3}/2$	$-1/\sqrt{2}$	$-1/2$
$\sqrt{3}/2$	$1/\sqrt{2}$	$1/2$	$-1/2$	$-1/\sqrt{2}$	$-\sqrt{3}/2$	$-\sqrt{3}/2$	$-1/\sqrt{2}$	$-1/2$	$1/2$	$1/\sqrt{2}$	$\sqrt{3}/2$
$1/\sqrt{3}$	1	$\sqrt{3}$	$-\sqrt{3}$	-1	$-1/\sqrt{3}$	$1/\sqrt{3}$	1	$\sqrt{3}$	$-\sqrt{3}$	-1	$-1/\sqrt{3}$

$-\pi/6$	$-\pi/4$	$-\pi/3$	$-2\pi/3$	$-3\pi/4$	$-5\pi/6$	$-7\pi/6$	$-5\pi/4$	$-4\pi/3$	$-5\pi/3$	$-7\pi/4$	$-11\pi/6$
$-1/2$	$-1/\sqrt{2}$	$-\sqrt{3}/2$	$-\sqrt{3}/2$	$-1/\sqrt{2}$	$-1/2$	$1/2$	$1/\sqrt{2}$	$\sqrt{3}/2$	$\sqrt{3}/2$	$1/\sqrt{2}$	$1/2$
$\sqrt{3}/2$	$1/\sqrt{2}$	$1/2$	$-1/2$	$-1/\sqrt{2}$	$-\sqrt{3}/2$	$-\sqrt{3}/2$	$-1/\sqrt{2}$	$-1/2$	$1/2$	$1/\sqrt{2}$	$\sqrt{3}/2$
$-1/\sqrt{3}$	-1	$-\sqrt{3}$	$\sqrt{3}$	1	$1/\sqrt{3}$	$-1/\sqrt{3}$	-1	$-\sqrt{3}$	$\sqrt{3}$	1	$1/\sqrt{3}$

Section 2.5

1. 0.559193, −0.829038, −0.674509, 1.78829, −1.20622, −1.48256
3. −0.777146, −0.629320, 1.23490, −1.28676, −1.58902, 0.809784
5. −0.515501, −0.856889, 0.601597, −1.93986, −1.16701, 1.66224
7. (−15.5899, 3.59922) **9.** (0.897513, −5) **11.** (−2, −6.05133) **13.** (4.12915, −3)
15. (0.813878, 0.581035) **17.** (−0.541642, 0.840609) **19.** −1 **21.** 6.87622 **23.** — **25.** 12.5634
27. 0.432189 **29.** −0.285093 **31.** 0.0816911 **33.** 0.669926 **35.** −1.09842 **37.** 567.886
39. 0.998342 **41.** 0.997053 **43.** −0.994210

Section 2.6

1, 3, 5. Graphs in text.

7.

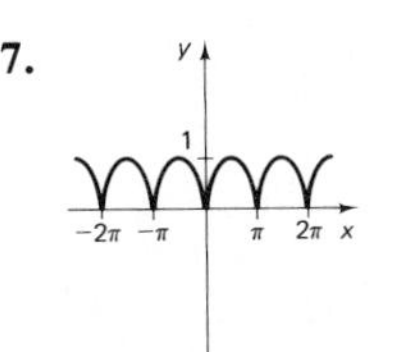

9.

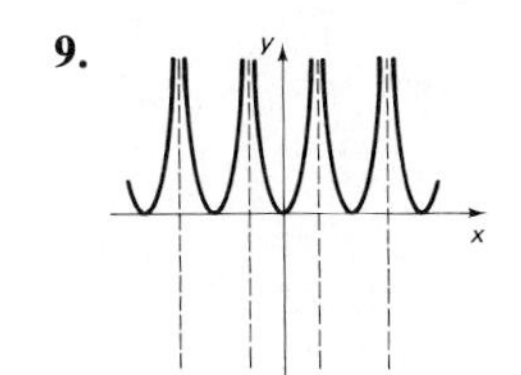

11.

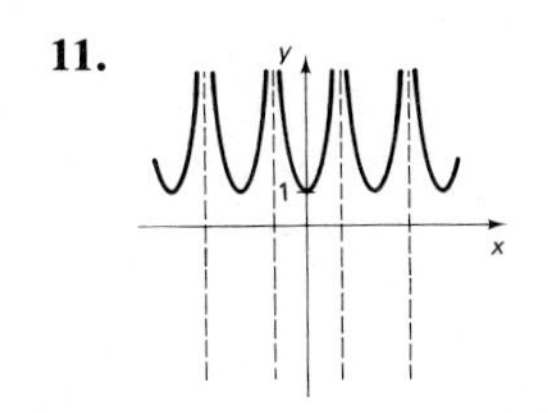

13.

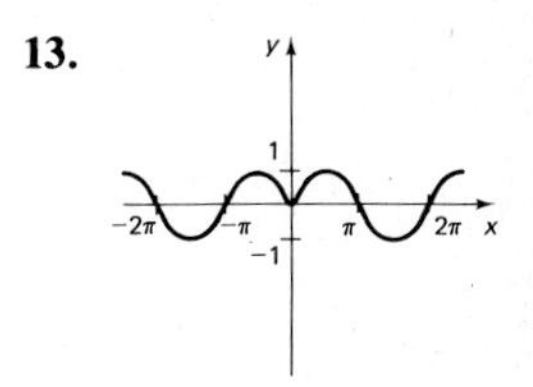

15.

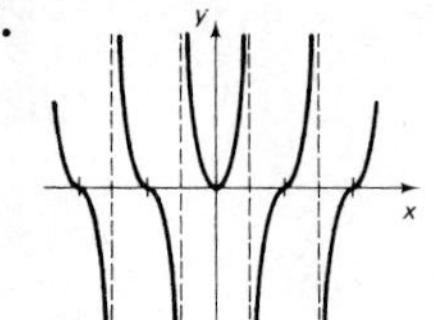

17.

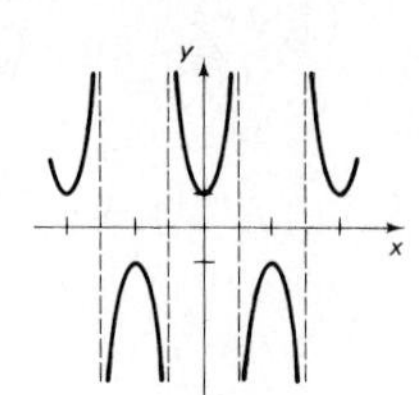

19.

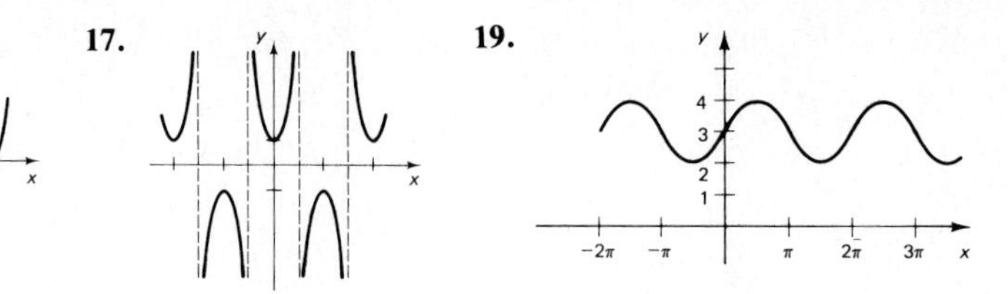

21.

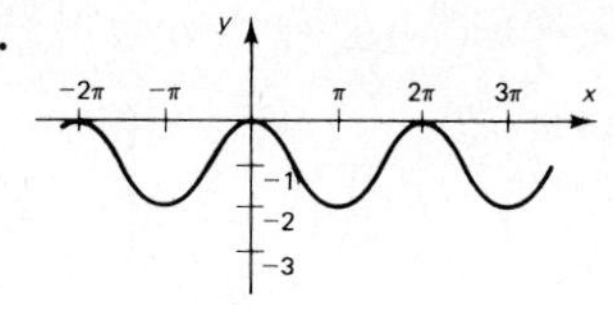

23.

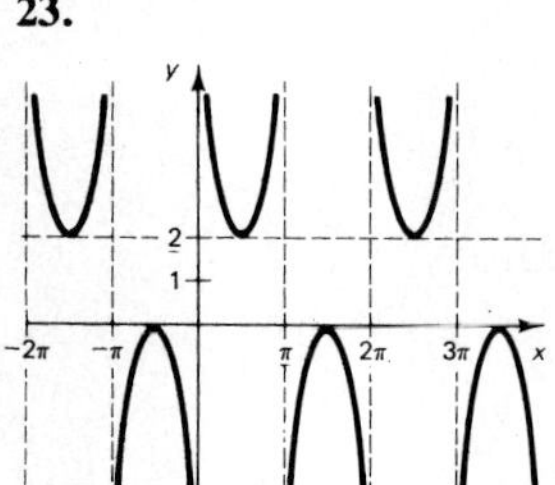

25.

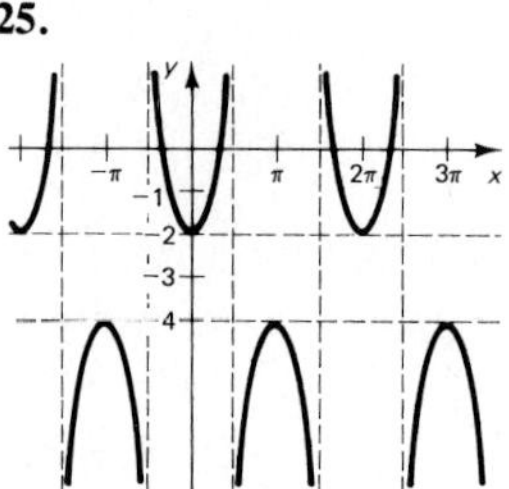

27.

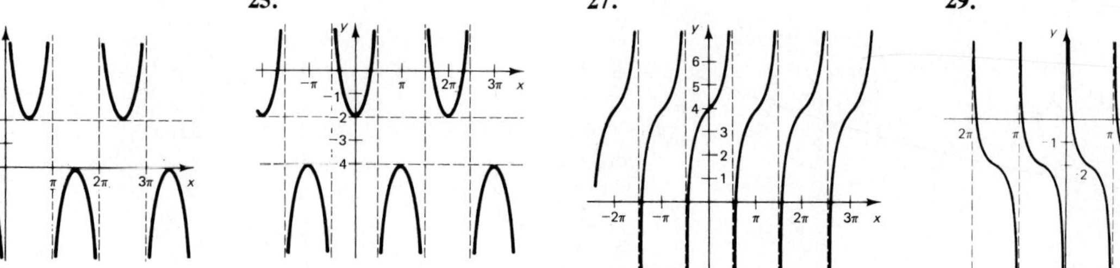

29.

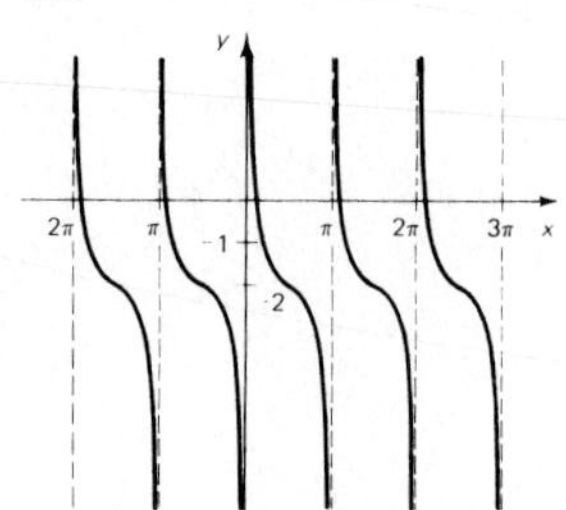

31. 2 **33.** 2 **35.** 3

Section 2.7

1. 2, 2π

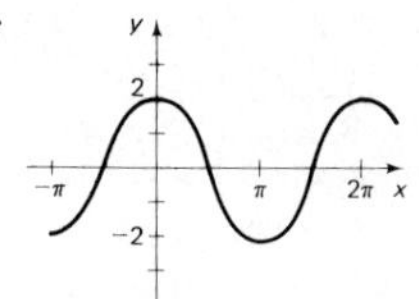

3. $\frac{1}{2}$, 2π

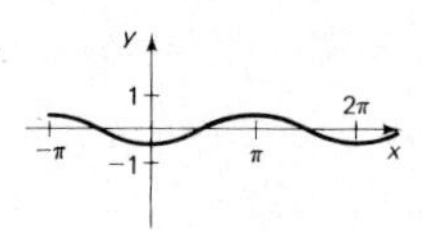

5. 3, 2π

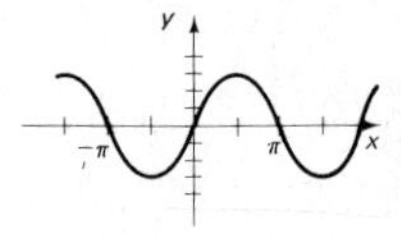

7. 1, 4π

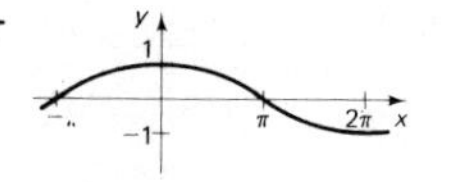

9. 1, $2\pi/3$

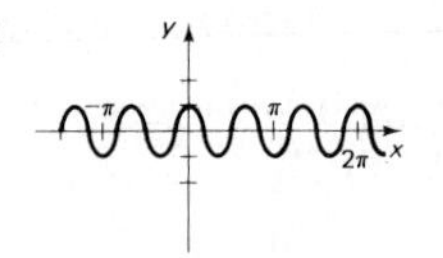

11. 1, 2π, $\pi/3$

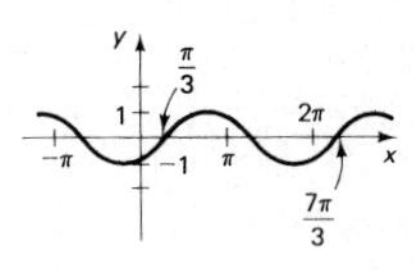

13. 1, 2π, -1

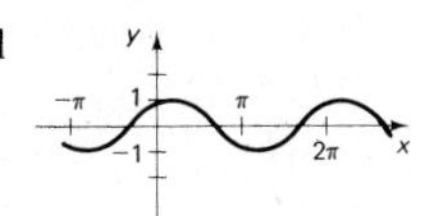

15. 1, 2π, π

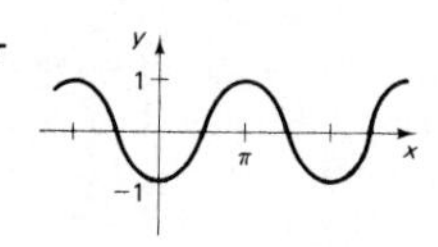

17. 1, $2\pi/3$, $\pi/3$

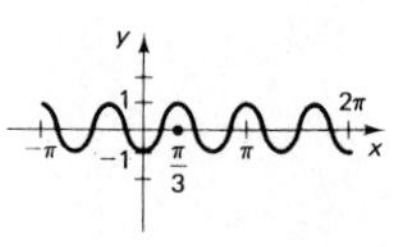

19. 1, $2\pi/3$, 2

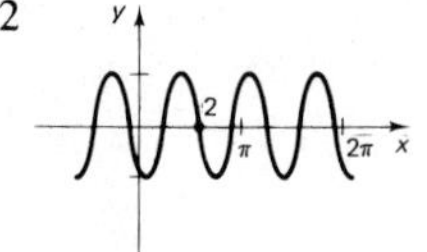

21. 3, 2, $-\frac{1}{2}$

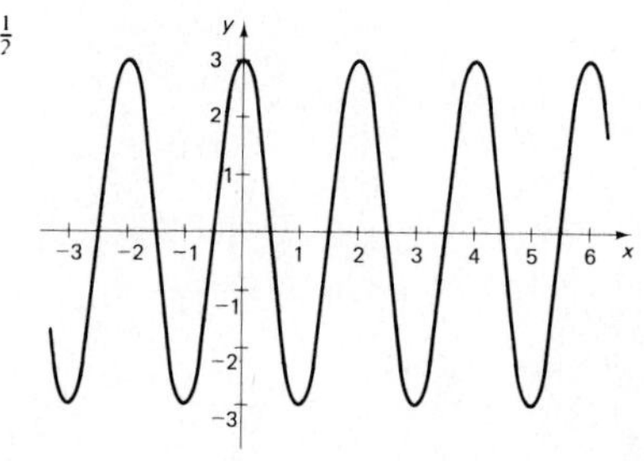

23. 4, $2\pi/3$, 1

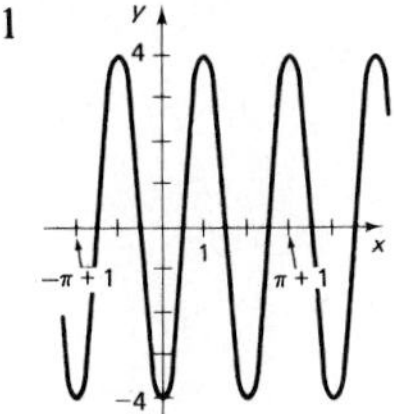

25. $\frac{1}{3}$, 2π, 2π

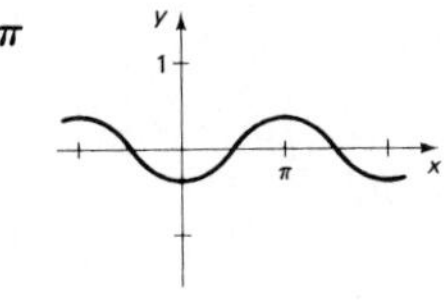

27.

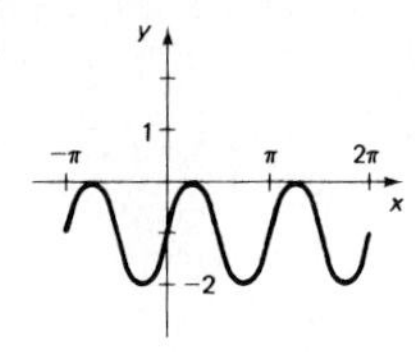

29.

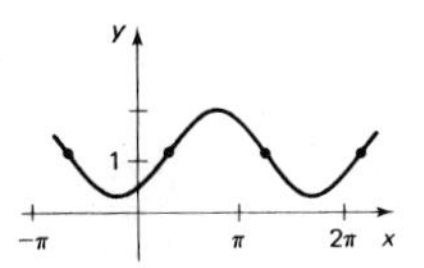

31.

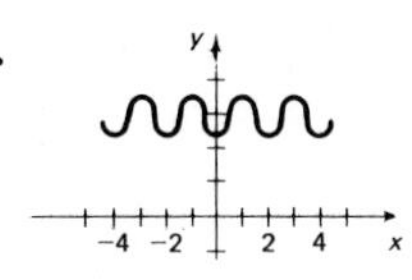

33.

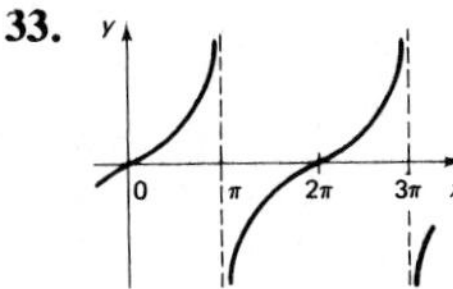

35.

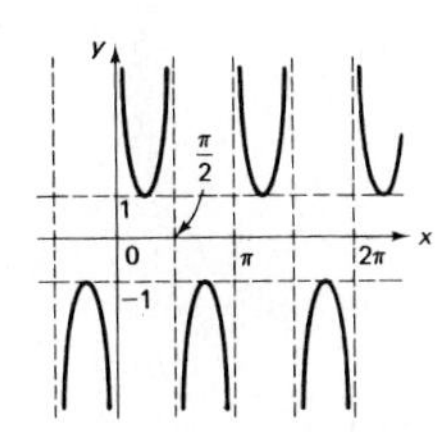

37.

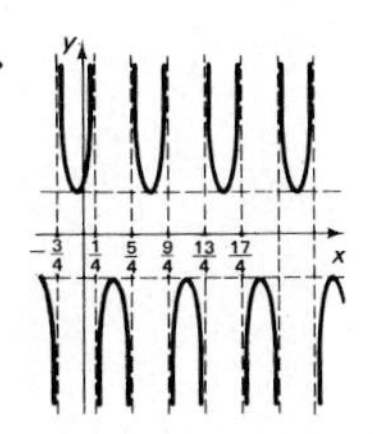

39.

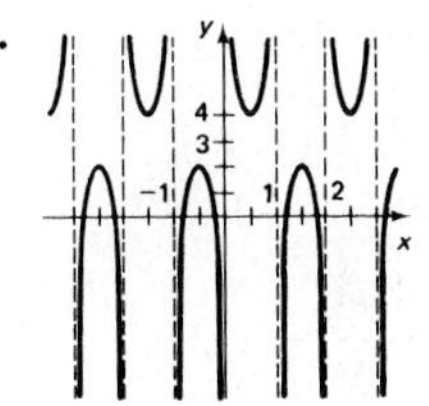

Section 2.8

1. 0.977317, 55.9962° **3.** 1.69875, 97.3309° **5.** −0.453759, −25.9985° **7.** 0.475604, 27.2501°
9. 0.980067 **11.** 0.730226 **13.** 0.615675 **15.** 4.44495 **17.** −35.0569 **19.** −0.00268026
21. $\pi/6$ **23.** $2\pi/3$ **25.** $-\pi/4$ **27.** $\pi/4$ **29.** $\pi/4$ **31.** $\pi/4$ **33.** $5\pi/6$ **35.** $5\pi/6$
37. *C* **39.** *C* **41.** *A* **43.** *A* **45.** *C* **47.** *C* **49.** 0.2 **51.** $-\pi/3$ **53.** 38° **55.** $5\pi/8$
57. −78° **59.** 57° **61.** 34° **63.** −479 **65.** −1, π **67.** $1/\sqrt{1-u^2}$
69. $\sqrt{u^2-1}$, $u \geq 1$; $-\sqrt{u^2-1}$, $u \leq -1$

71.

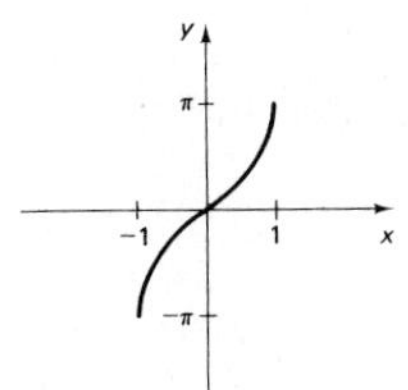

73.

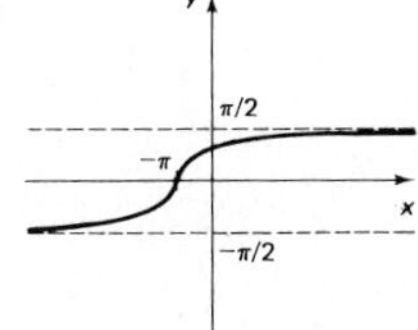

75.

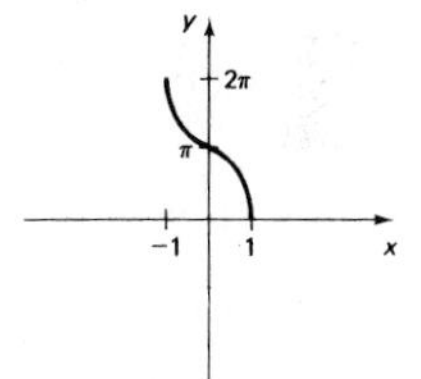

77.

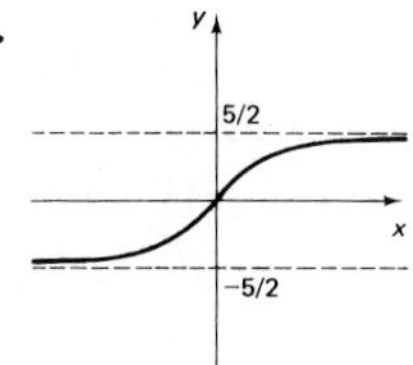

79.

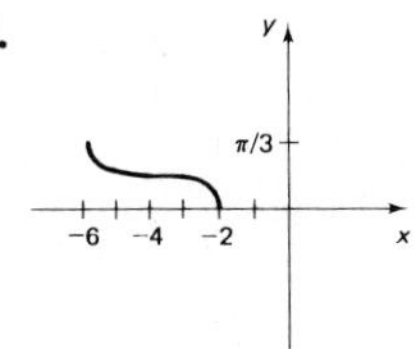

83.

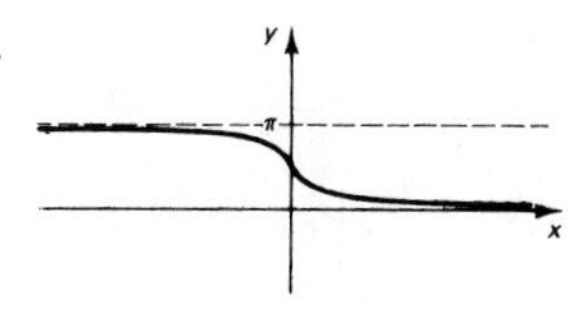

Section 2.9

1. 2.12 ft **3.** 880 hz **5.** 18.5 ft **7.** 1.25 ft, 0.625 ft **9.** 4.29×10^{14} hz **11.** 107 m **13.** 20%
15. 0.394791, $\sqrt{2}\sin(120\pi t + 0.394791)$ **17.** 0.363979, $4.71988\sin(120\pi t + 0.363979)$

Review

1. $4/\sqrt{17}, -1/\sqrt{17}, -4, \sqrt{17}/4, -\sqrt{17}, -1/4$ **3.** $\pm\sqrt{21}/5, 2/5, \pm\sqrt{21}/2, \pm 5/\sqrt{21}, 5/2, \pm 2/\sqrt{21}$
5. $\pm 3/\sqrt{10}, \mp 1/\sqrt{10}, -3, \pm\sqrt{10}/3, \mp\sqrt{10}, -1/3$ **7.** $\pm 12/13, \mp 5/13, -12/5, \pm 13/12, \mp 13/5, -5/12$
9. $4/5, 3/5, 4/3, 5/4, 5/3; 3/4$ **11.** $5/\sqrt{26}, -1/\sqrt{26}, -5, \sqrt{26}/5, -\sqrt{26}, -1/5$
13. $-\sqrt{15}/4, -1/4, \sqrt{15}, -4/\sqrt{15}, -4, 1/\sqrt{15}$ **15.** $(-2, 2.03788)$ **17.** $(-1, 0)$ **19.** $(-1/2, -\sqrt{3}/2)$
21. $(-0.981702, 0.190423)$ **23.** 0 **25.** 0.0936614 **27.** 2.5 **29.** 0 **31.** 3.78082
33. $1.09637, -0.889555$ **35.** $5°, -0.996195$
37. $3, 2\pi, \pi/3$ **39.** $2, 4\pi, \pi/2$ **41.** **43.** $-\pi/6$ **45.** $-62°$

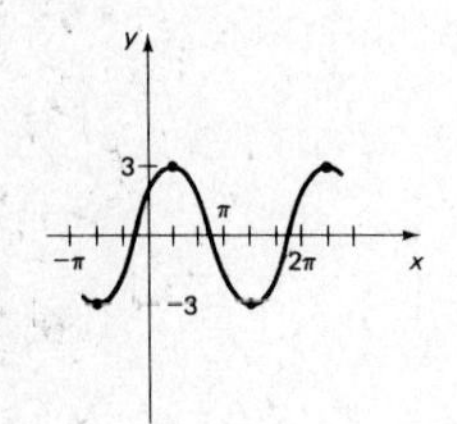

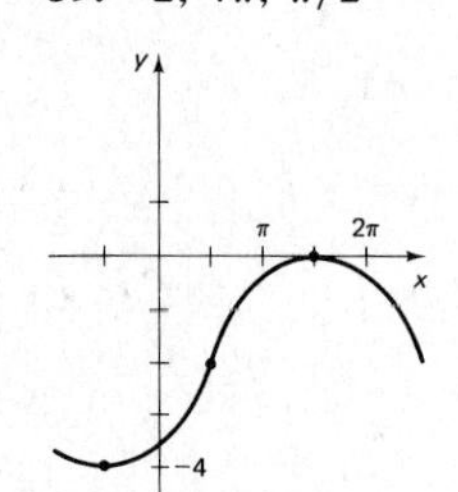

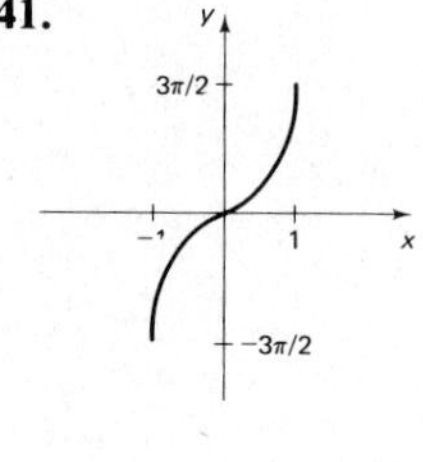

CHAPTER 3

Section 3.1

1. c **3.** f **5.** d **7.** d **9.** c **11.** b **19.** $\sec\theta$ **21.** $-\sin\theta$ **23.** $\cos x$ **25.** $1 - \sin\alpha$
27. $\cos\theta$ **29.** $-\frac{12}{13}, -\frac{5}{13}, \frac{12}{5}, -\frac{13}{12}, -\frac{13}{5}, \frac{5}{12}$ **31.** $1/2, -\sqrt{3}/2, -\sqrt{3}/3, 2, -2\sqrt{3}/3, -\sqrt{3}$

33. $\sin\theta, -\sqrt{1 - \sin^2\theta}, \dfrac{-\sin\theta}{\sqrt{1 - \sin^2\theta}}, \dfrac{1}{\sin\theta}, \dfrac{-1}{\sqrt{1 - \sin^2\theta}}, \dfrac{-\sqrt{1 - \sin^2\theta}}{\sin\theta}$

35. $\dfrac{1}{\csc\theta}, \sqrt{1 - \dfrac{1}{\csc^2\theta}}, \dfrac{-1}{\sqrt{\csc^2\theta - 1}}, \csc\theta, \dfrac{1}{\sqrt{1 - \dfrac{1}{\csc^2\theta}}}, -\sqrt{\csc^2\theta - 1}$ **37.** $2|\cos\theta|$ **39.** $3|\sec\theta|$

Section 3.2

1. $\pi/4 + 2n\pi, 3\pi/4 + 2n\pi$ **3.** $-\pi/6 + n\pi$ **5.** $\pm\pi/3 + 2n\pi$ **7.** $\pm\pi/3 + 2n\pi$ **9.** $\varnothing$ **11.** $n\pi$
13. $\pm\pi/3 + 2n\pi, \pi + 2n\pi$ **15.** $\pi/6 + 2n\pi, 5\pi/6 + 2n\pi, -0.339837 + 2n\pi, -2.80176 + 2n\pi$ **17.** $2n\pi$
19. $\varnothing$ **21.** $-\pi/4 + n\pi$ **23.** $\pi/6, 5\pi/6, 7\pi/6, 11\pi/6$; 30°, 150°, 210°, 330°
25. $\pi/3, \pi, 5\pi/3$; 60°, 180°, 300° **27.** $\pi/4, 3\pi/4, 5\pi/4, 7\pi/4$; 45°, 135°, 225°, 315° **29.** no solution
31. $\pi/2, 11\pi/6$; 90°, 330° **33.** $\pi/2, \pi$; 90°, 180°
35. $0, \pi/3, 2\pi/3, \pi, 4\pi/3, 5\pi/3$; 0°, 60°; 120°, 180°, 240°, 300° **37.** 0.401053, 2.74054, 3.83660, 5.58818

Section 3.3

41. Not an identity, e.g., $\pi/6$ **43.** Identity **45.** Not an identity, e.g., $\pi/6$ **47.** Not an identity, e.g., $\pi/4$

Section 3.4

1. $(\sqrt{3} - 1)/2\sqrt{2}$ **3.** $(\sqrt{3} - 1)/2\sqrt{2}$ **5.** $-2 - \sqrt{3}$ **7.** $\sqrt{2}(\sqrt{3} + 1)$ **9.** $1/\sqrt{2}$ **11.** 0.5 **13.** 1
15. 33/65, −63/65, 56/65, −16/65, 33/56, 63/16
17. $(\sqrt{6} + 1)/2\sqrt{3}, (\sqrt{6} - 1)/2\sqrt{3}, (\sqrt{3} - \sqrt{2})/2\sqrt{3}, -(\sqrt{3} + \sqrt{2})/2\sqrt{3}, (1 + \sqrt{6})/(\sqrt{3} - \sqrt{2}),$
$(1 - \sqrt{6})/(\sqrt{3} + \sqrt{2})$
45. $\sin 4\theta + \sin 2\theta$ **47.** $-3(\sin 7\theta + \sin 3\theta)/2$ **49.** $(\cos 6t - \cos 18t)/2$ **51.** $2\sin 4\theta\cos\theta$
53. $10\cos\frac{7}{2}\theta\cos\frac{1}{2}\theta$ **55.** $2\sin\frac{3}{2}x\cos\frac{1}{2}x$

Section 3.5

1. $\sqrt{3}/2, -1/2, -\sqrt{3}$ **3.** $-120/169, -119/169, 120/119$ **5.** $-24/25, 7/25, -24/7$ **7.** $\sqrt{2-\sqrt{3}}/2$
9. $-\sqrt{2-\sqrt{2}}/2$ **11.** $3/4$ **13.** $2/\sqrt{7}$ **33.** $7\pi/12, 7\pi/12+\pi, 11\pi/12, 11\pi/12+\pi$
35. $\pi/2, 3\pi/2, 7\pi/6, 11\pi/6$ **37.** $3\pi/2, \pi/6, 5\pi/6$ **39.** $\pi, \pi/3, 5\pi/3$ **41.** $0, \pi, 2\pi$

Section 3.6

1. $\alpha = 75°, b = 26.8973, c = 21.9615$ **3.** $\alpha = 120°, a = 33.6824, c = 13.3022$
5. $\beta = 5\pi/12, a = 36.6025, c = 44.8288$ **7.** $\beta = 44°, a = 50.1505, b = 37.5734$
9. $\gamma = 82.61°, a = 241.987, b = 359.974$ **11.** $\alpha = 0.991593, b = 115.167, c = 58.2914$
13. $\beta = 1.97159, a = 0.0320852, b = 0.0400342$ **15.** $\alpha = 44.61°, a = 4591.96, c = 5532.98$
17. $\gamma = 75.37°, a = 6.89226, c = 23.8889$ **19.** 168.511 ft **21.** 127.899 ft
23. 153.007 mi, 279.024 mi **25.** 762.510 nautical mi **27.** 263.507 km

Section 3.7

1. Unambiguous **3.** Ambiguous, 1 **5.** Ambiguous, 1 **7.** Unambiguous **9.** Ambiguous, 0
11. $\beta = 48.5904°, \gamma = 101.410°, c = 7.84190$, or $\beta = 131.410°, \gamma = 18.5904°, c = 2.55040$
13. $\beta = 43.8538°, \alpha = 16.1462°, a = 3.21110$ **15.** $\beta = 9.84655°, \gamma = 150.153°, c = 14.5512$
17. $\alpha = 90°, \gamma = 60°, c = 5\sqrt{3} \approx 8.66025$ **19.** No triangle **21.** $\beta = 33.3195°, \alpha = 90.2305°, a = 38.6484$
23. $\gamma = 1.11622, \beta = 1.52968, b = 18.9056$; or $\gamma = 2.02538, \beta = 0.620516, b = 11.0020$ **25.** No triangle
27. $\beta = 0.352464, \alpha = 2.47193, a = 121.594$; or $\beta = 2.78913, \alpha = 0.0352643, a = 6.90654$
29. No. No such triangle exists. **31.** 108.454 ft, 34.9642 ft, (No)

Section 3.8

1. $c = 2\sqrt{19} \approx 8.71780, \alpha = 96.5868°, \beta = 23.4132°$ **3.** $a = 70.6382, \beta = 36.8896°, \gamma = 21.1104°$
5. $b = 495.583, \gamma = 0.465726, \alpha = 2.29587$ **7.** $a = 7.25496, \beta = 27.8294°, \gamma = 138.171°$
9. $b = 8.89065, \alpha = 27.5157°, \gamma = 112.484°$ **11.** No triangle **13.** $\alpha = 50.2808°, \beta = 98.7143°, \gamma = 31.0049°$
15. $\alpha = 122.271°, \beta = 17.3422°, \gamma = 40.3866°$ **17.** 10.1214 hrs **19.** 36.3966 ft, 50.1332° **21.** 0.687161 km
23. 104.031° **25.** 33.1472 km

Section 3.9

1. 200 **3.** $18\sqrt{3} \approx 31.1769$ **5.** 73.2051 **7.** 92.8113 **9.** 182,490 **11.** 638.350
13. 719.687, 1966.64 **15.** 1.35897, 0.042500 **17.** 0.00775274 **19.** 146.654 **21.** $b = 35.7574, c = 5$
23. $a = 3.88247, b = 5.23083, c = 5.94828$ **25.** 2.47929 acres **27.** 4.81756, or 5 bags **29.** 16.5164 sq mi
31. $128.48 **33.** 178.928 cm

Section 3.10

1. 72.7328, 80.1039° **3.** 5.72242, 260.293° **5.** 34.9042, 3.54804° **7.** 0
9. **a.** 7 mph, S **b.** 1 mph, S **c.** 5 mph, 143.130° **d.** 2.83363 mph, 131.529° **e.** 6.47847 mph, 160.886°
11. **a.** 7.21110 mph, 56.3099° to *AB* **b.** 1.5 mi **c.** 15 min **d.** 15 min
13. **a.** 41.8103° upstream **b.** 8.94427 min **c.** No
15. 322.360°, 405.449 mph **17.** 128.176°, 398.765 kmph **19.** 38.4204 km, 4.54758° **21.** 17.2516 lbs
23. 24.8346° **25.** 13.2501 lbs

Section 3.11

1.–9.

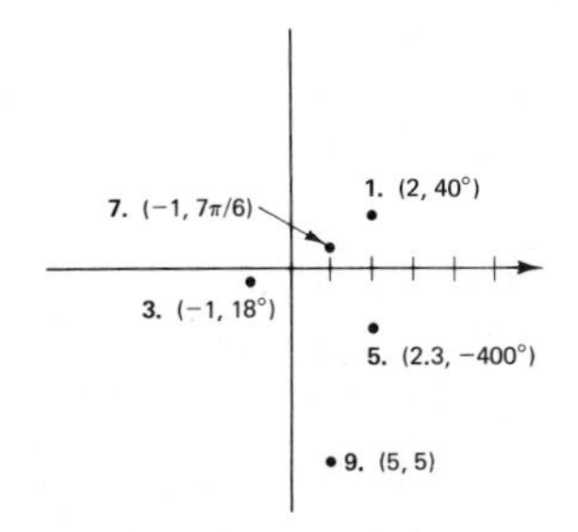

11. (3,180°) **13.** (4,60°) **15.** $(4\sqrt{2}, 135°)$ **17.** (13, −67.3801°) **19.** (10, 126.870°)
21. $(\sqrt{10}, -71.5651°)$ **23.** $(-1, \sqrt{3})$ **25.** (−5.86889, 1.24747) **27.** (0, 0) **29.** (0.832294, −1.81859)
31. $r \cos \theta = 7$ **33.** $\theta = \tan^{-1} 2$ **35.** $r^2(\cos^2 \theta + 4 \sin^2 \theta) = 4$ **37.** $x^2 + y^2 = 9$ **39.** $y = -x$
41. $y = 7$ **43.** $(x^2 + y^2)x = 3y - x$

45.

47.

49.

51.

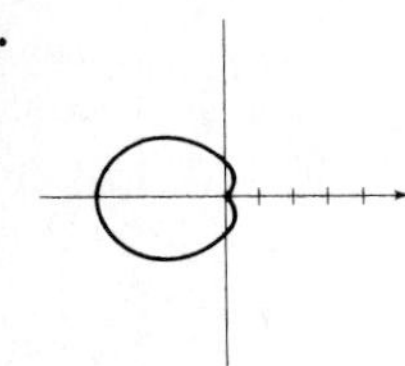

53.

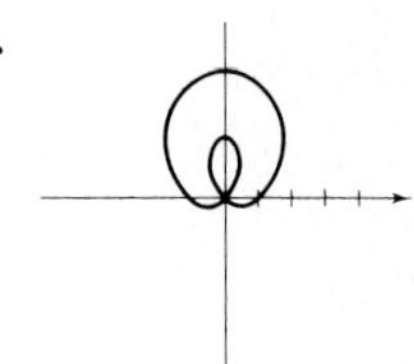

55.

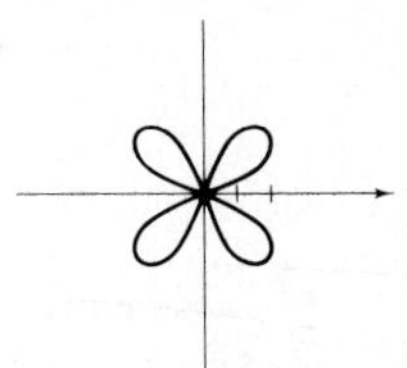

Review

11. $-\pi/6 + 2n\pi, -5\pi/6 + 2n\pi$ **13.** $n\pi, \pm 1.23096 + 2n\pi$ **15.** $2n\pi, \pm 2\pi/3 + 2n\pi$ **17.** $\frac{16}{65}, -\frac{63}{65}, -\frac{16}{63}$
19. $\gamma = 71°, a = 46.4587, b = 113.788$
21. $\alpha = 1.07941, \beta = 1.43387, b = 30.3368$ or $\alpha = 2.06219, \beta = 0.45109, b = 13.3501$
23. $a = 0.797409, \beta = 1.83834, \gamma = 0.323256$ **25.** 134.444 mi or 19.6074 mi **27.** 9.02936°
29. 123.796, 31.5351° **31.** 138.879 mph, 233.797°

CHAPTER 4

Section 4.1

1. $3 - i$ **3.** 5 **5.** $7 + 7i$ **7.** $15 - 7i$ **9.** 0 **11.** $6 - i$ **13.** $10 - 2i$ **15.** $3 + 3i$
17. $-3 - 2i$ **19.** $-8 - 5i$ **21.** $-4 - 11i$ **23.** $11 - 10i$ **25.** $2 - 9i$ **27.** $1 + 3i$ **29.** $9 + 15i$
31. $-5 - 12i$ **33.** i **35.** 1 **37.** $3i$ **39.** $\frac{8}{13} - \frac{1}{13}i$ **41.** $\frac{1}{17} + \frac{13}{17}i$ **43.** $\frac{1}{5} - \frac{2}{5}i$ **45.** $3 - 4i$
47. $-i$ **49.** $x = -2, y = \frac{7}{3}$ **51.** $x = -4, y = 7$ **53.** $x = 3, y = 3^{1/9}$ **55.** $x = \frac{3}{2}, y = \frac{1}{3}$
57. $x = \pi/2 + 2n\pi, y = \pm\pi/3 + 2n\pi$

Section 4.2

1. $2i$ **3.** $-2\sqrt{5}\, i$ **5.** $11i$ **7.** −90 **9.** 30 **11.** $\frac{13}{2}$ **13.** $12 - 5i$ **15.** $-20 - 20\sqrt{3}\, i$
17. $6 + 10i$ **19.** $-16\sqrt{2}\, i$ **25.** $-\frac{1}{3}i$ **27.** $1 + 2i$ **29.** $-\frac{5}{4}i$ **31.** $-3 + i$ **33.** $-\frac{6}{25} + \frac{17}{25}i$
35. $-\frac{11}{5} + \frac{2}{5}i$ **37.** $1 \pm i$ **39.** $\frac{3}{4} \pm \frac{1}{4}\sqrt{7}\, i$ **41.** $-\frac{2}{3} \pm \frac{1}{3}\sqrt{2}\, i$ **43.** $1, -\frac{1}{2} \pm \frac{1}{2}\sqrt{3}\, i$ **45.** $5, -\frac{5}{2} \pm \frac{5}{2}\sqrt{3}\, i$
47. $\pm 2, \pm 2i$ **49.** $\pm 2, \pm 1 \pm \sqrt{3}\, i$ **51.** $\pm i, \pm\frac{1}{2}\sqrt{2}\, i$ **53.** $-\frac{1}{2} \pm \frac{1}{2}\sqrt{3}\, i, 0, 0$ **55.** $i, -2i$
57. $-\frac{1}{2} + \frac{1}{2}(2 + \sqrt{3})i, -\frac{1}{2} + \frac{1}{2}(2 - \sqrt{3})i$ **59.** $-\frac{1}{2} + \frac{1}{2}i, -\frac{1}{2} + \frac{1}{2}i$ **61.** $0.833333 \pm 1.28019i$
63. $0.381532 \pm 0.339104i$ **65.** $x^2 + 4 = 0$ **67.** $x^2 + 4x + 13 = 0$ **69.** $x^2 + 3 = 0$
71. $x^3 + (-1 + 2i)x^2 + (3 - 2i)x - 3$ **73.** $x^3 - x^2 + 2$ **75.** $x^3 + (-4 - i)x^2 + (5 + 4i)x - 5i$

Section 4.3

1. $|z_n| = \sqrt{2}$, all n

$1 + i = \sqrt{2}\,(\cos \pi/4 + i \sin \pi/4) = \sqrt{2} \text{ cis } \pi/4$
$-1 + i = \sqrt{2}(\cos 3\pi/4 + i \sin 3\pi/4) = \sqrt{2} \text{ cis } 3\pi/4$
$-1 - i = \sqrt{2}(\cos 5\pi/4 + i \sin 5\pi/4) = \sqrt{2} \text{ cis } 5\pi/4$
$1 - i = \sqrt{2}(\cos 7\pi/4 + i \sin 7\pi/4) = \sqrt{2} \text{ cis } 7\pi/4$

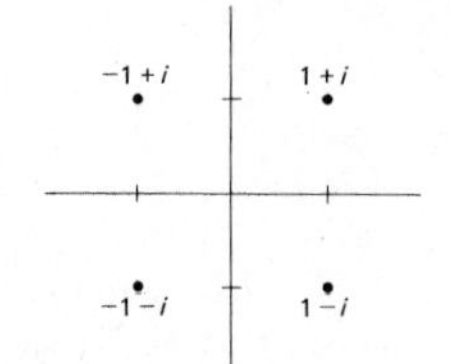

3. $|z_n| = 2$, all n

$\sqrt{3} + i = 2(\cos \pi/6 + i \sin \pi/6) = 2 \text{ cis } \pi/6$

$-\sqrt{3} + i = 2(\cos 5\pi/6 + i \sin 5\pi/6) = 2 \text{ cis } 5\pi/6$

$-\sqrt{3} - i = 2(\cos 7\pi/6 + i \sin 7\pi/6) = 2 \text{ cis } 7\pi/6$

$\sqrt{3} - i = 2(\cos 11\pi/6 + i \sin 11\pi/6) = 2 \text{ cis } 11\pi/6$

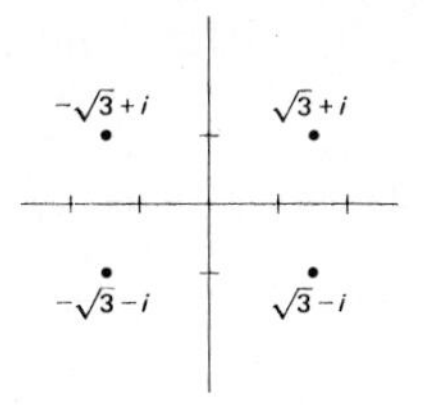

5. $|z_1| = 2\sqrt{10}$, $|z_2| = \sqrt{10}$, $|z_3| = 3\sqrt{10}$, $|z_4| = 2\sqrt{10}$

$6 + 2i = 2\sqrt{10}(\cos 18.4349° + i \sin 18.4349°) = 2\sqrt{10} \text{ cis } 18.4349°$

$-3 + i = \sqrt{10}(\cos 161.565° + i \sin 161.565°) = \sqrt{10} \text{ cis } 161.565°$

$-9 + 3i = 3\sqrt{10}(\cos 161.565° + i \sin 161.565°) = 3\sqrt{10} \text{ cis } 161.565°$

$6 - 2i = 2\sqrt{10}[\cos(-18.4349°) + i \sin(-18.4349°)] = 2\sqrt{10} \text{ cis}(-18.4349°)$

7. $-3\sqrt{3} - 3i$ **9.** $-0.317064 + 1.97471i$ **11.** $18 \text{ cis } 419°$ **13.** $740.479 \text{ cis } 10.12$ **15.** $0.5 \text{ cis}(-70°)$

17. $\frac{5}{3} \text{ cis}(\pi/5)$ **19.** $2 \text{ cis } 177°$ **21.** $1.31785 \text{ cis}(-2\pi/15)$ **23. a.** $16\sqrt{3} + 16i$ **b.** $-\frac{1}{4}$

Section 4.4

1. $3^6 \text{ cis } 612°$ **3.** $2^{-8} \text{ cis}(-1704°)$ **5.** $3.7^5 \text{ cis}(-5\pi/3)$ **7.** $(\frac{1}{2} + \frac{1}{2}\sqrt{3}\, i)$ **9.** $\frac{1}{2}\sqrt{2} - \frac{1}{2}\sqrt{2}\, i$

11. $154.686 - 606.795i$ **13.** $-3245.02 - 2574.04i$

15. $\text{cis } 0° = 1$, $\text{cis } 72° = 0.309017 + 0.951057i$
$\text{cis } 144° = -0.809017 + 0.587785i$
$\text{cis } 216° = -0.809017 - 0.587785i$
$\text{cis } 288° = 0.309017 - 0.951057i$

17. $2 \text{ cis}(-\pi/9) \approx 1.87939 - 0.684040i$
$2 \text{ cis } 5\pi/9 \approx -0.347296 + 1.96962i$
$2 \text{ cis } 11\pi/9 \approx -1.53209 - 1.28558i$

19. $2 \text{ cis } \pi/12 \approx 1.93185 + 0.517638i$
$2 \text{ cis } 5\pi/12 \approx 0.517638 + 1.93185i$
$2 \text{ cis } 9\pi/12 \approx -1.41421 + 1.41421i$
$2 \text{ cis } 13\pi/12 \approx -1.93185 - 0.517638i$
$2 \text{ cis } 17\pi/12 \approx -0.517638 - 1.93185i$
$2 \text{ cis } 21\pi/12 \approx 1.41421 - 1.41421i$

21. $1.29239 \text{ cis } 29.2620° \approx 1.12747 + 0.631726i$
$1.29239 \text{ cis } 101.262° \approx -0.252399 + 1.26751i$
$1.29239 \text{ cis } 173.262° \approx -1.28347 + 0.151636i$
$1.29239 \text{ cis } 245.262° \approx -0.540827 - 1.17379i$
$1.29239 \text{ cis } 317.262° \approx 0.949216 - 0.877078i$

23. a.

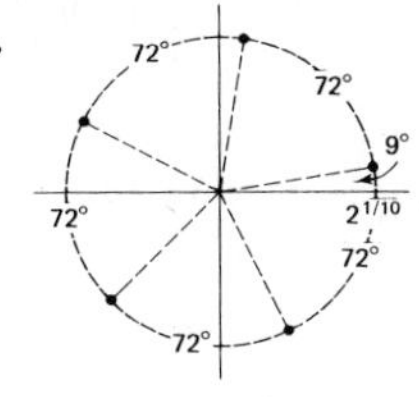

b.

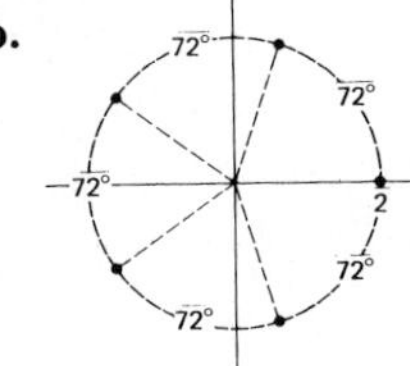

25. $\frac{1}{4}\sqrt{2}\, i, \frac{1}{2}\sqrt{2} - \frac{1}{4}\sqrt{2}\, i$ **27.** $-\frac{1}{4}i, \frac{1}{2}\sqrt{3} + \frac{1}{4}i$ **29.** $-4.67945 + 3.67945i, -0.320551 - 6.79449i$

31. $0.618034 - 0.618034i, -1.61803 + 1.61803i$

Review

1. $7 + 5i$ **3.** $-4 + 16i$ **5.** -1 **7.** $-\frac{1}{2} - \frac{1}{2}i$ **9.** $-81\sqrt{3}\, i$ **11.** $2\sqrt{2} + 2\sqrt{2}\, i$

13. $1.70197 - 5.75355i$ **15.** $29{,}524.5 + 29{,}524.5\sqrt{3}\, i$ **17.** $x = 0.5, y = 3.5$ **19.** $-\frac{17}{26} - \frac{7}{26}i$

21. $2, -1 \pm \sqrt{3}\, i$

23. $[-2 + (3 + \sqrt{5})i]/6, [-2 + (3 - \sqrt{5})i]/6$

25. $\sqrt[10]{17} \text{ cis }(-14.0362°)$, $\sqrt[10]{17} \text{ cis }(57.9638°)$, $\sqrt[10]{17} \text{ cis }(129.964°)$, $\sqrt[10]{17} \text{ cis }(201.964°)$, $\sqrt[10]{17} \text{ cis }(273.964°)$

CHAPTER 5

Section 5.1

1. 1330.45 **3.** 1248.88 **5.** 0.0214936 **7.** 1.08380 **9.** 1.48985 **11.** 480.540

13.

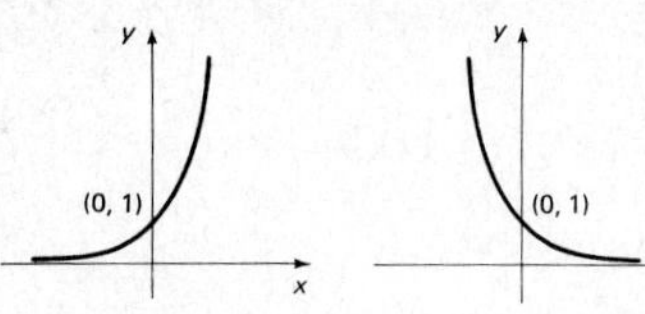

15. **a.** 1.7 **b.** 7 **c.** 0.8 **d.** 10 **17.** 2.2; 2.23607

19. 0.9; 0.870551 **21.** 1.1; 1.12246 **23.** 6; 6.24025 **25.** **27.**

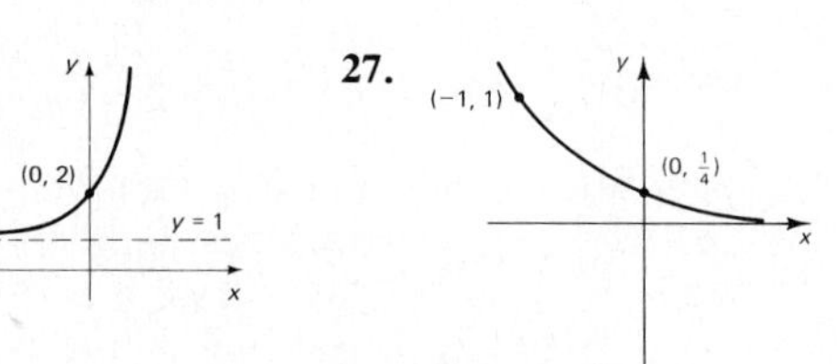

29.

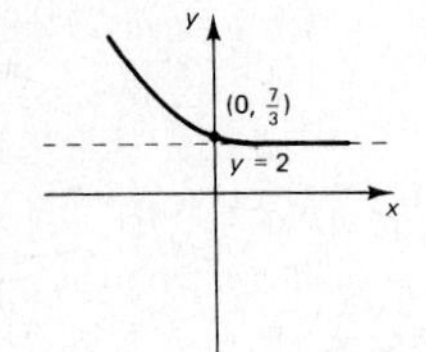

31.

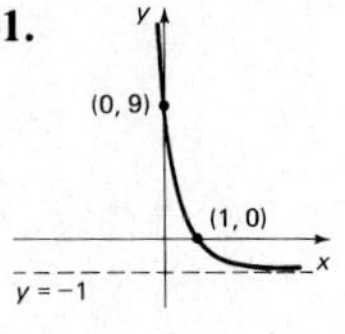

33.

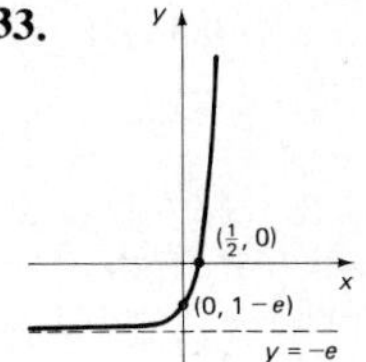

35.

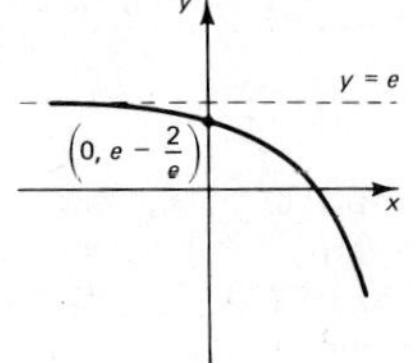

37.

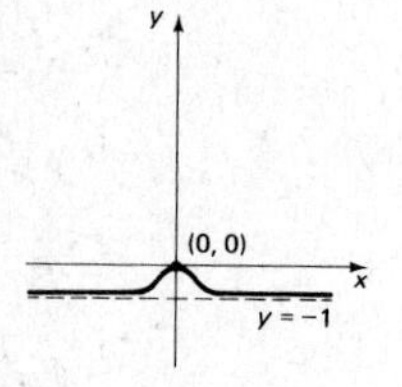

39.

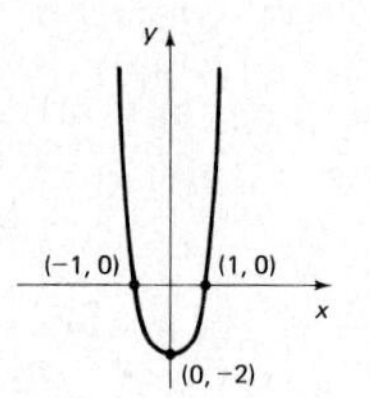

41.

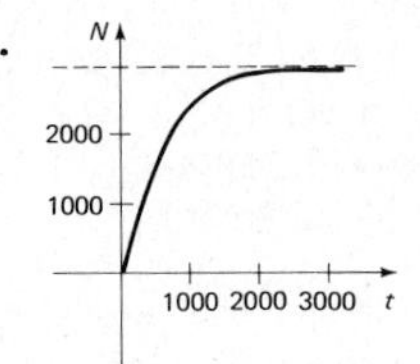

Section 5.2

1. 3 **3.** 0 **5.** 1 **7.** −2 **9.** 3/2 **11.** −12 **13.** −2

15. y (4, 1) (1, 0) x

17. y (1, 0) x (4, −1)

19.

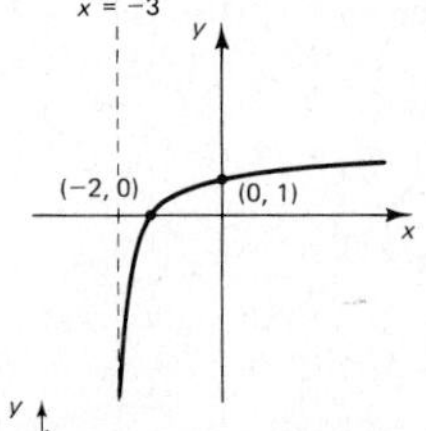

21.

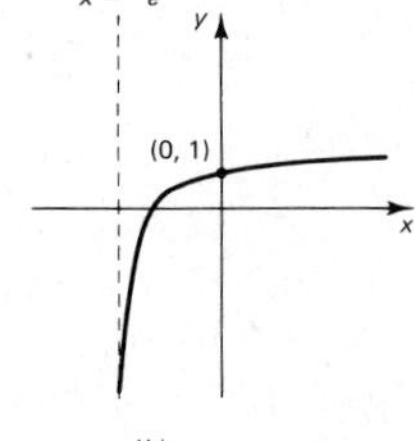

23.

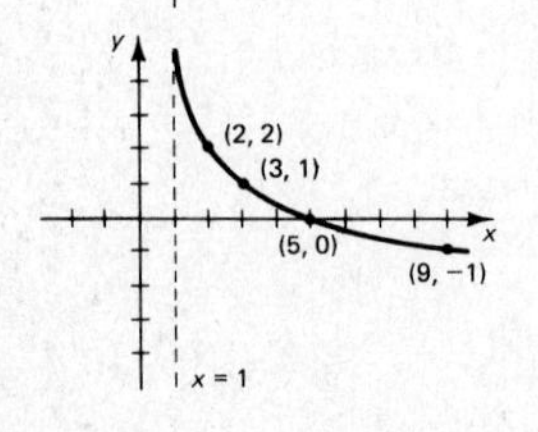

25.

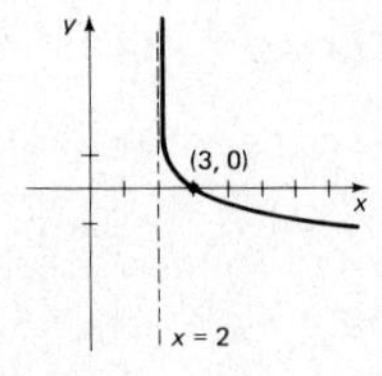

27.

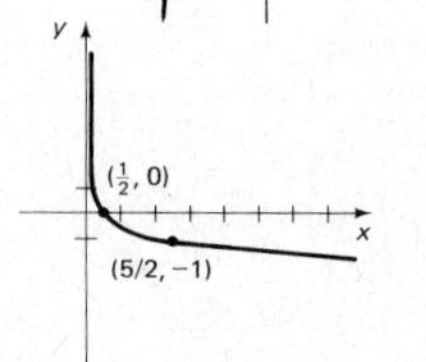

29.

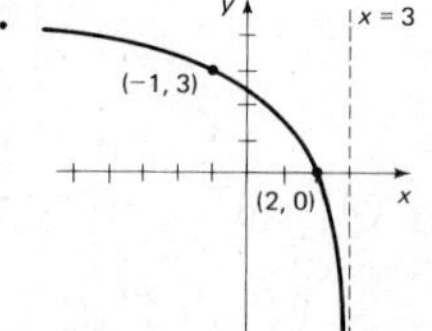

31.

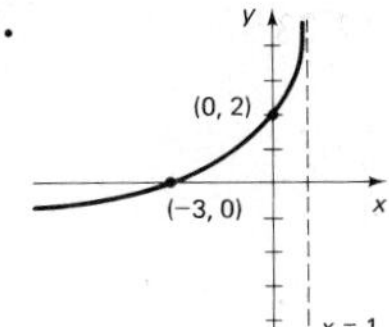

33.

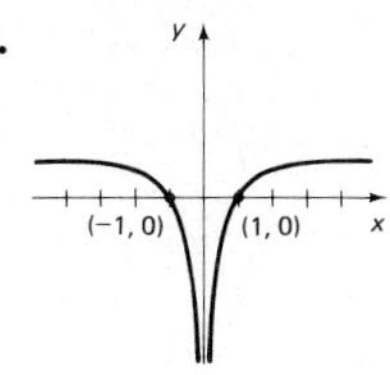

35.

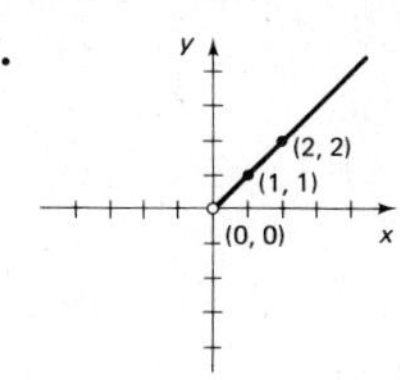

37.

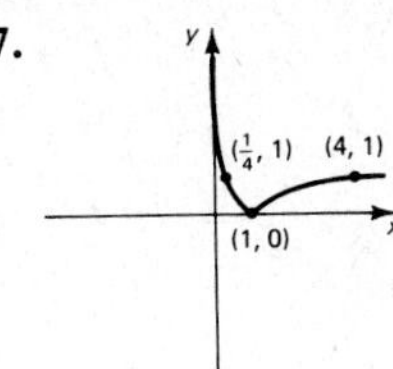

Section 5.3

1. 1.59331 **3.** 0.709440 **5.** 4.33021 **7.**

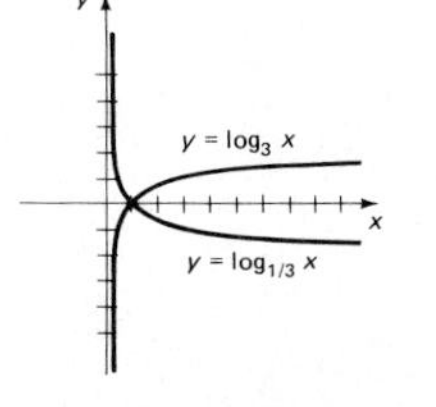

9. 1.26186; −1.26186

11. −0.203114; 0.203114 **13.** −2.72683; 2.72683 **15.** 1.26186 **17.** 2.77124 **19.** 5.28540
21. 1.35621 **23.** −2.09590 **25.** −0.748070 **27.** −0.827087 **29.** 16 **31.** 1,000,000
33. **a.** −0.287682; 0.792481; 2.07944; 0.333025 **b.** No, no **c.** Both = $\log_y x$

Section 5.4

1. 15 **3.** 4 **5.** 7/2 **7.** 2 **9.** $3^9 = 19{,}683$ **11.** $\sqrt{2}$ **13.** 1/2 **15.** 729 **17.** 11
19. 4 **21.** 5 **23.** 20 **25.** x **27.** −1/2 **29.** 36 **31.** $(23/4) \log_4 x$ **33.** $\log_{12}(x^2 - x + 1)$
35. 1/2 **37.** 1.11 **39.** 1.54 **41.** −0.57 **43.** 3 **45.** −1.43 **47.** 3 **49.** $8^{1/20} \approx 1.10957$
51. −0.138647 **53.** 24 **55.** **a.** 253.648 **b.** 2922.76 **c.** 0.0793484 **d.** 16.2376

Section 5.5

1. 3; 6.19174; 7.24465; 7.37431; 7.38906; 7.38906
3. −19; 1; 2.03704×10^{-10}; 1.68297×10^{-9}; 2.06111×10^{-9}; 2.06115×10^{-9}
5. **a.** $10,794.63 **b.** $11,040.20 **c.** $11,126.72 **d.** $11,127.70
7. **a.** 8.16334% **b.** 8.00321% **c.** 7.87294% **d.** 7.84723% **9.** b
11. **a.** $5882.80 **b.** $5663.77 **c.** $5594.44 **d.** $5588.62 **13.** **a.** $72,900 **b.** 15.2755 years
15. 20.5403%; 9.90308 years **17.** **a.** 1614 ants **b.** September 13
19. **a.** $P = 50e^{-0.0330070t}$ **b.** 69.7605 days **c.** 0.000292942 grams **21.** **a.** 26.6643 watts **b.** 1247.66 days

23. **a.** 2.20300; 2.39999; 2.40000 **b.**

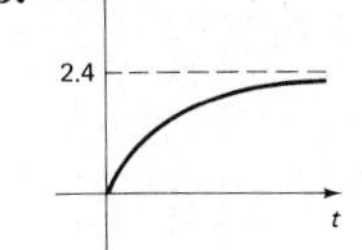

c. 1.43341; never **25.** 31.6228 times stronger

Section 5.6

1. **a.** $160.11 **b.** $11,527.92 **c.** $1,727.92
3. 20 years: **a.** $550.54 **b.** $132,129.60 **c.** $82,129.60
25 years: **a.** $526.61 **b.** $157,983.00 **c.** $107,983.00
30 years: **a.** $514.31 **b.** $185,151.60 **c.** $135,151.60
5. **a.** Think that you lend the bank money and it pays you back. **b.** $59.96 **c.** yes, $50 **d.** 11.5813 years
7. **a.** $353.39 **b.** $43,249.40 **9.** **a.** $4,492,760 **b.** 25.3073 years **11.** **a.** $18.08 **b.** $16.61
13. **a.** $205.58 **b.** $199.19

Section 5.7

1. 4 **3.** 2.52372 **5.** 17/2 **7.** $\pm\sqrt{10} \approx \pm 3.16228$ **9.** 1; −3 **11.** 8.43168 **13.** 10
15. 4^{-16}; 4 **17.** $-\log_4 3 \approx -0.792481$ **19.** 255; −3/4 **21.** 11/9 **23.** 1; 100 **25.** 1; 10,000
27. 0; $\log_3 8 \approx 1.89279$ **29.** $\log_4(8 \pm \sqrt{63}) \approx \pm 1.99717$ **31.** ±2 **33.** $\log_2 \sqrt{3} \approx 0.792481$
35. −0.684177 **37.** 1; $10^{\sqrt{3}} \approx 53.9574$; $10^{-\sqrt{3}} \approx 0.0185331$

Review

1. 0.00478656 **3.** 2; 1.62066 **5.** $\frac{1}{4}$; 4 **7.** 5 **9.** 2 **11.** ±7

13. **15.** **17.** **19.**

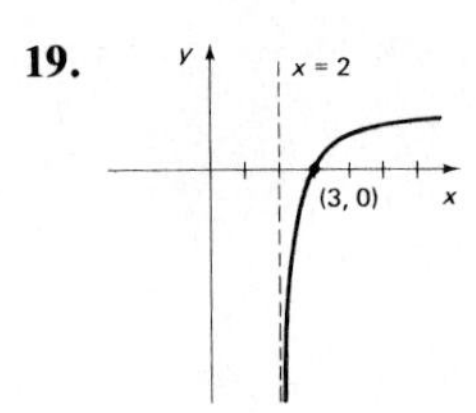

21. 2 **23.** $x^2 - 4x$ **25.** −0.25; 0.86; 2.54 **27.** 0.284065 **29.** 0.387904 **31.** 1/2
33. $\log_4 7 \approx 1.40368$ **35. a.** $7454.51 **b.** $7595.92 **c.** $7609.35 **d.** $7609.81 **37.** $4340.37; 3.49279 years
39. 32.9877 grams; 108.048 days **41. a.** $250.28 **b.** $16,016.80 **c.** $4,016.80

APPENDIX E

Section E.1

1. a. $(-2, 3)$ **b.** $(-2, 3]$

c. $[-2, 3)$ **d.** $[-2, 3]$

3. a. $(-7, -3)$ **b.** $\mathbb{R}$

c. $\emptyset$ **d.** $(-\infty, -7) \cup (-3, \infty)$

5.

7. 13 **9.** 7 **11.** $2\sqrt{a^2 + b^2}$
13. $s^{3/2} + s^{-3/2}$ **15.** (5, 2), (5, 6); (−3, 2), (−3, 6); (−1, 4), (3, 4) **17.** (0, 1), (0, 9)
19. (−2, −7), (−2, 17)
21. **23.** **25.**

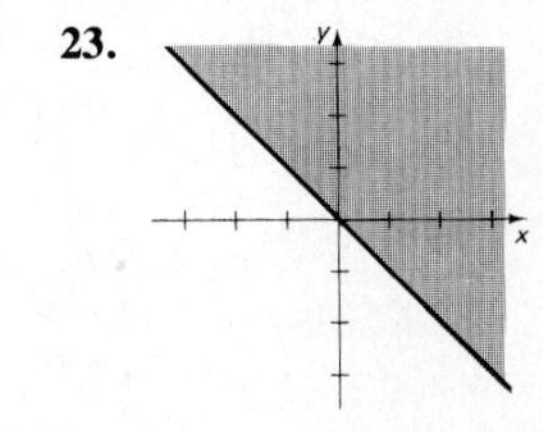

27.
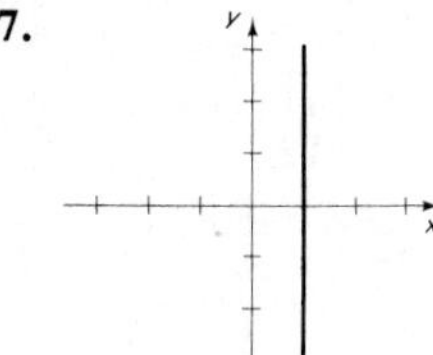

29.
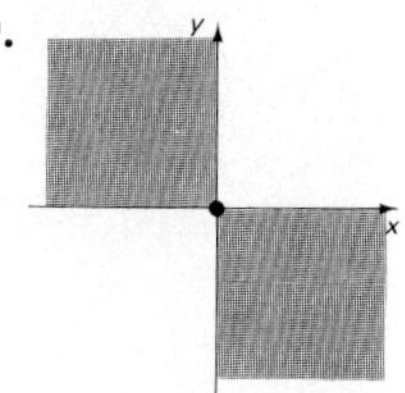

31.
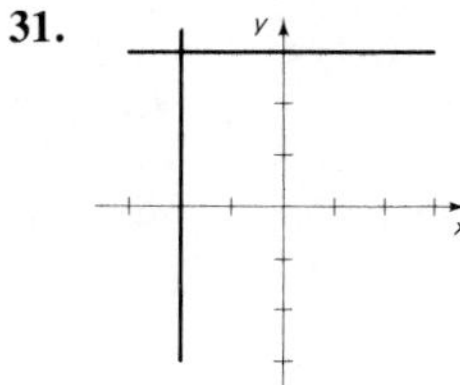

33.
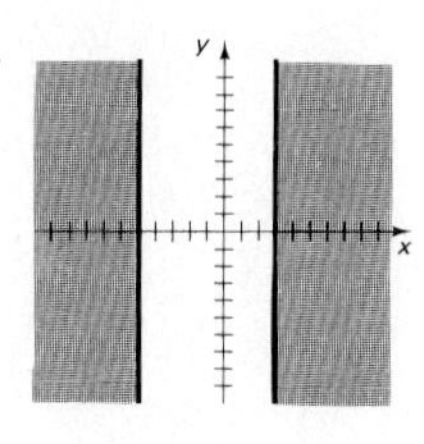

35.
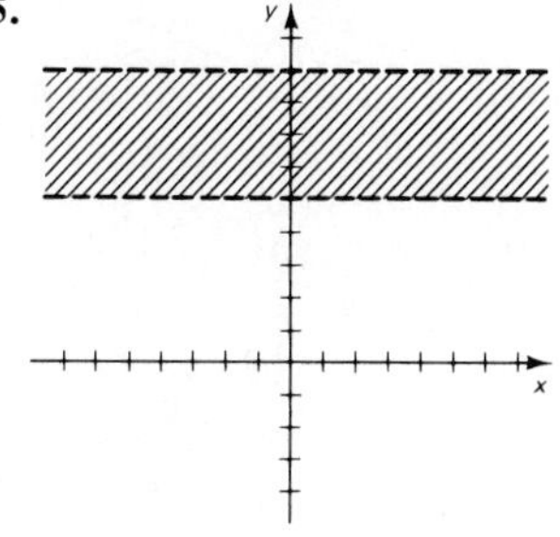

37.
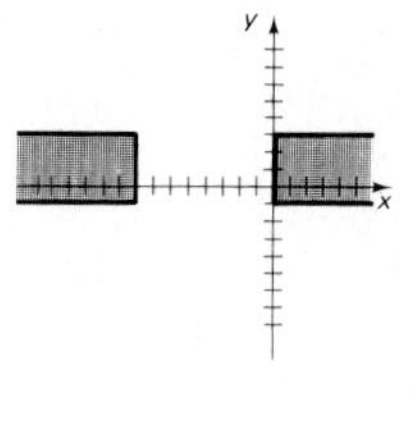

39.
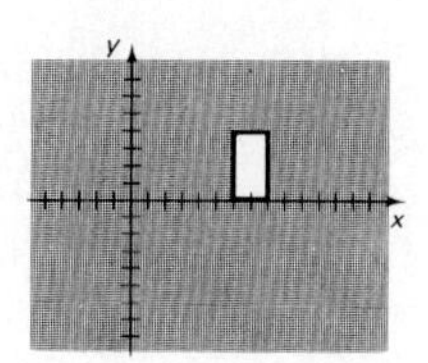

41.
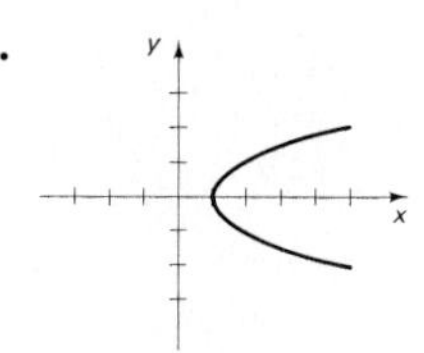

43.
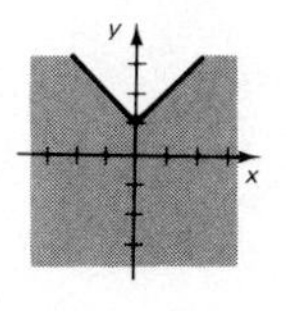

45.
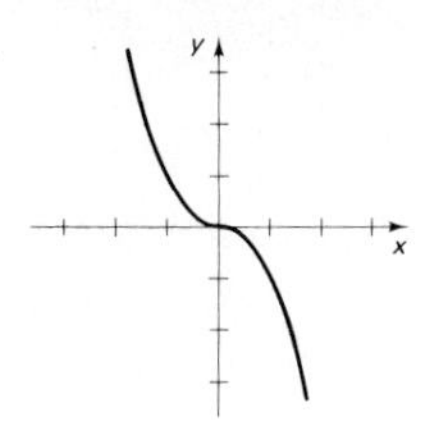

47.
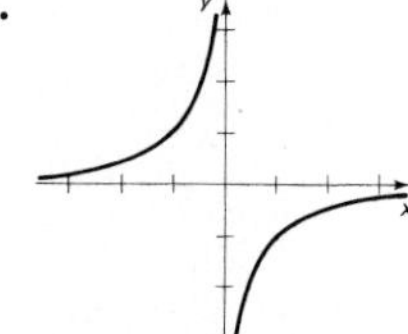

49.
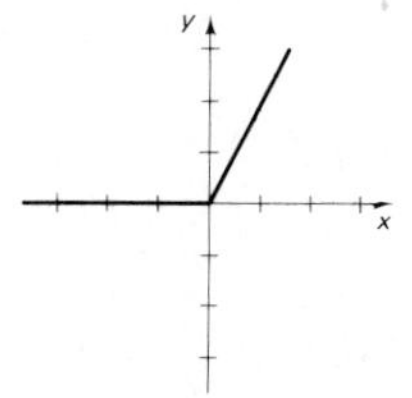

51.
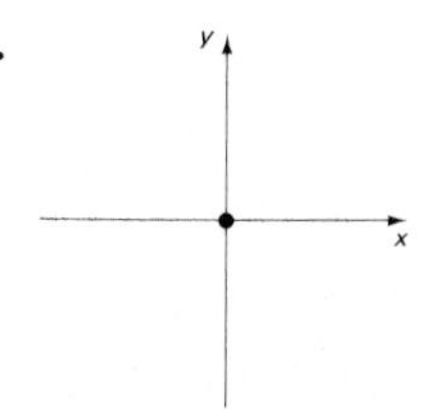

Section E.2

1. 34, 6, 1, 9 **3.** 3, 2, 1, 0 **5. a.** 3 **b.** -18 **c.** $7/a + 3$, **d.** $7a + 24$ **e.** 7
7. a. 2 **b.** 10 **c.** $5/2$ **d.** $5a^2$ **e.** 10 **9. a.** -0.8 **b.** 1.8 **c.** 1 **d.** 0 **f.** -1
11. a. 1.4 **b.** 2.1 **c.** 0.9 **d.** 1.8 **f.** 1.2 **13.** b, d **15.** b, d
17. $r = C/2\pi$ **19.** $A = 6V^{2/3}$ **21. a.** $V = (10 - 2x)(12 - 2x)x$ **b.** $0 < x < 5$

23. a. $T = \sqrt{1 + x^2}/3$ **b.** $T = \sqrt{5}/3$ hr **25.** $d = \begin{cases} 15t, \ 0 \le t \le 2 \\ \sqrt{400(t-2)^2 + 225t^2}, \ 2 < t \end{cases}$

27. $C = \begin{cases} 20n, \ 0 \le n < 25 \\ (22.40 - 0.10n)n, \ 25 \le n \le 110 \end{cases}$

Section E.3

1. $8x - 1; 8x - 11$ **3.** $12x^2 + 1; 6x^2 + 2$ **5.** $x + 1, x \geq -2; \sqrt{x^2 + 1}$ **7.** $8x - 1; 2\sqrt[3]{x^3 - 1}$
9. $x/(1 - 3x), x \neq 0, 1/3; x - 3, x \neq 3$ **11.** $1/x^4, x \neq 0; 1/x^4, x \neq 0$ **13.** $f(x) = x^2, g(x) = x + 1$
15. $f(x) = x + 1, g(x) = x^2$ **17.** $f(x) = \sqrt[3]{x}, g(x) = 1/x$ **19.** $f(x) = 1/x, g(x) = \sqrt[3]{x}$ **21.** 1.25, 2.25
23. 0.2, −1.8 **25.** 0.7, 0.17 **27.** 0.84, −1.16 **29.** Yes **31.** No **33.** No **35.** Yes **37.** Yes

39.

$f^{-1}(x) = x + 1$

$f(x) = x - 1$

41.

$h(x) = -\frac{1}{2}x$

$h^{-1}(x) = -2x$

43.

$F^{-1}(x) = \frac{1}{2}x + \frac{3}{2}$

$F(x) = 2x - 3$

45.

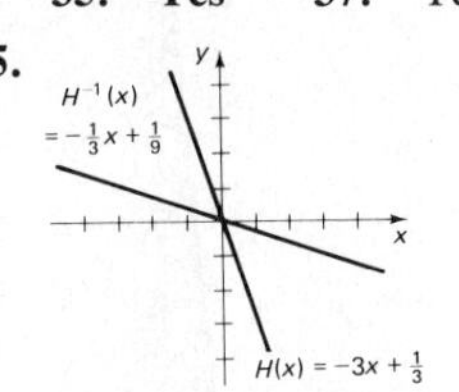

47.

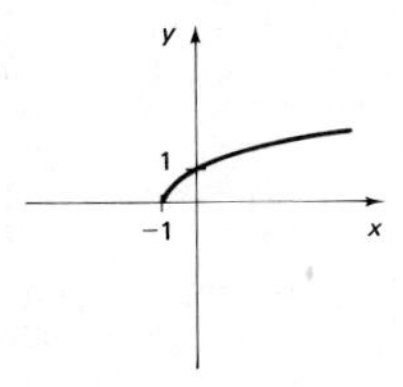

49.

$h(x) = (x + 1)^2, x \geq -1$

$h^{-1}(x) = \sqrt{x} - 1, x \geq 0$

51.

$F(x) = (x - 2)^2 - 1, x \leq 2$

$F^{-1}(x) = -\sqrt{x + 1} + 2, x \geq -1$

53. $f^{-1}(x) = x - 1$ **55.** $f^{-1}(x) = \frac{1}{2}x - \frac{3}{2}$ **57.** $f^{-1}(x) = x/(x - 1), x \neq 1$ **59.** $f^{-1}(x) = \sqrt{x} + 1, x \geq 0$

67.

$D_f = [0, \infty) = R_{f^{-1}}$
$R_f = [-1, \infty) = D_{f^{-1}}$

69.

$D_h = (0, \infty) = R_{h^{-1}}$
$R_h = (-2, \infty) = D_{h^{-1}}$

71.

$D_F = \{x \neq -3\} = R_{F^{-1}}$
$R_F = \{x \neq -1\} = D_{F^{-1}}$

Index

A

B

C

D

E

F

G

H

I

L

M

N

O

P

Q

R

S

T

U

V

W

X

Y

Z

TRIGONOMETRIC FUNCTIONS

Of Acute Angles

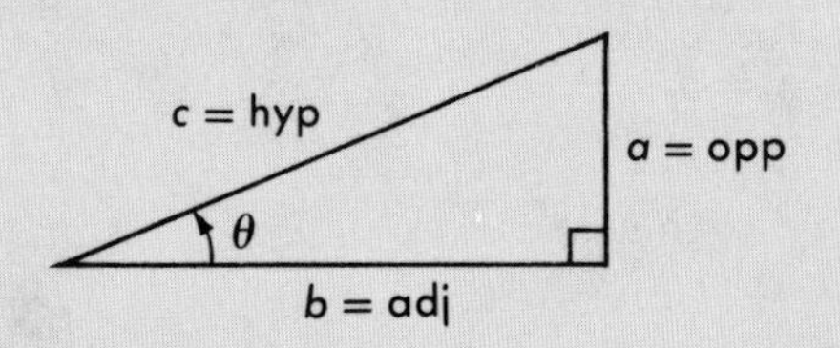

$$\sin\theta = \frac{a}{c} = \frac{\text{opp}}{\text{hyp}} \qquad \csc\theta = \frac{c}{a} = \frac{\text{hyp}}{\text{opp}}$$

$$\cos\theta = \frac{b}{c} = \frac{\text{adj}}{\text{hyp}} \qquad \sec\theta = \frac{c}{b} = \frac{\text{hyp}}{\text{adj}}$$

$$\tan\theta = \frac{a}{b} = \frac{\text{opp}}{\text{adj}} \qquad \cot\theta = \frac{b}{a} = \frac{\text{adj}}{\text{opp}}$$

Of Arbitrary Angles

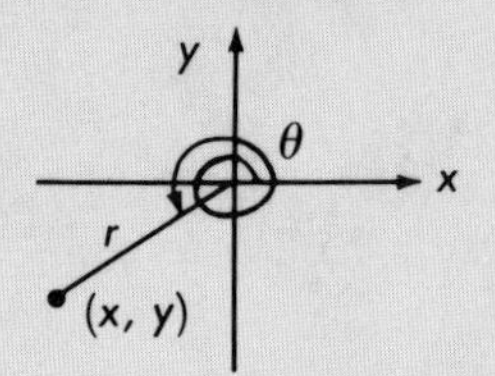

$$\sin\theta = \frac{y}{r} \qquad \csc\theta = \frac{r}{y}, \quad y \neq 0$$

$$\cos\theta = \frac{x}{r} \qquad \sec\theta = \frac{r}{x}, \quad x \neq 0$$

$$\tan\theta = \frac{y}{x}, \quad x \neq 0 \qquad \cot\theta = \frac{x}{y}, \quad y \neq 0$$

FUNDAMENTAL IDENTITIES

$$\csc x = \frac{1}{\sin x} \qquad \sec x = \frac{1}{\cos x} \qquad \cot x = \frac{1}{\tan x}$$

$$\tan x = \frac{\sin x}{\cos x} \qquad \cot x = \frac{\cos x}{\sin x}$$

$$\sin^2 x + \cos^2 x = 1 \qquad \tan^2 x + 1 = \sec^2 x \qquad 1 + \cot^2 x = \csc^2 x$$

IDENTITIES FOR NEGATIVES

$$\sin(-x) = -\sin x \qquad \cos(-x) = \cos x \qquad \tan(-x) = -\tan x$$

ADDITION FORMULAS

$$\sin(x + y) = \sin x \cos y + \cos x \sin y \qquad \sin(x - y) = \sin x \cos y - \cos x \sin y$$

$$\cos(x + y) = \cos x \cos y - \sin x \sin y \qquad \cos(x - y) = \cos x \cos y + \sin x \sin y$$

$$\tan(x + y) = \frac{\tan x + \tan y}{1 - \tan x \tan y} \qquad \tan(x - y) = \frac{\tan x - \tan y}{1 + \tan x \tan y}$$